DRUG DESIGN RESEARCH PERSPECTIVES

DRUG DESIGN RESEARCH PERSPECTIVES

STANLEY P. KAPLAN
EDITOR

Nova Biomedical Books
New York

For permission to use material from this book please contact us:
Telephone 631-231-7269; Fax 631-231-8175
Web Site: http://www.novapublishers.com

Library of Congress Cataloging-in-Publication Data

Available upon request.

ISBN: 978-1-60021-823-1
ISBN: 1-60021-823-7

Published by Nova Science Publishers, Inc. ✢ New York

Contents

Preface

Drug design is the approach of finding drugs by design, based on their biological targets. Typically a drug target is a key molecule involved in a particular metabolic or signalling pathway that is specific to a disease condition or pathology, or to the infectivity or survival of a microbial pathogen.

Some approaches attempt to stop the functioning of the pathway in the diseased state by causing a key molecule to stop functioning. Drugs may be designed that bind to the active region and inhibit this key molecule. However these drugs would also have to be designed in such a way as not to affect any other important molecules that may be similar in appearance to the key molecules. Sequence homologies are often used to identify such risks.

Other approaches may be to enhance the normal pathway by promoting specific molecules in the normal pathways that may have been affected in the diseased state.

The structure of the drug molecule that can specifically interact with the biomolecules can be modeled using computational tools. These tools can allow a drug molecule to be constructed within the biomolecule using knowledge of its structure and the nature of its active site. Construction of the drug molecule can be made inside out or outside in depending on whether the core or the R-groups are chosen first. However many of these approaches are plagued by the practical problems of chemical synthesis.

Newer approaches have also suggested the use of drug molecules that are large and proteinaceous in nature rather than as small molecules. There have also been suggestions to make these using mRNA. Gene silencing may also have therapeutical applications.

This new book presents leading-edge research from around the world.

Chapter I - Development of drugs is a risky, expensive and time intensive process. The pharmaceutical industry spends billions of dollars on various trials of numerous molecules to prove their suitability. In silico drug development can save a tremendous amount of effort and make drug development a much less expensive venture. To do that efficiently, the mechanisms whereby drugs interact with the relevant organs needs to be studied and this must be done in consort with related effects. A systems approach is necessary to simulate both the target area with which the drug is designed to interact as well as the surrounding areas where feedback mechanisms may alter the expected effect. This is particularly important since side effects are to be included in the simulation. In addition both the physics and chemistry of the processes must be included so that all mechanisms at a broad range of length scales are included in the model. This requires multi scale models. Recent developments in computer technology involving multi core processors have greatly expanded the possibilities of solving increasingly complex problems with multithreading so that computers can process the various mechanisms simultaneously. The example related to Propranolol considered here brings out the importance of research into the details of neuronal response in studying heart research. The primary example system described here is that of the influence of angiotensin (AngII) on the development of hypertrophy. Effects of AngII include complex system events about which some details are still unknown. The present study describes the overall system and gives details of two components of the AngII system. One component is the interaction with the effects of Amlodipine, a calcium channel blocker, interacting with the arterioles. One of the results here relates the dosage of the channel blocker Amlodipine to the fraction of Calcium ions absorbed. The other component considered is the action of Ang II in causing gene upregulation in the left ventricle of the heart and the effect of the angiotensin receptor blocker Losartan. This in turn studies the effect of AngII on the increase in left ventricular mass and development of hypertrophy. A suggested method for the action of Losartan is deduced. Speculation on the action of the drug Propranolol on mood changes is presented. In silico simulations not only provide important data for the pharmaceutical industry, they also highlight areas where further important research is required.

Chapter II - Regulation of a target gene at will is one of the most prospective themes in the post-genomic era. An artificial transcription factor with desired DNA binding specificity could work as a powerful tool and drug to control target genes. The C_2H_2-type zinc finger motif is one of the most typical DNA binding motifs found in a lot of transcription factors. The zinc finger domain is extremely attractive for design of artificial transcription factors with desired DNA binding specificities because of its characteristic DNA binding properties: (1) recognition of 3-4 bp per one motif, (2) tandemly connected modular structure, and (3) binding to non-palindrome sequences as an independent monomer. In order to target a specific gene by using a protein-based drug, it is required for a protein to bind to any desired DNA sequences. By randomizing amino acid residues important for DNA recognition, a number of successful zinc finger peptides have been selected to bind to the specific DNA sequence and to control gene expression. For the purpose of selective targeting in the huge genome (3×10^9 bp), long DNA sequence (18 bp) should be specifically recognized by a protein-based drug. The authors have shown that 6-zinc finger peptides in which two 3-zinc finger domains are connected with the consensus linker bind to the expected 18 bp sequences containing continuous two target sites for the original 3-zinc finger peptides. Artificial 6-zinc finger peptides with various kinds of linkers have also been developed and shown to skip DNA sequences. Artificial zinc finger domains can be fused with functional domains such as transcriptional regulatory domains or DNA cleavage domains for therapeutic use. Such zinc finger proteins have been shown to control specific gene expression or to induce homologous recombination. Herein, effects of linker design of artificial 6-zinc finger peptides on DNA binding are mainly described based on our experimental results. the author also review the latest trends in artificial zinc finger proteins toward gene therapy.

Chapter III - Rho-kinase is a leading player in the regulation of cytoskeletal events involving smooth muscle contraction and neurite growth-cone collapse and retraction, and is a promising drug target in the treatment of both vascular and neurological disorders. Recent crystal structures of Rho-kinase (ROCK 2) and ROCK 1 have revealed structural details of the ATP-binding site, which represents the target site for conventional small-molecule kinase inhibitors. Four series of kinase inhibitors are known which differ in scaffold chemical structures and includes isoquinolines, 4-aminopyridines, indazoles, and others derivatives containing amide and urea. The crystal structures of the fasudil- and Y27632-bound forms have provided valuable clues to compound development of the first two derivative series with improved selectivity. Especially, both kinase and drug conformational pliability and stability are underlined for conferring selectivity. In this study, the authors set out to predict the binding-modes of two series of the inhibitors, indazole and amide/urea series, in Rho-kinase by docking study in order to understand the structure-activity relationship of the Rho-kinase inhibitors from structural point of view. One isoquinoline inhibitor was also included in the study to examine our hypothesis that was previously addressed concerning the pliability of the P-loop. In the docking studies, three Rho-kinase structures with different P-loop conformations were served to examine the importance of the P-loop pliability in the binding mode and selectivity of all four derivatives including the indazoles and amide/urea derivatives. The authors show that the extended P-loop conformation is suitable to accommodate the indazoles and amide/urea derivatives that possess an aromatic pendant ring so as to bind a pocket beneath the phosphate-binding region which favors binding of a small aromatic ring, whereas the folded P-loop conformation accommodates the smaller compounds of the isoquinoline and 4-aminopyridine series. In the latter case, the conserved P-loop phenylalanine occupies the pocket, resulting in an increase of protein-ligand contacts. Thus, the P-loop pliability plays an important role in inhibitor binding affinity and specificity.

Chapter IV - The 1,3-dipolar cycloaddition reaction (1,3-DCR) of a 1,3-dipole to a dipolarophile for the synthesis of heterocycles is a ubiquitous transformation in synthetic organic chemistry. Recently, the Sharpless and Meldal groups have reported the dramatic rate enhancement (up to 10^7 times) and improved regioselectivity of the Huisgen 1,3-DCR of an organic azide to a terminal acetylene to afford, specifically a 1,4-disubstituted-1,2,3-triazole in the presence of a Cu^I catalyst. This Cu^I catalysed 1,3-DCR has successfully fulfilled the requirements of “click chemistry” as prescribed by Sharpless and within the past few years has become a premier component of the click chemistry paradigm. Click chemistry has proven to be of remarkable utility and broad scope, not only in organic synthesis, but in chemical biology and drug discovery. Click chemistry is highly modular and simplifies difficult syntheses. The biocompatibility of the reaction, tolerance towards a broad range of pH and relative inertness of acetylenes and azides within highly functionalised biological milieus has allowed click chemistry to become a viable bioconjugation strategy both for labelling biomolecules and for *in situ* lead discovery applications. More recently, click chemistry has emerged as a powerful conjugation strategy for the preparation of structurally diverse neoglycoconjugates of biomedical interest. This chapter aims to highlight recent developments appearing within the literature concerning the use of click chemistry in carbohydrate based drug discovery and glycobiology. Topics range from small molecule probes and drug leads, multivalent neoglycoconjugates acting as lectin inhibitors and potential vaccines, to bioconjugation strategies for labelling of engineered cell surface glycans. The chapter aims to be comprehensive with commentary on future perspective.

Chapter V - In computer-aided drug design, molecular similarity study plays an important role in finding and designing new drugs from the known drugs or inhibitors for a given protein target. A list of compounds can be compared according to the similarity of their structures or properties. With regards to the comparison of several molecular structures, some similarity calculation methods based on 2D structural characteristics, 3D structural characteristics or 3D pharmacophore characteristics have been developed; as to the comparison of the similarity of the property of a set of compounds, some chemoinformatics tools for clustering compounds have been developed.

In addition, in the combinatorial library design, more attention is given to design target focusing library along with diversity and drug-like criteria. On the basis of the compounds similarity comparison study, the similarity of the two compounds library can be compared; furthermore, in high-throughput-screening, high quality compounds library, in particular, the focusing library of drug-likeness (or lead-likeness) for a given target could be built based on the results of molecular similarity study such as the 3D pharmacophore characteristics.

Molecular similarity study will continue to occupy an important role in drug discovery and library design.

Chapter VI - An important trend in drug design is the evaluation of the *in vivo* pharmacokinetics and pharmacodynamics of new chemical entities (NCE). Although the pharmacokinetic properties of NCE can be studied by means of traditional blood sampling, regulatory authorities emphasize the evaluation of drug distribution at the target site. Another issue in the assessment of NCE is that the discovery of the existence of a large number of orphan receptors will increase the number of NCE to be tested. Therefore, a more rigorous selection process will be necessary at early stages of drug development. Integration of pharmacokinetic-pharmacodynamic (PK-PD) concepts through PK-PD modeling is a potential tool to enhance information gain and efficiency of drug selection during development. Therefore, the evaluation of drug concentrations at the biophase and their relationship with the pharmacological effect are two important trends in modern drug discovery.

New sampling techniques, including imaging techniques and microdialysis, are now available for the study of drug distribution at the target site. However, microdialysis seems to be the most adequate methodology for the evaluation of drug levels at the biophase in both preclinical and clinical research, considering that this technique samples extracellular free concentrations of drugs. Finally, microdialysis sampling might become a reference technique for PK-PD modeling by the fact that this technique allows the simultaneous determination of drug concentrations in one and more tissues and their effect on biochemical and clinical markers in the same subject and with good temporal resolution.

The aim of the chapter is to review the methodological and practical aspects of the microdialysis technique and its application for the study of drug distribution and PK-PD modeling in basic and clinical research of drugs and NCE.

Chapter VII - Stroke is the third most common cause of death worldwide and the most common cause of adult disability in industrialized countries. Stroke alone absorbs 6% of all heath costs. To date the only approved therapy for acute ischemic stroke is thrombolysis within 3 hours of stoke onset in well-selected cases with only 5% of all ischemic stroke patients currently receiving thrombolytic therapy. Therefore, new therapies must be developed urgently. One major obstacle in the development of effective therapies for ischemic stroke has been the lack of versatile imaging techniques. Magnetic resonance imaging (MRI) in stroke permitted major pathophysiological insights into the understanding of acute cerebral ischemia, but now is also considered of major importance in patient management and can serve as a surrogate end point. Because of its noninvasive nature, versatility, patient compatibility with few contraindications and its potential for imaging with relatively high spatial and temporal resolution, MRI serves as an indispensible tool in stroke evaluation for two major purposes: to ensure a definite diagnosis and to follow-up the ischemic lesion size over time. Besides, its availability both in clinical and preclinical settings allows direct bridging of findings in humans and animals. More than a decade, new MRI modalities, especially diffusion- and perfusion-weighted MRI (DWI and PWI, respectively) have been used in experimental studies with great success. DWI first has been performed and developed in animals and rapidly applied to clinical practice. Multimodal stroke MRI provides numerous advantages: DWI enables detecting very early ischemic lesion(s), allows for observation of lesion evolution with repeated measurements and gives clues for etiology; PWI shows where and to what degree cerebral blood flow is compromised; PWI and DWI can together approximate penumbra (the salvageable brain tissue); MR angiography discloses the occluded artery and other vascular pathologies; T2*-weighted images helps to exclude new and old hemorrhagic lesions; and conventional images show the final infarct volume. Thus, MRI improves patient triage to clinical therapy trials. The authors propose that implementation of a similar MRI protocol we used in animal models to human therapy trials will deliver quick answers to whether a novel therapy is effective or not, in fairly small and homogenous patient groups unnecessitating large, expensive, and time-consuming old-fashioned clinical trials.

Chapter VIII - Molecular docking and 3D-QSAR analyses were performed to understand the interaction between a series of efavirenz derivatives with WT and K103N HIV-1 RT. To model the potential binding modes of efavirenz derivatives in the binding pocket of WT and K103N HIV-1 RT, molecular docking approaches by using GOLD and Autodock 3.05 programs were performed. The results show that the docking results obtained from both methods reveal a good ability to reproduce the X-ray bound conformation with rmsd less than 1.0 Å for both WT and mutant enzymes. The docking calculations of all efavirenz derivatives in the data set were, consecutively, performed to elucidate their orientations in the binding pockets. The results derived from docking analysis give additional information and further probes the inhibitor-enzyme interactions. The correlation of the results obtained from docking models and the inhibitory activities validate each other and lead to better understanding of the structural requirements for the activity. Therefore, these results are informative to improve the development of more efficient HIV-1 RT inhibitors, especially, active against mutant enzyme. Based on the molecular alignment of conformations obtained from molecular docking procedures, the high predictive 3D-QSAR models were produced by using CoMFA and CoMSIA approaches. The CoMFA models reveal the importance of steric and electrostatic interactions through contour maps. The resulting CoMSIA models enhance the understanding of steric, electrostatic, hydrophobic, electron donor and acceptor requirements for ligands binding to the K103N HIV-1 RT. Moreover, quantum chemical calculations were carried out to analyze the interaction energies of the selected inhibitors with the individual amino acids in the binding pocket. The obtained results show the important interactions of inhibitors with the enzyme at the residue level. Consequently, the results obtained from structure-based, ligand-based design approaches and quantum chemical calculations can be integrated to identify the structural requirements of HIV-1 RT inhibitors in the class of efavirenz compounds. The principle derived from the present study provides a gainful guideline to design and predict new and more potent compounds active against WT and K103N HIV-1 RT.

Chapter IX - Drug discovery is a complex process that involves a combination of classical methods, modern technologies and innovative strategies addressed to the design of new chemical entities as drug candidates against a variety of biological targets. Drug discovery has moved toward more rational strategies based on our increasing understanding of the fundamental principles of protein-ligand interactions. The combination of available knowledge of several 3D protein structures with hundreds of thousands of small-molecule drug candidates have attracted the attention of scientists from all over the world for the application of structure- and ligand-based drug design approaches. The investigation of new potential biological targets is often limited by the absence of initial compounds that bind to a specific site in the protein, possessing reasonable chemical and biological profiles for further lead optimization. Biological screening requires testing large numbers of compounds in standard models to assess potency, selectivity and affinity. Virtual screening (VS) is a computational high-performance approach to analyze large databases of chemical compounds in order to prioritize novel potential hits for biological screening. Molecular docking is used to predict the binding affinities of each compound in the active site of a biological target. Similarity searching is another important component of VS methods that involves finding molecules in a database having the strongest similarity to a user-defined target structure. Typically, the target will be known to have some biological activity of interest and the objective of the search is to prioritize a dataset to optimize the pharmacodynamic and pharmacokinetic properties. The contribution in this book chapter will briefly outline our view on the present scope and future advances of VS based on some instructive examples from recent research describing this emerging strategy that is providing new valuable and effective routes for the traditional high-throughput screening (HTS) in drug discovery.

Chapter X - The fact that protein sequences largely outnumber their structures gives rise to a steadily increasing demand for homology-based, atom-scale modeling of protein structures applying computational methods. Pertaining to the three-dimensional (3D) model generation of proteins, low amino acid sequence identity of the target sequence to known template structures imparts a twilight zone with uncertainties about the either homologous (evolutionary related) or coincidental (by chance) nature of relationships between target (known only by its sequence) and templates (known 3D structures). As a consequence, fully automated 3D model predictions tend to suffer from inappropriate candidate selections and/or sequence alignments. As an introduction for beginners the authors present a technical report on mainly web-based computational approaches how to convert protein sequences into 3D models. Using the gene registered at *GenBank* (Acc. No. AY364634), this chapter illustrates the necessary steps to obtain an atomic-scale model: Upon domain recognition (280 residues out of 898 total sequence length) which belongs to the alpha/beta hydrolase fold superfamily exploit the highly conserved fold similarities to the cytosolic aminopeptidase family (EC 3.4.11.x) to find template structures. The resulting peptidase is theoretically involved in protein processing, a very plausible function in the light of the experiments in rat ovaries during cellular differentiation. In order to fill in the structural gap for this protein class the new aminopeptidase was deposited at PDB model repository under entry code 2B62 by AutoDep at RCSB (http://pdb.rutgers.edu/adit/).

In: Drug Design Research Perspectives
Editor: Stanley P. Kaplan, pp. 1-8
ISBN: 978-1-60021-823-1

Expert Commentary A

Astroglial Inwardly Rectifying Potassium Channel Kir4.1 as a Potential Target for the Novel Antidepressant Agents

Yukihiro Ohno and Yoshihisa Kurachi*
Division of Molecular and Cellular Pharmacology;
Department of Pharmacology; Osaka University,
Graduate School of Medicine

Introduction

It is well documented that antidepressant agents commonly inhibit the reuptake of monoamines (e.g., serotonin and noradrenaline) into the nerve terminals and elevate extracellular levels of the monoamines at synaptic regions (Sanchez and Hyttel, 1999; Frazer, 2001). These actions have long been proposed as the principle mechanism of actions for antidepressants to normalize the deficits of monoaminergic neurons in depression. It should be, however, pointed out that this "monoamine mechanism" cannot solely account for clinical features of antidepressants such as a delayed onset of the actions, differences between the high levels of drug concentration in the brain and their affinities to the monoamine transporters, and diverse efficacies of antidepressants for various disorders (e.g., anxiety disorders and chronic pain) (Briner and Dodel, 1998; Sanchez and Hyttel, 1999; Pollack and Marzol, 2000; Ninan, 2003, Henry et al., 2005). Therefore, actions of antidepressants other than that on monoaminergic neurons are to be elucidated to clarify the mechanisms of the drug-action (Fuchs et al., 2004; Malberg and Blendy, 2005).

* Corresponding author: Yoshihisa Kurachi, M.D., Ph.D. Division of Molecular and Cellular Pharmacology; Department of Pharmacology; Graduate School of Medicine, Osaka University; 2-2 Yamada-oka, Suita, Osaka 565-0871, Japan; TEL: +81-6-6879-3512; FAX: +81-6-6879-3519; e-Mail: ykurachi@pharma2.med.osaka-u.ac.jp

We have recently demonstrated that various antidepressants, including the tricyclic antidepressants (TCAs) and the selective serotonin reuptake inhibitors (SSRIs), can block the inwardly rectifying potassium Kir4.1 channels (Su et al., 2007; Ohno et al., 2007), which are predominantly expressed on cerebral astrocytes and responsible for their spatial potassium (K^+) buffering action (Poopalasundaram et al., 2000; Walz, 2000; Kofuji and Newman, 2004; Simard and Nedergaard, 2004). Our findings strongly suggest that antidepressants alter the Kir4.1-mediated astroglial K^+ buffering currents in the brain, which may be involved in pharmacological actions of these agents. Here we summarize the interactions of astroglial Kir4.1 channels with various antidepressants and discuss their potential as a target of actions for the antidepressant treatment.

Astroglial Kir4.1 Channels and Spatial K^+ Buffering

The inwardly rectifying potassium (Kir) channel subunits comprise more than fifteen members that can be functionally classified into four subfamilies (Isomoto et al., 1997; Kubo et al., 2005). They play crucial roles in determining the resting membrane potential of neurons (Kir2.x), in regulating the neuronal excitability via the G protein-mediated pathways (Kir3.x) or the intracellular ATP-dependent signaling (Kir6.x), and in transporting K^+ in glial and epithelial cells (Kir1.x, Kir4.x and Kir5.x). Among these subunits, Kir4.1 and Kir5.1 are predominantly expressed in brain astrocytes as well as retinal Müller cells, where they are involved in buffering the excessive extracellular K^+ elevated during neuronal activity (referred to as 'spatial potassium buffering' in the brain or 'potassium siphoning' in the retina) (Takumi et al., 1995; Ishii et al., 1997, 2003; Poopalasundaram et al., 2000; Walz, 2000; Higashi et al. 2001; Hibino et al., 2004; Kofuji and Newman, 2004; Simard and Nedergaard, 2004).

Astroglial Kir channels are composed of two types, the homo-tetramer of Kir4.1 (Kir4.1 channels) or the hetero-tetramer of Kir4.1 and Kir5.1 (Kir4.1/Kir5.1 channels), both of which constitutively allow large inward K^+ currents at potentials negative to the K^+ equilibrium potential (E_K) and small, but significant, outward K^+ currents at those positive to E_K (figure 1) (Takumi et al., 1995; Ishii et al., 1997; Tanemoto et al, 2000; Higashi et al. 2001; Hibino et al., 2004). Thus, depending on the difference between local E_K and the membrane potential of astrocytes, they can absorb the excessive extracellular K^+ locally elevated near the synaptic sites and transport it to the regions of low K^+ such as blood vessels (Walz, 2000; Ishii et al., 2003; Kofuji and Newman, 2004; Simard and Nedergaard, 2004; Kucheryavykh et al., 2007). In addition, the water channels, aquaporin-4, are co-localized with Kir4.1 channel in certain membrane domains of brain astrocytes. This suggests that spatial K^+ buffering may couple with water movement across the astroglial membrane (figure 1) (Nagelhus et al., 1999; Amiry-Moghaddam et al., 2003), which should be indispensable for homeostasis of ionic and osmorality surrounding synapses.

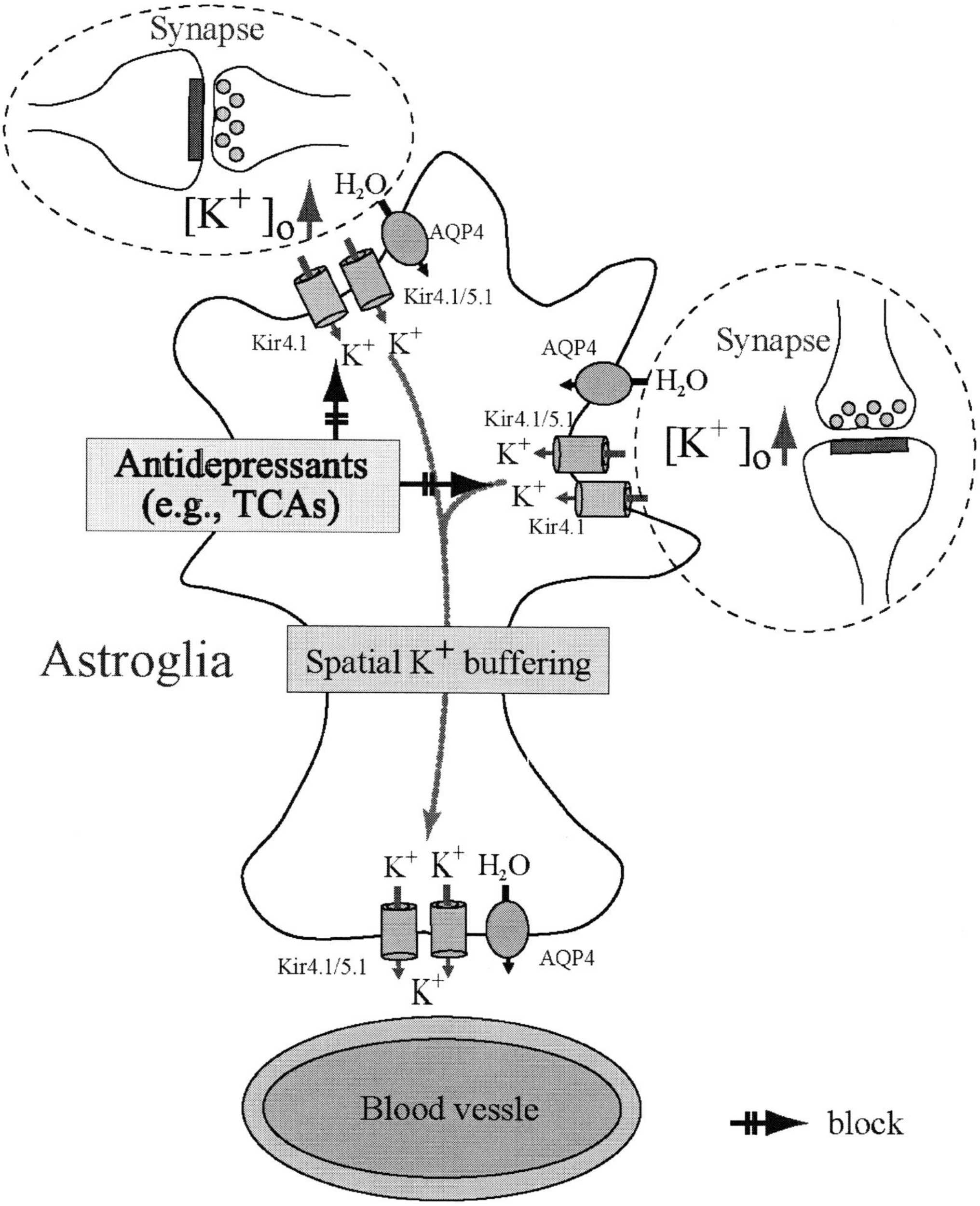

Figure 1 Astroglial Kir4.1 channels and the spatial K^+ buffering. The astroglial Kir channels are composed of two subtypes, the homo−tetramer of Kir4.1 (Kir4.1 channels) or the hetero-tetramer of Kir4.1 and Kir5.1 (Kir4.1/Kir5.1 channels). The both channels function as intermediate inwardly rectifying channels by conducting a large inward K^+ current at membrane potentials negative to E_K and a relatively small outward K^+ current at those positive to E_K. Therefore, they can absorb the excessive extracellular K^+ locally elevated near the synaptic sites and transport it to the regions of low K^+ such as blood vessels (spatial K^+ buffering). Several antidepressants can negatively regulate the astroglial spatial K^+ buffering by inhibiting Kir4.1 channels.

Inhibition of Astroglial Kir4.1 Channels by Antidepressants

The typical TCA, nortriptyline, and the SSRI, fluoxetine, both reversibly inhibited Kir4.1 channel currents in a concentration-dependent manner. The inhibition of Kir4.1 channels by nortriptyline depended on membrane potentials (i.e., the voltage difference between the membrane potential and the E_K) with greater potency at more positive potentials (figure 2). The dissociation constant (K_D) of nortriptyline for Kir4.1 inhibition was 28.1 μM at E_K, which suggests the therapeutic importance of the action because the concentrations of TCAs during the chronic treatment can reach several tens of μM in the brain (Glotzbach and Preskorn, 1982; Baumann et al., 1983; Miyake et al., 1990). It is conceivable, therefore, that TCAs at clinical doses can inhibit the Kir4.1-mediated astroglial K^+-buffering action. Kinetic analysis of the nortriptyline blockade of Kir4.1 has revealed that the agent acts as a pore blocker with an electrical distance parameter of 0.62, implying that the agent might pass 62% of the electrical field across the plasma membrane from inside of the cell to reach its binding site (figure 2) (Su et al., 2007).

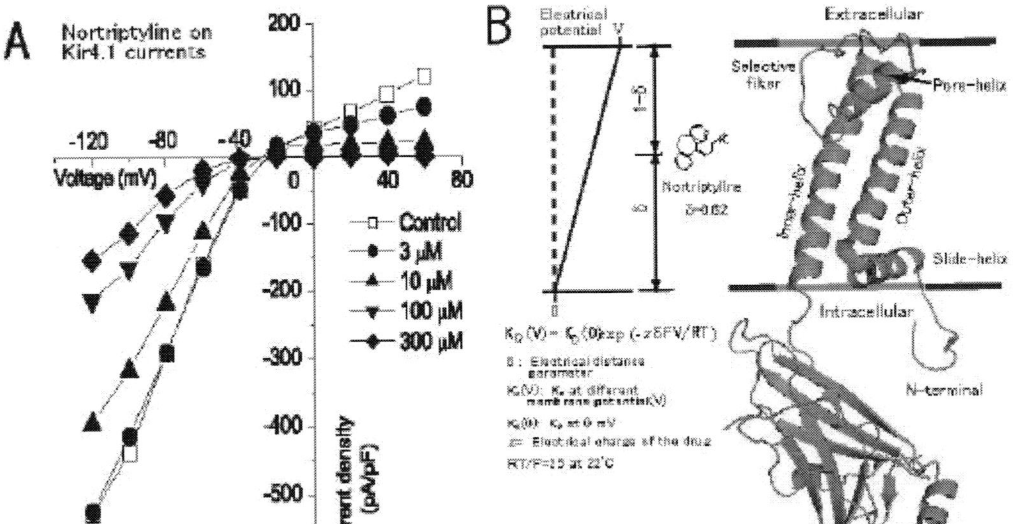

Figure 2 Inhibition of Kir4.1 channel currents by nortriptyline. Nortriptyline inhibits the astroglial Kir4.1 channels in voltage- and time-dependent fashions with greater inhibition at more positive membrane potentials (A). Blocking kinetics of the drug could be described by first order kinetics, where dissociation of the drug slowed down and association accelerated as the membrane was depolarized. In addition, kinetic analysis of the nortriptyline blockade revealed that the agent acts as a deep pore blocker with an electrical distance parameter (δ) of 0.62, implying that the agent might pass 62% of the electrical field across the plasma membrane from inside of the cell to reach its binding site (B). The Kir subunit model based on the structure of Kirbac1.1 (Kuo et al., 2003) is also shown as a structural reference.

The inhibitory effect of fluoxetine, a typical SSRI, on Kir4.1 channels was more potent than that of nortriptyline, and was virtually voltage-independent (Ohno et al., 2007). The Hill's coefficients for the concentration–response curves of SSRIs were more than 2 whereas the value for nortriptyline was nearly unit (Su et al., 2007). Thus it seems likely that the SSRIs, unlike TCAs, bind multiple sites within Kir4.1 channels. On the other hand, the tetracyclic antidepressant (i.e., mianserin) or 5-HT_{1A} receptor partial agonist (i.e., buspirone) only weakly or negligibly affected Kir4.1 channel activity (Ohno et al., 2007).

It is of interest that fluoxetine selectively inhibited Kir4.1 channel currents with barely affecting Kir2.1 or Kir1.1 channel currents (Ohno et al., 2007). Nortriptyline also showed a selectivity to Kir4.1 (*vs.* Kir2.1) channels in its inhibition (Su et al., 2007). The Kir2.1 channels are dominantly expressed on neurons and are critical in determining the resting membrane potentials of neurons (Horio, Y, 1996; Kubo et al., 2005). The Kir1.1 channels were originally identified as the K^+-transporter channels in the outer medulla of the rat kidney (Kubo et al., 2005), and also shown expressed in some brain regions (e.g., cerebral cortex and hippocampus) (Kenna et al., 1994). Thus, antidepressants such as fluoxetine and nortriptyline seem to preferentially block the astroglial Kir4.1 channels without affecting the neuronal type Kir2.1 channels or the K^+-transporter type Kir1.1 channels in the brain. Furthermore, a previous study using *Xenopus* oocytes (Kobayashi et al., 2003, 2004) demonstrated that these agents act as only a weak partial antagonist at Kir3.x channels. Therefore, the blocking actions of fluoxetine and nortriptyline are relatively specific to Kir4.1 among various Kir channels.

Hypothesis on Astroglial Kir4.1 Channels as a Potential Target for Antidepressants

The cerebral astrocytes are now known as one of the target sites for the antidepressants' actions (Fuchs et al., 2004; Mercier et al., 2004; Malberg and Blendy, 2005; Czeh et al., 2006). They form the blood-brain barrier by surrounding capillaries, maintain water and extracellular ion homeostasis, and mediate neural plasticity and/or neurogenesis through secretion of the neurotrophic factors (e.g., BDNF and GDNF) (Malberg and Blendy, 2005). In addition, it was reported that morphological alterations (e.g., reduced volume and atrophy) of astrocytes in the limbic structures are identified in animal models of chronic stress and also in patients with depression (Fuchs et al., 2004; Czeh et al., 2006).

Our results provide the evidence that various antidepressants inhibit the Kir4.1-mediated astroglial K^+ buffering currents, implying that astroglial Kir4.1 channels may be a common target potentially involved in the pharmacological actions of antidepressants. The inhibition of astroglial K^+-buffering by antidepressants may enhance neuronal activity indirectly by increasing $[K^+]_o$ surrounding neurons. Interestingly, it was reported that various K^+ channel blockers such as tetraethylammonium (for all K^+ channels), glibenclamide and glyburide (for ATP-sensitive K^+ channels), apamine and charybdotoxin (for Ca^{2+}-activated K^+ channels), which were supposed to commonly increase neuronal excitability by inhibiting the activity of different types of K^+ channels, produce a significant antidepressant action in animal models (Guo et al., 1996; Kaster et al., 2005). Thus, the inhibition of astroglial Kir4.1 channels might also be involved in the therapeutic actions of antidepressants. On the other hand, antidepressants at very high concentrations or under sustained treatments may lead to neuronal suppression, since an excessive elevation of the extracellular K^+ over the level of 10-12 mM $[K^+]_o$ is known to induce spreading suppression (Walz, 2000; Kofuji and Newman, 2004).

In conclusion, we demonstrated that the antidepressants at clinical concentrations can inhibit the Kir4.1-mediated astroglial K^+ buffering currents, which may influence the neuronal excitability and/or ion-water homeostasis in the brain. Our findings may give a novel possibility that the astroglial Kir4.1 channels may serve as a potential target of actions for the novel antidepressants, which may account for at least partly diverse actions of the antidepressants in patients with depression and related mood disorders. Further studies such as *in vivo* assessment of neural activity under suppression of astroglial Kir channels are required to delineate detailed mechanisms and pharmacological relevance of the actions of antidepressants in the brain.

References

Amiry-Moghaddam, M., Otsuka, T., Hurn, P.D., Traystman, R.J., Haug, F.M., Froehner, S.C., Adams, M.E., Neely, J.D., Agre, P., Ottersen, O.P., Bhardwaj, 2003. Analpha-syntrophin-dependent pool of AQP4 in astroglial end-feet confers bidirectional water flow between blood and brain. *Proc. Natl. Acad. Sci. USA.* 100, 2106-2111.

Briner, K., Dodel, R.C., 1998. New approaches to rapid onset antidepressants. *Curr. Pharm. Des.* 4, 291-302.

Baumann, P., Gaillard, J.M., Perey, M., Justafre, J.C., Le, P., 1983. Relationships between brain concentrations of desipramine and paradoxical sleep inhibition in the rat. *J. Neural. Transm.* 56, 105-116.

Czeh, B., Simon, M., Schmelting, B., Hiemks, C., Fuchs, E., 2006. Astroglial plasiticity in the hippocampus is affected by chronic psychosocial stress and concomitant fluoxetine treatment. *Neuropsychopharmacology.* 31, 1616-1626.

Frazer, A., 2001. Serotonergic and noradrenergic reuptake inhibitors: Prediction of clinical effects from in vitro potencies. *J. Clin. Psychiatry,* 62, 16-23.

Fuchs, E., Czeh, B., Kole, M.H.P., Michaelis, T., Lucassenm P.J., 2004. Alterations of neuroplasiticity in depression: the hippocampus and beyond. *Eur. Neuropsychpharmacol.* 14, S481-S490.

Glotzbach, R.K., Preskorn, S.H., 1982. Brain concentrations of tricyclic antidepressants: single-dose kinetics and relationship to plasma concentrations in chronically dosed rats. *Psychopharmacology.* (Berl), 78, 25-27.

Guo, W., Todd, K., Bourin, M., Hascoet, M., Kouadio, F., 1996. Additive effects of glyburide and antidepressants in the forced swimming test: evidence for the involvement of potassium channel blockade. *Pharmacol. Biochem. Behav.* 54, 725-730.

Henry, M.E., Schmidt, M.E., Hennen, J., Villafuerte, R.A., Butman, M.L., Tran, P., Kerner, L.T., Cohen, B., Renshaw, P., 2005. A comparison of brain and serum pharmacokinetics of R-fluoxetine and racemic fluoxetine: A 19-F MRS study. *Neuropsychopharmacology,* 30, 1576-1583.

Hibino, H., Fujita, A., Iwai, K., Yamada, M., Kurachi, Y., 2004. Differential assembly of inwardly rectifying K+ channel subunits, Kir4.1 and Kir5.1, in brain astrocytes. *J. Biol. Chem.* 279, 44065-44073.

Higashi, K., Fujita, A., Inanobe, A., Tanemoto, M., Doi, K., Kubo, T., Kurachi, Y., 2001. An inwardly rectifying K(+) channel, Kir4.1, expressed in astrocytes surrounds synapses and blood vessels in brain. *Am. J. Physiol. Cell. Physiol.* 281, C922-C931.

Horio, Y., Morishige, K., Takahashi, N., Kurachi, Y., 1996. Differential distribution of classical inwardly rectifying potassium channel mRNAs in the brain; comparison of IRK2 with IRK1 and IRK3. *FEBS Let.,* 379, 239-243.

Ishii, M., Horio, Y., Tada, Y., Hibino, H., Inanobe, A., Ito, M., Yamada, M., Gotow, T., Uchiyama, Y., Kurachi, Y., 1997. Expression and clustered distribution of an inwardly rectifying potassium channel, KAB-2/Kir4.1, on mammalian retinal Muller cell membrane: their regulation by insulin and laminin signals. *J. Neurosci.* 17, 7725-7735.

Ishii, M., Fujita, A., Iwai, K., Kusaka, S., Higashi, K., Inanobe, A., Hibino, H., Kurachi, Y., 2003. Differential expression and distribution of Kir5.1 and Kir4.1 inwardly rectifying K+ channels in retina. *Am. J. Physiol. Cell. Physiol.* 285, C260-C267.

Isomoto, S., Kondo, C., Kurachi, Y., 1997. Inwardly rectifying potassium channels: their molecular heterogeneity and function. *Jan. J. Physiol.* 47, 11-39.

Kaster, M.P., Ferreira, P.K., Santos, A.R., Rodrigues, A.L., 2005. Effects of potassium channel inhibitors in the forced swimming test: possible involvement of L-arginine-nitric oxide-soluble guanylate cyclase pathway. *Behav. Brain Res.* 165, 204-209.

Kenna, S., Ropert, S., Ashcroft, S.J.H., Ashcroft, F.M., 1994. Differential expression of the inwardly-rectifying K-channel ROMK1 in rat brain. *Mol. Brain. Res.* 24, 353-356.

Kobayashi, T., Washiyama, K., Ikeda, K., 2003. Inhibition of G protein-activated inwardly rectifying K+ channels by fluoxetine (Prozac). *Brit. J. Pharmacol.* 138, 1119-1128.

Kobayashi T, Washiyama K and Ikeda K (2004) Inhibition of G protein-activated inwardly rectifying K+ channels by various antidepressant drugs. *Neuropsychopharmacology.* 29:1841-1851.

Kofuji, P., Newman, E.A., 2004. Potassium buffering in the central nervous system. *Neuroscience,* 129, 1045-1056.

Kubo, Y., Adelman, J.P., Clapham, D.E., Jan, L.Y., Karschin, A., Kurachi, Y., Lazdunski, M., Nichols, C.G., Seino, S., Vandenberg, C.A., 2005. International Union of Pharmacology. LIV. Nomenclature and molecular relationships of inwardly rectifying potassium channels. *Pharmacol. Rev.* 57, 509-526.

Kuo, A., Gulnis, J.M., Antcliff, J.F., Rahman, T., Lowe, E.D., Zimmer, J., Cuthbertson, J., Ashcroft, F.M., Ezaki, T., Doyle, D.A., 2003. Crystal structure of the potassium channel Kirbac1.1 in the closed state. *Science,* 300, 1922-1926.

Kucheryavykh, Y.V., Kucheryavykh, L.Y., Nichols, C.G., Maldonado, H.M., Baksi, K., Reichenbach, A., Skatchkov, S.N., Eaton, M.J., 2007. Downregulation of Kir4.1

inward rectifying potassium channel subunits by RNAi impairs potassium transfer and glutamate uptake by cultured cortical astrocytes. *Glia,* 55, 274-281.

Malberg, J.E., Blendy, J.A., 2005. Antidepressant action: to the nucleus and beyond. *Trends Pharmacol. Sci.* 26, 631-638.

Mercier, G., Lennon, A.M., Renouf, B., Dessouroux, A., Ramauge, M., Courtin, F., Pierre, M., 2004. MAP kinase activation by fluoxetine and its relationship to gene expression in cultured rat astrocytes. *J. Mol. Nurosci.* 24, 207-216.

Miyake, K., Fukuchi, H., Kitaura, T., Kimura, M., Sarai, K., Nakahara, T., 1990. Pharmacokinetics of amitriptyline and its demethylated metabolite in serum and specific brain regions of rats after acute and chronic administration of amitriptyline. *J. Pharm. Sci.* 79, 288-291.

Nagelhus, E.A, Horio, Y., Inanobe, A., Fujita, A., Haug, F.M., Nielsen, S., Kurachi, Y., Ottersen, O.P., 1999. Immunogold evidence suggests that coupling of K+ siphoning and water transport in rat retinal Muller cells is mediated by a coenrichment of Kir4.1 and AQP4 in specific membrane domains. *Glia,* 26, 47-54.

Ninan, P.T., 2003. Obsessive-compulsive disorder: implication of the efficacy of an SSRI, paroxetine. *Psychopharmacol. Bull.* 37 (Suppl 1), 89-96.

Ohno, Y., Hibino, H., C. Lossin, Inanobe, A., Kurachi, Y., 2007. Inhibition of astroglial Kir4.1 channels by selective serotonin reuptake inhibitors. *Brain Res., in press.*

Pollack, M.H., Marzol, P.C., 2000. Panic: course, complications and treatment of panic disorder. *J. Psychopharmacol.* 14(2 Suppl 2), S25-30.

Poopalasundaram, S., Knott, C., Shamotienko, O.G., Foran, P.G., Dolly, J.O., Ghiani , C.A., Gallo, V., Wilkin, G.P., 2000. Glial heterogeneity in expression of the inwardly rectifying K(+) channel, Kir4.1, in adult rat CNS. *Glia,* 30, 362-372.

Sanchez. C., Hyttel, J., 1999. Comparison of the effects of antidepressants and their metabolites on reuptake of biogenic amines and on receptor binding. *Cell. Mol. Neurobiol.* 19, 467-489.

Simard, M., Nedergaard, M., 2004. The neurobiology of glia in the context of water and ion homeostasis. *Neuroscience,* 129, 877-896.

Su, S., Ohno, Y., Lossin, C., Hibino, H., Inanobe, A., Kurachi, Y., 2007. Inhibition of astroglial inwardly rectifying Kir4.1 channels by a tricyclic antidepressant, nortriptyline. *J. Pharmacol. Exp. Ther.* 320, 573-580.

Takumi, T., Ishii, T., Horio, Y., Morishige. K,, Takahashi, N., Yamada, M., Yamashita, T., Kiyama, H., Sohmiya, K., Nakanishi, S., Kurachi, Y., 1995. A novel ATP-dependent inward rectifier potassium channel expressed predominantly in glial cells. *J. Biol. Chem.* 270, 16339-16346.

Tanemoto, M., Kittaka, N., Inanobe, A., Kurachi, Y., 2000. In vivo formation of a proton-sensitive K+ channel by heteromeric subunit assembly of Kir5.1 with Kir4.1. *J. Physiol.* 525, 587-592.

Walz, W., 2000. Role of astrocytes in the clearance of excess extracellular potassium. *Neurochem. Int.* 36, 291-300.

In: Drug Design Research Perspectives
Editor: Stanley P. Kaplan, pp. 9-24

ISBN: 978-1-60021-823-1

Chapter I

In Silico Drug Design Systems

A.K. Macpherson[1], *S. Neti*[1] *and P.A. Macpherson*[2]
[1] Institute for Biomedical Engineering and Mathematical Biology
Lehigh University, Bethlehem, PA, USA
[2] Department of Applied Technology,
Rogers State University, OK, USA

Abstract

Development of drugs is a risky, expensive and time intensive process. The pharmaceutical industry spends billions of dollars on various trials of numerous molecules to prove their suitability. In silico drug development can save a tremendous amount of effort and make drug development a much less expensive venture. To do that efficiently, the mechanisms whereby drugs interact with the relevant organs needs to be studied and this must be done in consort with related effects. A systems approach is necessary to simulate both the target area with which the drug is designed to interact as well as the surrounding areas where feedback mechanisms may alter the expected effect. This is particularly important since side effects are to be included in the simulation. In addition both the physics and chemistry of the processes must be included so that all mechanisms at a broad range of length scales are included in the model. This requires multi scale models. Recent developments in computer technology involving multi core processors have greatly expanded the possibilities of solving increasingly complex problems with multithreading so that computers can process the various mechanisms simultaneously. The example related to Propranolol considered here brings out the importance of research into the details of neuronal response in studying heart research. The primary example system described here is that of the influence of angiotensin (AngII) on the development of hypertrophy. Effects of AngII include complex system events about which some details are still unknown. The present study describes the overall system and gives details of two components of the AngII system. One component is the interaction with the effects of Amlodipine, a calcium channel blocker, interacting with the arterioles. One of the results here relates the dosage of the channel blocker Amlodipine to the fraction of Calcium ions absorbed. The other component considered is the action of Ang II in causing gene upregulation in the left ventricle of the heart and the effect of the angiotensin receptor blocker Losartan. This in turn studies the effect of AngII on the increase in left ventricular mass and development of hypertrophy. A suggested method for the action of Losartan is deduced. Speculation on the action of the drug Propranolol on mood changes is presented. In silico simulations not only provide important data for the pharmaceutical industry, they also highlight areas where further important research is required.

Introduction

The development of new medicines is a very expensive and time consuming task. Many potentially useful chemicals are considered before tests are performed on animals and of those that pass, many fail when tested on human subjects. It would be very beneficial to have a tool and a protocol that is capable of ranking proposed medicines and identify those, which have the most likely chances of success. The aim here is to compare different potential drugs and in silico (computer) modelling of drugs can play a very important role. With judicious technical choices, great savings in development costs would be possible for the pharmaceutical industry.

The proposition underlying this chapter is that for computer drug design (often known as in silico biology) to play a significant role in the future it is necessary to take a systems approach to the modeling of processes. As a drug often acts not only on the part of the body for which it is designed but may have other systemic effects sometimes leading to harmful contra indications. A recent example is the testing of TGN 1412 an immunomodulator medication which made it clear that it is necessary to simulate all parts of the body which may be affected. This will require the development of a large data base containing models of various parts of the body. In addition, as shown here, there is a requirement to undertake in depth research in areas where information is very limited. However this will be much less expensive than wet bench development of new drugs and possible side effects being found later.

In silico drug design has had limited application due to present day limitations in computer power and the development of appropriate data bases necessary for this purpose. There are some detailed models such as the neuron [1], virtual cell [2]. In such cases the model is very detailed but the computational power is not available to use them as part of an overall drug design system analysis. This limitation will very likely be less of hurdle in the future with the wider use of multi-core computer processors and other improvements. The Intel Company has suggested in a number of presentations that processors with hundreds of cores could be available in the next ten years. Expanded computational power is presently available with some super computers. Such facilities are useful for running programs once the program has been developed. Generally the procedures necessary to administer such large facilities do not lend themselves to the trial and error methods necessary to develop the analysis of intricate interdependent systems. Use of large core processors will make it possible to define threads in the program so that different aspects of the drug effects can be simulated at the same time on moderate size machines. A thread is a segment of computer code which can interact with other code but operate on its own while the main code continues. For example a print command can be assigned a thread and control the printer while the main computations continue.

Specific examples of present day system design will be discussed in the main body of this chapter. It will be shown that an important aspect of such studies is that multiple scales of both time and length are important. These are run on different threads and it will be shown how the process can be controlled so that one part of the simulation does not proceed till all the data is available. The JAVA language was used to achieve the control and the processes are dynamic. Consider for example if calcium is required to activate a muscle. If the rate that the calcium is consumed is greater than the diffusion rate into the area, then the process will slow down so that both the processes are compatible.

In any such system there will be a number of unknowns since the available experimental data is rarely exhaustive. It is possible to consider a range of concentrations of variables and observe the result. In addition it demonstrates where experimental studies can be most effectively undertaken in drug design. However consider a hypothetical example which would be possible using a computer with hundreds of cores. The drug could be a new heart drug which would interact with the receptors on the myocardium and alter the internal structure of the myocardium muscle cells. The drug may interact with the alveoli in the lungs, interact with the neurons and dendrites, the pancreas, the adipose tissue, the retina, the pituitary gland, and finally have an interaction with the liver. Threads could be established to study the blood flow linking the components. They would be used to examine the effects that the drug would have on the heart wall endothelium and receptors. Threads would be established to examine the overall flow in the heart, the interaction with the receptors, the effect on the internal structure of the myocardium cells and subsequent gene changes. Threads would also be established to examine the penetration of the blood brain barrier, the external and internal interaction with the neurons and subsequent effect some systems acting on the axons. A version of the neuron model, for example, could be used here. At the same time the blood flow through the lungs would be studied to determine if there would be an effect on the alveoli. Finally threads would be established to examine the influence on the other components of the system which may influence the heart due to the presence of the drug in the blood. With experience it would be possible to examine the chemical reactions involved in different parts of the system to predict undesirable side effects. Such a study would require careful design to optimize the time spent in each core (processor) and how efficiently the available threads could be used.

The Renin-Angiotensin System Effect on the Heart

The Renin-Angiotensin System (RAS) affects many parts of the body [3]. If the large multi-core computer was available this would be an ideal candidate for simulation. In the present discussion the affect of RAS will be restricted to the heart. The development of the important heart antagonist angiotensin II (AngII) usually begins with its precipitated by the enzyme Renin. Angiotensin II is generated from a peptidergic precursor, named angiotensinogen, which is cleaved by the aspartyl protease Renin to form the decapeptide angiotensin I. The angiotensin converting enzyme (ACE), a dipeptidyl carboxypeptidase, acting on the angiotensin I hydrolyzes it at its carboxyterminus, leading to the generation of the octapeptide angiotensin II.

The effects of AngII are wide ranging. The renin-angiotensin system can be found [4], for example, in the liver [5], gonades [6], placenta [7], pancreas [8,10], adipose tissue [10, 11], heart [12], retina [13-17], pituitary [18, 19], and brain. It is clear that AngII is generated in various parts of the body and that a system to take into account all possible influences would be very large, The analysis here will be restricted to the effect of AngII on the heart due to three routes only.

The three systems to be considered where AngII is developed independently are the peripheral blood, the neuronal system behind the blood brain barrier and in the ventricular tissue of the heart. There are two principal types of receptors for AngII, known as AT (1) and AT (2). It is known that there is an increase in the AT (1) receptors located in cerebrovascular endothelial cells and results in a change in cerebrovascular flow [20]. However it is unclear if the AngII results in an increase in blood pressure directly due to the interaction with the cerebrovascular flow. It would appear to be due to the diameters of the cerebrovascular system that the effect could be a related influence. There are many other effects of AngII being attached to the cerebrovascular cells as described in [21]. AngII is known not to cross the blood brain barrier [22].

The system considered here will start after the formation of AngII by the ACE. As a model for the system is being considered there will be several aspects of the AngII considered. These are, (a) the effect on the heart by direct interaction with the receptors in the left ventricle tissue [2] resulting in molecular changes in the signal transduction pathways and protein expressions that lead to hypertrophy. The AngII comes from both the arterial flow and by generation within the ventricular tissue. (b) There is an interaction of the AngII with the neurons in the brain [4] which results via the axons in the constriction of the muscles around the arterioles. In this example it is shown how the drugs Losartan and Amlodipine act to relieve high blood pressure by blocking the receptors on the myocardium in the heart and the calcium channels on the alveoli muscles which restrict the blood flow into the capillaries.

The System Overview

There are many reasons for hypertrophy. It is known from animal tests that aortic banding which increases the blood pressure in the left ventricle will induce hypertrophy. Thus in this study pressure overload will be considered the only cause. One mechanism whereby pressure overload can be generated is by an action potential traveling from the neurons down through the axons to signal the alveoli muscles to contract and increase the pressure. It is known [3] that AngII can cause an increase in neuronal firing and the generation of an action potential. Thus it appears likely that AngII is the trigger to set this process in motion. As AngII is always present behind the blood brain barrier, then an important consideration for drug design is the determination of the conditions under which the action potential can be generated. Although the arterial AngII cannot cross the blood-brain barrier it can influence the brain at particular sites, e.g., the circumventricular organs

(CVO) [5]. That there is a relation between AngII activity in the brain and the control of cardiovascular functions has been established [23]. As discussed below, in connection with the left ventricle, it is necessary for the receptors to be activated. The mechanism for activation in the heart is presently unknown although it is thought that the metabolites AngII and AngIV may be instrumental in activation [24]. Possibly an increase in AngII around the neurons from the arterial AngII could activate the receptors and thus result in neuronal firing. Further, it is shown in connection with the left ventricle that once activated the AngII receptors are quickly blocked.

Presently there is insufficient data about the neuronal receptors and the relation between the AngII system behind the blood-brain barrier and the arterial AngII to attempt a simulation of the neural system. Thus it will be assumed that the neuron will fire and send a signal to the axon. Another important unknown relation is that between the AngII presence and the magnitude of the action potential. Although Losartan has been shown to have an effect on this behavior it does not appear to have been shown to influence the interaction of AngII with the neuronal receptors [4] as it does in the heart. There is an influence of Losartan via the brain on the heart as norepinephrine-mediated neuromodulation involves AT 1 -receptor-mediated Losartan-dependent rapid norepinephrine release resulting in the inhibition of K^+ channels [3]. Some of these effects may be related to or mediated by the effects of Propranalol on the neural Renin-angiotensin system. A brief discussion pertaining to Propranolol is considered below.

Propranolol

It has been observed that Propranolol has a significant influence on secondary prevention of post-traumatic stress disorder. Propranolol was developed as a beta-adrenergic receptor blocking agent but has been found to have a useful effect on post-traumatic stress disorder. There are at present no fully explainable suggested mechanisms for this action. The following is a suggestion of a possible mechanism. It is known that AngII is involved with the regulation of emotional responses [4]. Both AngII and Propranolol are involved with the neurotransmitter norepinephrine so that it may be conjectured that the Propranolol does not directly influence post-traumatic stress but blocks the effect of AngII on the neurons. Another possible candidate for performing the memory changes would be AngIV. More research is needed to look into details of the effect of the various angiotensin forms on the brain.

Effects of AngII

Thus there are two events related to the AngII system which can be simulated at this time. These are the contraction of the arterioles and the Effect of the AngII directly on the ventricle.

Simulation 1 - Contraction of the Arteriole and Anti-Hypertensive Drugs

To enable an appreciation of the complexity of a systems simulation the two processes will be described in some detail. This simulation starts with the arrival of the sodium signal (the action potential) at the nerve ending known as the axon terminal [25]. The axon terminal contains chemical messengers (neurotransmitters) that are released into the cleft between the axon and the muscle. The neurotransmitters activate receptors on the muscles around an arteriole. Once a receptor is activated it opens a channel that allows calcium to enter the muscle. The calcium subsequently causes the muscle to constrict around the arteriole, thus increasing the blood pressure. There are four stages in the simulation. The calculations were undertaken using different length scales, a continuum scale, a Monte Carlo scale and a molecular scale. In addition, the calculation was undertaken on two inter-connected computers simulating a multi core computer. The first stage simulated the motion of the sodium ions along the axon using a dynamic Monte Carlo method. The second stage was the diffusion of the neurotransmitters across the cleft. This involved a continuum calculation of the diffusion process. The third stage was the interaction of the neurotransmitters and drug molecules with the receptor on the muscle. If the drug molecule attaches to the receptor then the neurotransmitter molecules could not open the channel. The interactions with the receptor were undertaken using a direct simulation method known as molecular dynamics. The final stage involves the entry of the calcium molecules into the receptor. Again a Monte Carlo solution method was used. Stages one , two and four were undertaken on one computer, stage three was undertaken concurrently on a second computer and the results sent to the other computer when required.

The ionic fluid can be considered to be contained in a cylindrical shell between two cylinders. The lower end is closed to ions and only neurotransmitter vesicles can pass through. Ions are reflected from the axon ending with a coefficient of restitution of zero, i.e., all kinetic energy is given up. The free ionic concentrations can be estimated from [26]. The diameter of the axon at the nerve ending was estimated to be 0.3 μm a value between that of the rabbit vagus which is 0.7 μm and that of the garfish olfactory nerve which is 0.2 μm^2. The upper end was open and if an ion passed out a new ion was placed at the x (along the axon) and r (radial location) locations that the departing ion would have attained if it were reflected from the end. The angular position (r) of the new ion was chosen at random. The inner cylinder was permeable to ions and if an ion passes through the inner cylinder it was replaced by a randomly chosen ion positioned as if reflected from the inner cylinder. Ions entering through channels at the axon wall are bare ions and located at the x and r position of the colliding ion. The angular position was also selected at random. Examination of the ion velocities indicated that they must be restricted to a region near the axon wall to process in less than a msec of the arrival of the action potential and the release of the transmitters. One possibility is that the ions are held to the wall by an electrostatic force. This would require the simulation of the proteins that compose the wall. For the present simulation it was assumed the ions are restricted to a region near the wall. The diameter of this inner core was initially chosen as 2700 Å.

The Monte Carlo solution method [27] was used in two regions of the solution. One was in the ion motion in the axon and the other was in the region close to the muscle where the neurotransmitters activate the receptors on the muscle. The two different regions require different applications of the technique. In the axon motion, the ion densities were sufficiently large that consideration of the ion interaction with each other must be considered. In the region near the muscle, the density of the neurotransmitters is so low that the neurotransmitters will not interact and the solution was obtained using the motion of a test particle in a fluid background first proposed by Chandrasekhar [28].

The initial assignment of ion positions was random in the radial and angular positions. Initially the x positions were assigned at random. It was found that some cells would have concentrations of K^+ lower than the average value. This generated a higher potential than expected in regions of the axon and opened sodium channels ahead of the action potential. It was necessary to use an initial uniform distribution of all ions along the axon. The velocities of the ions were set with a Maxwellian distribution appropriate to the temperature. The velocity of the action potential ions were again chosen from a Maxwellian distribution with an average velocity four times the average velocity of the normal axon ions.

The details of the procedure are presented in [25]. However the type of problem to be overcome was in the final stage involving collisions of the calcium ions with the receptor and a probability of entry used. The calcium was uniformly distributed throughout the volume. The concentration of calcium ions in skeletal muscle [26], would not produce a significant number of calcium ions near the muscle. There is some evidence that the concentration of calcium increases near the active sight [29]. Thus the concentration of Ca^{++} was increased. Again the test particle Monte Carlo method was used. The Ach is released from the receptor in 2 msec and the muscle is activated every 250 msec, [30] and thus release is not an impediment.

A typical type of result that could be useful to the pharmaceutical industry is shown in Figure 1. It can be seen that if the average dose of Amlodipine was distributed uniformly throughout the body (uniform dosage) then it would be ineffective in reducing the calcium ions absorbed. Thus it is necessary for the local concentration of Amlodipine in the vicinity of the arteriole to greatly increase. This may be achieved by an extra ingredient in the dose.

In the other example involving the left ventricle of the heart there is considerably more data available so that the simulation can be produced with more confidence. It is known that [20] the effect of the hormone AngII on the heart and vascular system is to initially cause hypertension and subsequently pressure overload [32]. Ang II has the effect of changing the heart shape, size and function (remodeling) [33]. The changes are due to hemodynamic load and neurohormonal activation. The AngII interacts with receptors (known as ATR_1) on the ventricle and vascular walls. The interaction occurs at certain docking sites [34, 35] on the receptor. The effect of the AngII is to instigate reprogramming of cardiac gene expressions resulting in molecular, cellular and interstitial changes [36, 32]. Although the mechanism whereby the gene reprogramming occurs is unknown, proposed mechanisms [37, 32] are discussed below. There are a number of AngII antagonists [38] (known as ARBs or ARAs), which act to block the effects of AngII by docking at sites of the receptors different to the AngII sites. As well as reducing blood pressure, the ARBs are known to reverse reprogramming of the myocardium [32]. Only the diastole phase of the blood flow in the left ventricle will be discussed here although the full cycle analysis has been presented [39]. The initial flow simulation was at a macroscopic level and involved the whole ventricle. The macroscopic flow provided input to the mid-scale flow. The mid-scale flow simulated flow at the albumin, platelet or blood cell size. Finally the microscopic level simulated the interaction of the Losartan and AngII with the ATR_1. The latter calculation has properties similar to the calculation of affinity parameters. Affinity parameters are equilibrium values whereas in the present calculation the values are time dependent. The results of the present calculations indicated that for dynamic situations then time dependent values are required.

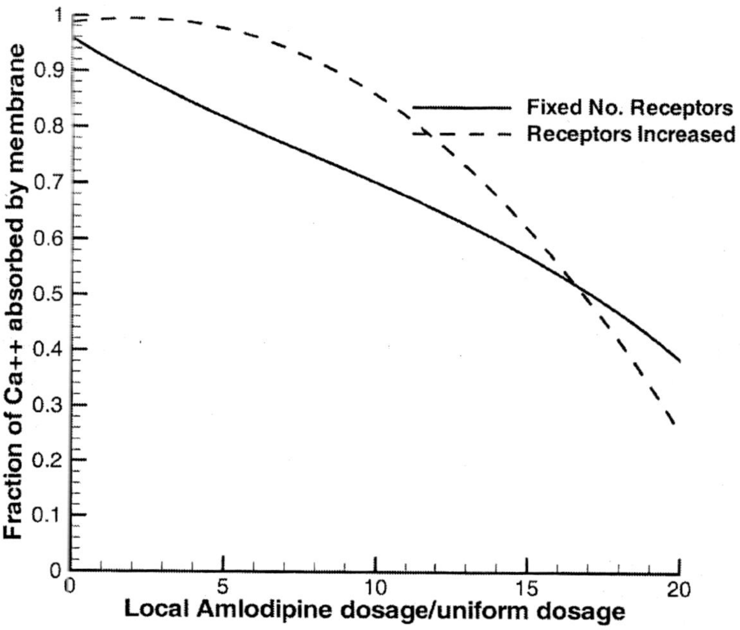

Figure 1 Reduction in Calcium entering Muscle due to Amlodipine Increase.

Simulation 2 - Effect of the Angii on the Ventricle and Hypertrophy Mitigation

The simulation is undertaken on three length scales as mentioned above and described in [25]. The scheme is as shown on the top row of figure 2.

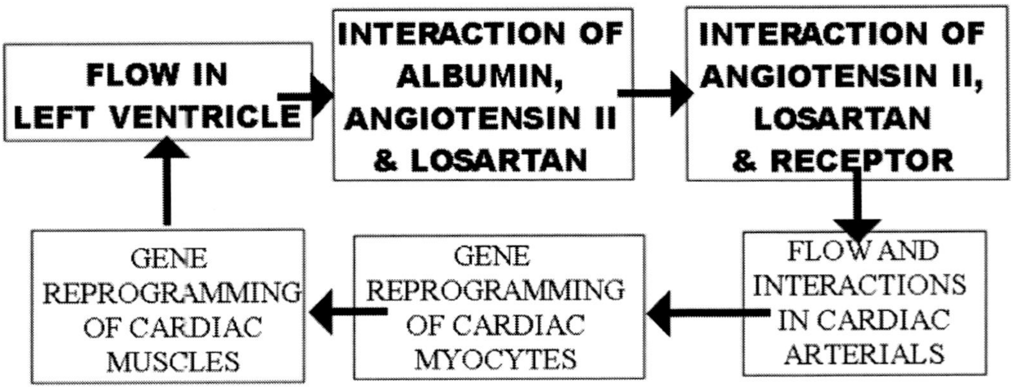

Figure 2. Flow chart of the three stages of simulation.

The lower row of figure 2 shows how the system results in changing the wall structure of the genes in the heart.

Continuum Scale

The flow in the left ventricle is calculated using a macroscopic simulation to obtain the velocity of the blood interacting with the ventricle walls. The state of the ventricle is at the beginning of diastole with the pulmonary and aortic valves closed. The inflow to the atrium is simulated by distributed sources within the atrium. The conservation of mass is maintained by sinks distributed outside the ventricular membrane. The solution follows methods given in [40, 41] using different boundary conditions.

Blood Cell Scale

The Monte Carlo method has been described previously in [25] and in the present case the blood is considered to be composed of water, erythrocyte, albumin, angiotensin II and Losartan. This acts as an intermediate scale between the continuum and molecular level.

Molecular Scale

The AngII and the drug Losartan interact with the receptors on the heart cell muscles. The extra cellular and upper membrane portion of the ATR_1 is shown in figure 3.

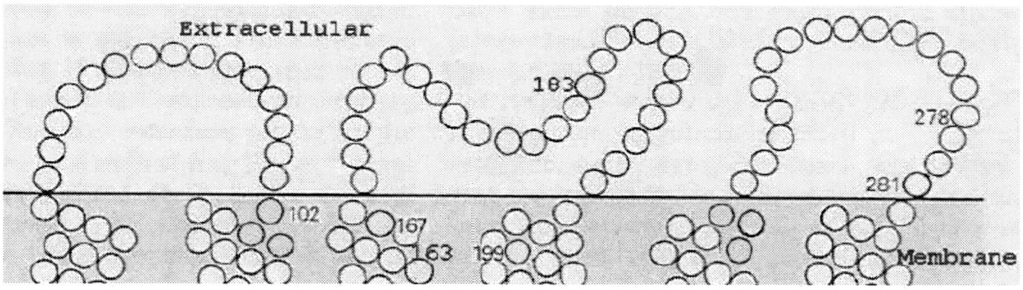

Figure 3. Binding sites for AngII and Losartan on ATR_1.

The numbers are the appropriate amino acid binding sites. The binding sites for AngII are 183,199,278 and 281. The amino acid sites on AngII were chosen as Asp^1, Arg^2 and Phe^8. The binding sites for Losartan were 102,163,167 and 199. As the active metabolite EXP3174 of Losartan differs only in the carboxyl group, it was decided to assume that it was this group that bound to the receptor. As mentioned above the receptors normally are not in the active state (R*) and may be in a state of flux between the inactive state (R and R*) [35]. The mechanism for activation is presently unknown although it is thought that the metabolites AngII and angIV may be instrumental in activation [24].

The procedure followed for the interaction was similar to that in [25]. If a Losartan or AngII molecule would hit the lower boundary in the next time step then the probability of a collision was calculated using the density of ATR1 which in a normal heart is 7.75 fmol/mg protein [42]. A random number was used to decide if a collision would occur. This random selection of deciding if a collision would occur provides provision for activation of the receptor. The normal level of AngII in the blood appears to be about 50 fmol/ml [43]. As shown below, the AngII quickly binds to the receptor and even if it would be necessary for activation to occur before binding, there is ample time for this to occur. If a collision was to be considered then the AngII or Losartan molecule was setup in the computer at a random position a short distance from the receptor. The molecule was given a random rotation in three dimensions appropriate to the temperature. Molecular dynamics [25] was used to determine if the appropriate atoms of the interacting molecule would come within 1.5 Angstroms of an appropriate docking site. Docking was then assumed to occur and the receptor was removed from the system.

There are approximately one hundred times more drug molecules in the domain of the left ventricle compared to the number of receptors. If the number of drug or AngII molecules fell below the initial uniform distribution due to the sufficiently large number of these molecules in the left ventricle it was assumed that when a molecule was removed by a receptor it would be replaced by diffusion. If due to convection the number of molecules was greater than uniform distribution then it was not replaced. A typical result from the calculation of drug interaction with the receptors is shown in figure 4. When a drug molecule docks with a receptor then it is removed as shown by the circles figure 4.

There are 15 receptors blocked due to Losartan and 6 due to angiotensin II at the end of the computational period. Thus the drug (ARB) docks more readily than the angiotensin II. Additional drug molecules can enter the Monte Carlo region by diffusion and convection. As mentioned above, there are approximately 100 times as many drug and AngII molecules as receptors in the left ventricle, myocardium (MC). Thus if the concentration in the MC region is less than in the main fluid and a molecule docks with the receptor, then that molecule is replaced by a new molecule near the top of the region. These are the delta shaped data points representing the molecules convected in due to the bulk fluid motion. There is also transport by diffusion calculated from the continuum calculation and these are the square data points in the graph. Thus it is possible for the number of molecules to have a greater concentration in the left ventricle due to convection than in the bulk fluid (diffusion). Under these conditions the molecules are not replaced when they dock with the membrane.

Once it had been established that gene changes have been initiated it is necessary to follow the development of the disease. Drugs such as Losartan do not ‘block’ receptor channels as with calcium channel blockers. As shown above drug molecules are attached to various sites on the cell wall and inhibit the up regulation of genes by the AngII. It is unknown exactly how drugs like Losartan (ARB) work. A recent survey [44] indicates that there are probably several hundred genes up regulated during hypertrophy. Specific changes in cardio-myocytes during the development of cardiac hypertrophy include the induction of ‘immediate-early’ genes (heat shock protein genes and proto-oncogenes), changes in gene expression (re-expression of fetal genes) and increased protein synthesis [5]. Thus it was necessary to use a computation system that used the experimental data concerning only four typical genes and allow for the possible effects of the neglected genes. Further, there must be a certain probability that the interaction of an AngII molecule with a receptor will result in the growth of a cell. One technique used to satisfy both these requirements was a supervised linear Gaussian state-space model known as a linear dynamical system [45]. This enabled the calculation of the spread of the disease. If the chemical and biological mechanisms involved

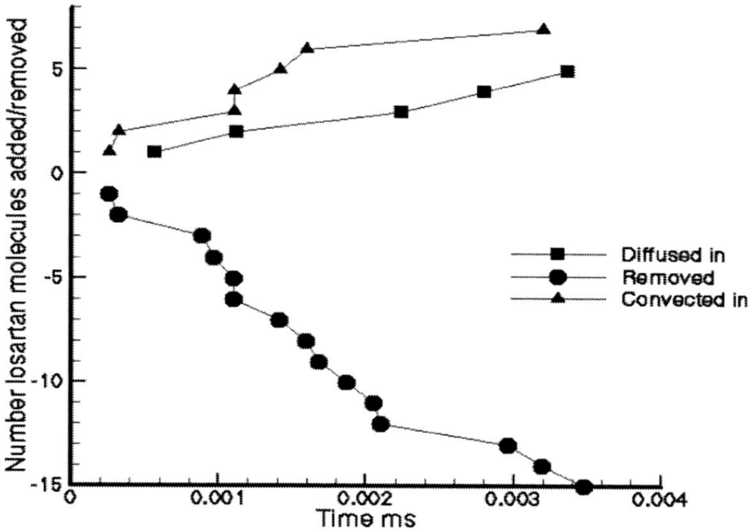

Figure 4. The docking of Losartan with a convective velocity 1cm/sec.

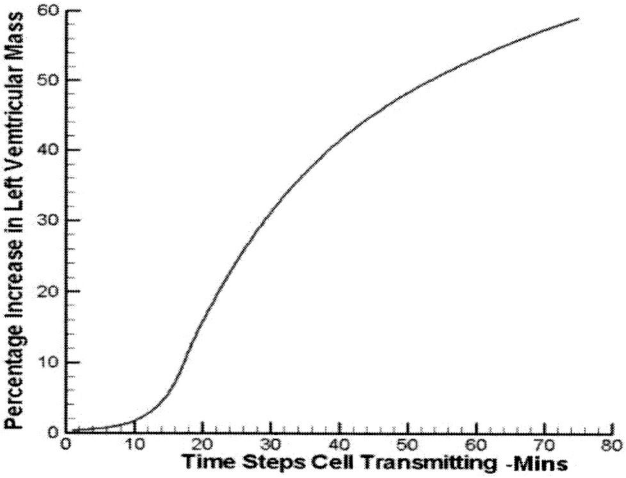

Figure 5. The predicted percentage increase in left ventricular mass.

with the different gene up regulations were known, then the effect of the various gene expression changes on the appropriate tissues could be determined. At present these mechanisms are unknown. Lacking such data it would seem that a probabilistic approach would be the most profitable. A technique called cellular automata credited to John von Neumann [46] might be suitable for the present purpose. Although the term at first might seem to indicate a system that is apt to act by itself (automata), in the present context the meaning is that discrete (particulate) cells respond to a stimulus according to a predetermined set of rules. These rules are usually local in nature and a recent description of the method is presented in [47].

Using these techniques it was possible calculate for example the relation between the cell transmitting time and the 12 month increase in left ventricular mass [48] as shown in Figure 5. As the normal cell transmission time is about 12 minutes [49] it is suggested that the effect of drugs such as Losartan is to reduce the cell transmission time to the normal value and thus help the medical treatment.

Conclusions

In silico drug development can save a tremendous amount of effort and make drug development a much less expensive venture. To achieve that goal, the mechanisms whereby drugs interact with the relevant organs needs to be studied and this must be done in consort with related effects. The potential and capabilities of in silico drug design has been growing along with computational capabilities that now routinely include multi-core processing that can handle multi-threading using JAVA and other means. The complexity and nature of biological systems that can be modeled using in silico techniques has improved to the point where it can now generate detailed and useful biology data. Such modeling and drug design must be aware of the fact that drug molecules act in a systemic manner and thus always have the potential for generating contra indications. The present chapter has described the complexities involved in modeling drug design and the need for multi-scale modeling. The Renin-Angiotensin system has been described and details and results pertaining to the modeling of two specific effects of angiotensin II, namely the contraction of the arteriole and Effect of the angiotensin II on the ventricle, are presented. In Silico biology and drug design have great potential and will be used more extensively and routinely as more and more biochemistry and cell biology data become available.

References

[1] Hines M, Moore J.W., Carnevale, T. Neuron: for empirically-based simulations of neurons and networks of neurons, 1984-present, Available from *www.neuron.yale.edu/neuron/*

[2] Loew L, The virtual cell program (1999), University of Connecticut.

[3] von Bohlen O, Halbach O, Albrecht D (2006) The CNS renin-angiotensin system. *Cell Tissue Res.* Nov;326(2):599-616.

[4] O. von Bohlen and Halbach, (2005) The Renin-Angiotensin System in the Mammalian Central Nervous System, *Current Protein and Peptide Science*, 6 (In press)

[5] C Sernia, L Sinton, WG Thomas, and W Pascoe (1985) Liver angiotensin II receptors in the rat: binding properties and regulation by dietary Na+ and angiotensin II *Journal of Endocrinology,* 106, 1, 103-111

[6] Speth, R.C., Daubert, D.L., and Grove, K.L. (1999) Angiotensin II: a reproductive hormone too? *Regul. Pept., 79*, 25-40.

[7] Nielsen, A.H., Schauser, K.H., and Poulsen, K. (2000) Current topic: the uteroplacental renin-angiotensin system. *Placenta, 21*, 468-477.

[8] Regoli, M., Bendayan, M., Fonzi, L., Sernia, C., and Bertelli, E. (2003) Angiotensinogen localization and secretion in the rat pancreas. *J. Endocrinol., 179*, 81-89.

[9] Leung, P.S., Chan, W.P., Wong, T.P., and Sernia, C. (1999) J. Expression and localization of the renin-angiotensin system in the rat pancreas. *Endocrinol.*, 160, 13-19.

[10] Engeli, S., Schling, P., Gorzelniak, K., Boschmann, M., Janke, J., Ailhaud, G., Teboul, M., Massiera, F., and Sharma, A.M. (2003) The adipose-tissue renin-angiotensin-aldosterone system: role in the metabolic syndrome? *Int. J. Biochem. Cell Biol., 35*, 807-825.

[11] Engeli, S., Negrel, R., and Sharma, A.M. (2000) Physiology and pathophysiology of the adipose tissue renin-angiotensin system. *Hypertension, 35*, 1270-1277.

[12] Bader, M., Peters, J., Baltatu, O., Muller, D.N., Luft, F.C., and Ganten, D. (2001) Tissue renin-angiotensin systems: new insights from experimental animal models in hypertension research. *J. Mol. Med., 79*, 76-102.

[13] Sjolie, A.K. and Chaturvedi, N. (2002) The retinal renin-angiotensin system: implications for therapy in diabetic retinopathy. *J. Hum. Hypertens., 16*, S42-S46.

[14] Wagner, J., Jan Danser, A.H., Derkx, F.H., de Jong, T.V., Paul, M., Mullins, J.J., Schalekamp, M.A., and Ganten, D. (1996) Demonstration of renin mRNA, angiotensinogen mRNA, and angiotensin converting enzyme mRNA expression in the human eye: evidence for an intraocular renin-angiotensin system. *Br. J. Ophthalmol., 80*, 159-163.

[15] Jurklies, B., Eckstein, A., Jacobi, P., Kohler, K., Risler, T., and Zrenner, E. (1995) The renin-angiotensin system--a possible neuromodulator in the human retina? *Ger. J. Ophthalmol., 4*, 144-150.

[16] Kohler, K., Wheeler-Schilling, T., Jurklies, B., Guenther, E., and Zrenner, E. (1997) Angiotensin II in the rabbit retina. *Vis. Neurosci, 14*, 63-71.

[17] Zhang, X., Lassila, M., Cooper, M.E., and Cao, Z. (2004) Retinal Expression of Vascular Endothelial Growth Factor Is Mediated by Angiotensin Type 1 and Type 2 Receptors. *Circulation.* 109: 1536-1542

[18] Ganong, W.F., Deschepper, C.F., Steele, M.K., and Intebi, A. (1989) Renin-angiotensin system in the anterior pituitary of the rat. *Am. J. Hypertens., 2*, 320-322.

[19] Ganong, W.F. (1993) Blood, pituitary, and brain renin-angiotensin systems and regulation of secretion of anterior pituitary gland. *Front Neuroendocrinol., 14*, 233-249.

[20] Saavedra JM, Benicky J, Zhou J. *Mechanisms of the Anti-Ischemic Effect of Angiotensin II AT(1) Receptor Antagonists in the Brain. Cell Mol. Neurobiol.* 2006 Nov;26(7-8):1097-109.

[21] Mark G, Luciano, David J. Skarupa, Amy M Booth, Arcangela S Wood, Christine L Brant and Martha J Gdowski, *Cerebrovascular Adaptation in Chronic Hydrocephalu, Journal of Cerebral Blood Flow and Metabolism.* (2001) 21, 285–294.

[22] Harding JW, Sullivan MJ, Hanesworth JM, Cushing LL, Wright JW., Inability of [125I] Sar1, Ile8-angiotensin II to move between the blood and cerebrospinal fluid compartments, *J. Neurochem. 1988 Feb;50(2):554-7.*

[23] Pan HL (2004) Brain angiotensin I1 and synaptic transmission. *Neuroscientist.* 10:422-431

[24] De Gasparo M, Catt KJ, Ing T, Wright JW and Unger TH. 2000 International Union of Pharmacology. XXIII. *The Angiotensin II Receptor Pharma. Revs.* 52, 415-472

[25] A.K.Macpherson , S.Neti and P.A.Macpherson (2001) A rapid procedure for initial drug evaluation, *Phys. In Med. And Biol.*46,6.

[26] Hille,B, 1991 *Ionic Channels of Excitable Membranes*, 2nd edition, Sinauer Assoc., Mass.

[27] Macpherson,A.K. (1971) Rotational Temperature Profiles of Shock Waves in Diatomic Gases, *J.F.M* 49,337,1971.

[28] Chandrasekhar,S. 1942 *Principles of Stellar Dynamics*, Uni. Of Chicago Press, Chicago

[29] Nicholls,J,G, Martin, A.R. and Wallace,B.G. 1992 *From Neuron to Brain,* Sinauer Assoc., Sunderland, Ma.

[30] Bennett M.R.1972, Autonomic *Neuromuscular Transmission,* C.U.P. Cambridge.

[31] Silverman, R.B.1992, *The Organic Chemistry of Drug Design and Drug Action* Academic Press, New York 162-170.

[32] Dostal DE, (2000), The cardiac Renin-angiotensin system: novel signaling mechanisms related to cardiac growth and function, *Reg. Peptides*, 91, 1-11.

[33] Cohn JN, Ferrari R, Sharpe N 2000, Cardiac remodeling-concepts and clinical implications: a consensus paper from an international forum on cardiac remodeling. Behalf of an International Forum on Cardiac Remodeling , *J. Am. Coll Cardiology*, 35(3):569-82.

[34] Boucard AA,Wilkes BC,Laporte SA, Escher E,Guillemette G and Leduc R 2000, Photolabeling Identifies Position 172 of the Human AT_1 Receptor as a Ligand Contact Point: Receptor-Bound Angiotensin II Adopts and Extended Structure BioChem 39,9662-9670.

[35] Noda K,Feng Y-H,Liu X-p,Saad Y, Husain A and Karnik SS 1996 *The Active State of the AT_1 Angiotensin Receptor Is Generated by Angiotensin II Induction.* 35,16435-16442(574.19205 B6152)

[36] Paradis P, Dali-Youcef N, Paradis FW, Thibault G and Nemer M. 2000 Overexpression of angiotensin II type I receptor in cardiomyocytes induces cardiac hypertrophy and remodeling. PNAS. 97,2 931-936.

[37] Kim S and Iwao H 2000, Molecular and Cellular Mechanisms of Angiotensin II mediated Cardiovascular and Renal Diseases. Pharma. Revs, 52(1) 11-34.

[38] Burnier M and Brunner HR 2000, Angiotensin II receptor antagonists. The Lancet,355, 637-645.

[39] Macpherson, A.K. Neti, S. and Macpherson, P.A.(2005) Mechanical Stress and Hypertrophy , *in Modelling in Medicine and Biology VI.*, Eds W.Ursino, C.A. Brebba, G.Pontrelli and E.Magosso,WIT press, p 171-181.

[40] Peskin CS, 1977, Numerical Analysis of Blood Flow in the Heart, J. Comp. Phys 25,220-252.

[41] Peskin CS and Printz BF, 1993 Improved Volume Conservation in the Computation of Flows with Immersed Elastic Boundaries J.Comp. Phys 105, 33-46.

[42] Zisman Ls., Asano K., Dutcher D.L., Ferdensi A., Robertson A.D., Jenkin M., Bush E.W., Bohlmeyer T., Perryman M.B. and Britow M.R. 1998 *Differential Regulation of Cardiac Angiotensin Converting Enzyme Binding Sites and* AT_1 *Receptor Density in the Failing Human Heart Circulation.* 98,(17) 1735-41.

[43] Schuijt MP, van Kats JP, de Zeeuw S, Dunker DJ, Verdouw PD, Schalekamp MA, Danser AH 1999, *Cardiac interstitial levels of angiotensin I and II in the pig* , *J. Hypertens.* 17 (12pt2) 1885-91

[44] Liew C.C. Heart Failure a Genomics Approach in Proteomic and Geonomic Analysis of Cardiovascular Disease Eds J.E. Van Eyk and M.J.Dunn Wiley ,45-60,2003

[45] Rangel C., Wild D.L. and Falciani F. 2002, Modelling biological responses using gene expression profiling and linear dynamical systems, Proceedings of the 2nd International Conference on Systems Biology. Madison, WI: OmniPress, 248-56.

[46] Poundstone W. 1987 The Recursive Universe: *Cosmic Complexity and the Limits of Scientific Knowledge.* Oxford: OUP..

[47] Wolfram S.2002 *A New Kind of Science*, Champagne, Illinois, Wolfram Media,Inc

[48] Macpherson, A.K., Macpherson, P.A., and Crosby,L (2006) Application of Gene Expression Data and Cellular Automata To Predict Disease Progress in a Cardiac Hypertrophy Model, presented at 2nd Modern Drug Discovery and Development Summit,

[49] Morré D.J., Chueh P-J, Pletcher J., Tang X., Wu L-Y and Morré DM. 2002, *Biochemical Basis for the Biological Clock, Biochemistry*,41,40, 11941-45.

In: Drug Design Research Perspectives
Editor: Stanley P. Kaplan, pp. 25-37
ISBN: 978-1-60021-823-1

Chapter II

Artificial Zinc Finger Proteins: Desired DNA Sequence-Specific Function Toward Gene Regulation

Miki Imanishi*
Institute for Chemical Research, Kyoto University, Uji, Kyoto 611-0011, Japan

Abstract

Regulation of a target gene at will is one of the most prospective themes in the post-genomic era. An artificial transcription factor with desired DNA binding specificity could work as a powerful tool and drug to control target genes. The C_2H_2-type zinc finger motif is one of the most typical DNA binding motifs found in a lot of transcription factors. The zinc finger domain is extremely attractive for design of artificial transcription factors with desired DNA binding specificities because of its characteristic DNA binding properties: (1) recognition of 3-4 bp per one motif, (2) tandemly connected modular structure, and (3) binding to non-palindrome sequences as an independent monomer. In order to target a specific gene by using a protein-based drug, it is required for a protein to bind to any desired DNA sequences. By randomizing amino acid residues important for DNA recognition, a number of successful zinc finger peptides have been selected to bind to the specific DNA sequence and to control gene expression. For the purpose of selective targeting in the huge genome (3×10^9 bp), long DNA sequence (18 bp) should be specifically recognized by a protein-based drug. We have shown that 6-zinc finger peptides in which two 3-zinc finger domains are connected with the consensus linker bind to the expected 18 bp sequences containing continuous two target sites for the original 3-zinc finger peptides. Artificial 6-zinc finger peptides with various kinds of linkers have also been developed and shown to skip DNA sequences. Artificial zinc finger domains can be fused with functional domains such as transcriptional regulatory domains or DNA cleavage domains for therapeutic use. Such zinc finger proteins have been shown to control specific gene expression or to induce homologous recombination. Herein, effects of linker design of artificial 6-zinc finger peptides on

* Miki Imanishi: Tel: +81-774-38-3212, Fax: +81-774-32-3038, E-mail: imiki@scl.kyoto-u.ac.jp

DNA binding are mainly described based on our experimental results. I also review the latest trends in artificial zinc finger proteins toward gene therapy.

Introduction

Zinc finger proteins are one of the most ubiquitous DNA binding proteins. There are more than 10 different classes of zinc finger motifs depending on the way of coordinating zinc ions, for example, C_2H_2, C_3H, C_4, and C_6 (Matthews and Sunde, 2002). Some of zinc finger proteins are known to interact not only with DNA but also with RNA, proteins and lipids. Among them, the C_2H_2-type zinc finger motif is the most major DNA binding motif in human transcription factors. It is estimated that the C_2H_2-type zinc finger proteins are corded in about 700 genes in human genome (Tupler et al., 2001). The motif commonly contains the sequence of the type, (Tyr, Phe)-X-Cys-$X_{2\text{-}4}$-Cys-X_3-Phe-X_5-Leu-X_2-His-$X_{3\text{-}5}$-His, where X represents relatively non-conserved amino acids (Miller et al., 1985). Each finger is approximately 30 amino acid residues long and folds into a compact $\beta\beta\alpha$ structure by coordination of a zinc ion with the conserved two cystein and two histidine residues in a tetrahedral fashion (figure 1). This fold is stabilized by hydrophobic interactions (Lee et al., 1989; Klevit et al., 1990; Omichinski et al., 1990).

The X-ray crystal structures of the C_2H_2-type zinc finger protein Zif268-DNA complexes clearly revealed the characteristic DNA binding mode of C_2H_2-type zinc finger proteins as follows. (1) One finger typically recognizes contiguous 3 or 4 base pairs (bp) of DNA sequence. The α-helix of a zinc finger motif is called the "recognition helix". The specific three amino acid residues on the α-helix (positions -1, 3 and 6) are responsible for contacting with contiguous three DNA bases. In addition, the amino acid residue at the position 2 is also revealed to contact with a DNA base on the other strand of the DNA. The base contacts are mediated by their side chains (Elrod-Erickson et al., 1996, 1998) (figure 2). (2) Multiple zinc finger motifs are tandemly repeated by simple covalent linkage, namely connected by the consensus linker region, the TGEK(R)P sequence. In general, more than two tandem zinc fingers are required to obtain DNA binding affinity. (3) Zinc finger domains bind to DNA as a monomer. Therefore, they recognize sequences of asymmetric base pairs unlike other nucleic acid recognition motifs such as basic leucine-zipper and helix-turn-helix. In addition, it is possible to create sequence-specific functional zinc fingers by fusing zinc finger domain with various effector domains.

These properties of C_2H_2-type zinc finger proteins strongly indicate possible strategy to obtain proteins with novel DNA binding specificities and functions by "mix, match, and/or link" zinc fingers (Desjarlais and Berg, 1992, 1993). Based on the first property, novel zinc fingers have been selected by randomizing specific amino acid residues to bind to desired DNA triplets. Artificial 6-zinc finger proteins that bind to longer DNA sequences have been developed taking advantage of the second property. Linking with other functional domains, sequence specific transcription factors and nucleases have been created. To develop design strategies of zinc finger proteins is important for gene regulation by artificial sequence specific DNA binding proteins.

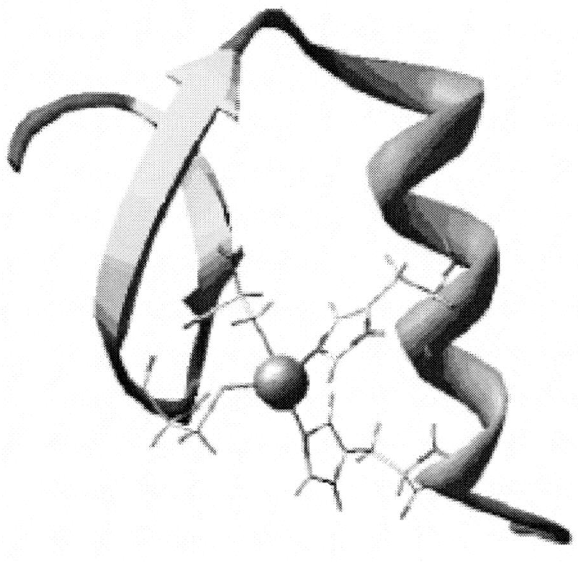

Figure 1. Structure of a C_2H_2-type zinc finger motif, Sp1 finger 3 (Narayan et al., 1997). By coordination of a zinc ion, the peptide folds into a ββα structure.

1. Design of Zinc Finger Domains: Recognition of Desired DNA Triplets by Zinc Finger Motifs

As described above, one zinc finger recognizes contiguous 3 or 4 bp and the base recognitions are mediated by interactions with side chains of amino acid residues at the specific contacting positions on the α-helix (figure 2). Such a concept, "recognition code", makes it possible to arrange amino acid residues in order to contact with any desired DNA bases (Desjarlais and Berg, 1992, 1993, Choo and Klug, 1994, Rebar and Pabo, 1994). Selection methods such as phage display have been used to create novel sequence specific zinc finger proteins. In these selections, the key DNA-contacting residues on the α-helix are randomized while leaving all other regions of the structure unchanged. Sequential selection strategy was developed by Pabo's group (Greisman and Pabo, 1997, Wolfe et al., 2001). In the method, the terminal finger of 3-fingers is randomized and selected at first, while other two fingers are unchanged. The same steps are repeated for the other fingers. In parallel method, the middle finger of 3-fingers is randomized to form a library of variants towards desired DNA triplets (Segal et al., 1999). The selected and optimized fingers are assembled to form 3- or 6-zinc finger proteins that bind to contiguous DNA triplets. So far, many zinc finger motifs for recognition of most of 5'-GNN-3', 5'-ANN-3' and 5'-CNN-3' triplets have been reported (Segal et al., 1999, Dreier et al., 2000, 2001, 2005, Mandell and Barbas, 2006). As another strategy, Klug et al. introduced a bipartite library approach, in which one and a half fingers are randomized and selected at a time (Isalan et al., 2001). Other screening systems using bacteria, yeast, or ribosomal display, et al. have been also developed to create zinc fingers with novel DNA binding specificities (Joung et al., 2000, Cheng et al., 1997, Hughes et al., 2005, Ihara et al., 2006).

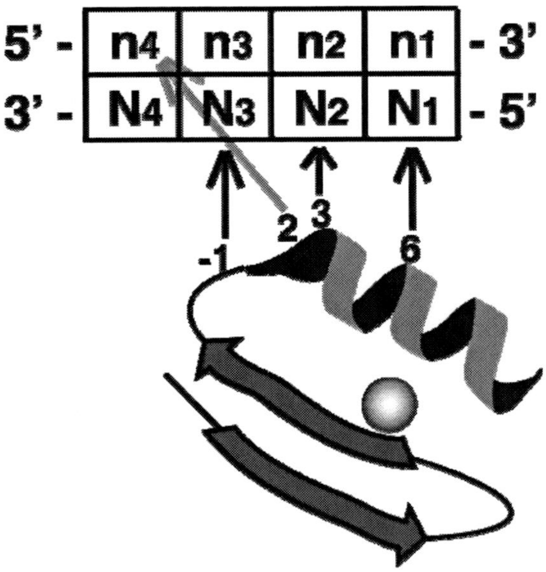

Figure 2. A typical DNA recognition mode by a C_2H_2-type zinc finger motif. Amino acid residues in positions of -1, 3 and 6 on α-helix interact with continuous three DNA bases. The amino acid in position 2 interacts with a DNA base in the complementary strand.

2. Creation of Multiple Zinc Finger Proteins: Recognition of Long DNA Sequences and Gain of Function

Considering the size of the human genome (3 x 10^9 bp), it is required for sequence selective DNA binding proteins to recognize more than 16 bp (4^{16} = 4 x 10^9) in order to target a unique site. However, recognition of a long DNA sequence is a rare function in most DNA-binding proteins with such domains as bZIP, helix-turn-helix, or helix-loop-helix. On the other hand, a natural C_2H_2-type zinc finger protein generally consists of multiple zinc finger motifs recognizing DNA triplet and typically binds to continuous 5-10 bp using its two or three continuous zinc finger motifs. Therefore, it would be possible to create artificial zinc finger proteins that recognize more than 18 bp by connecting more than six independent zinc finger motifs or multiple naturally-occurring 3-zinc finger domains corresponding to the length of target DNA (Liu et al., 1997; Kamiuchi et al., 1998; Hirata et al., 2005). We have created 6-zinc finger proteins by connecting two 3-zinc finger domains derived from the DNA binding domain of human transcription factor Sp1. The connection region, namely "linker" between two Sp1 DNA binding domains, has been changed to be long, flexible, cationic, structural, or functional (figure 3). In addition, the effects of the artificial linker on the characteristics of DNA binding of the 6-zinc finger proteins have been examined from the point of view of DNA structural change, DNA binding kinetics, selectivity to discontinuous target DNA sequence. It has been clarified that not only a DNA recognition helix but also a linker region would be an important target for design of novel DNA binding zinc finger proteins as described below (Imanishi et al., 2000, 2002, 2005; Nakatsukasa et al., 2005; Yan et al., 2007).

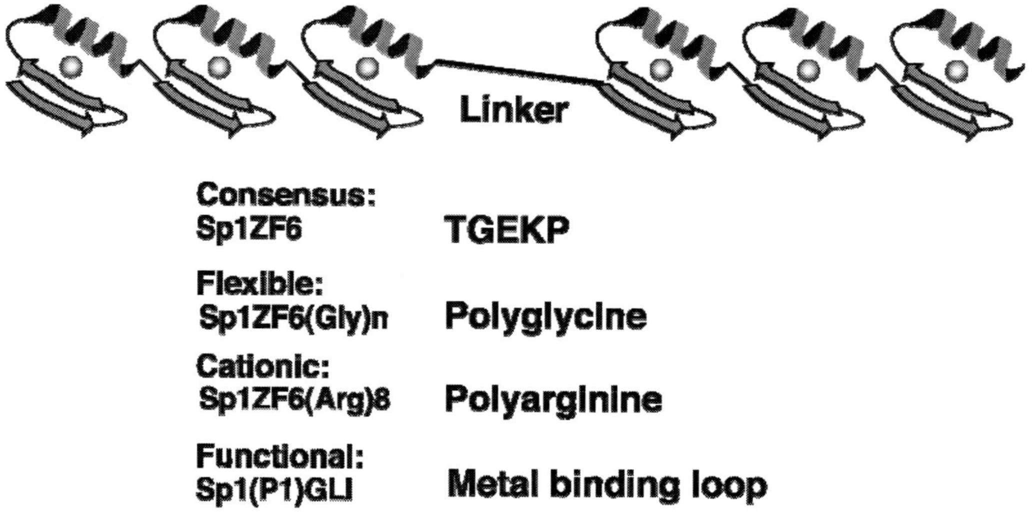

Figure 3. Artificial multi zinc finger proteins with various linkers.

2.1. Targeting Continuous Binding Sequences

Human transcription factor Sp1 is well known to be a ubiquitous transcription factor (Kadonaga et al., 1987). Sp1 contains three zinc finger motifs (Sp1ZF3) at C-terminal region as a DNA binding domain. Sp1ZF3 binds to a GC-rich DNA sequence (5'- GGGGCGGGGC-3') called GC-box. The three zinc finger motifs are tandemly connected by a highly conserved linker, TGEK(R)P. This linker functions in controlling the orientation and spacing of adjacent zinc fingers (Laity et al., 2000). To examine the possibility of artificial zinc finger proteins to recognize more than 16 bp, we have connected two, three, and five Sp1ZF3s by the consensus linker and created Sp1ZF6, Sp1ZF9 (Kamiuchi et al., 1998) and Sp1ZF15 (Hirata et al., 2005). The multiple zinc finger proteins bind to the expected long DNA sequences, 2GC, 3GC and 5GC, respectively. It is indicated that the length of DNA binding sequence increases in proportion to the number of zinc finger motifs. Other designed 6-zinc finger peptides based on the Zif268 framework have also been reported to bind to expected long DNA sequences (Liu et al., 1997). Regarding binding affinity, the multiple zinc finger proteins do not show much cooperativity comparing from the original 3-zinc finger unit, maybe because the conserved TGEK(R)P linkers can put an excessive strain on the complex. Kim and Pabo extended the linker length between two 3-zinc finger units by inserting Gly and Ser residues and demonstrated to greatly raise DNA binding affinity (Kim and Pabo, 1998). Neuteboom et al. examined the differences of linker regions between two 3-zinc finger units on targeting to genomic DNA. They concluded that the consensus linker could properly target a contiguous 18 bp sequence in genome (Neuteboom et al., 2006).

2.2. DNA Structural Change by Artificial Zinc Finger Proteins

DNA structural change is one of the most important factors for gene regulation. Therefore, an artificial protein that induces a DNA conformational change is interesting as a transcriptional regulator. We created 6-zinc finger proteins, Sp1ZF6(Gly)$_7$ and Sp1ZF6(Gly)$_{10}$, by connecting two 3-zinc finger units of Sp1 with flexible poly-glycine linkers (Imanishi et al., 2000). As the target DNA, we used two GC-box sequences separated by one helical turn of DNA (10 bp). Sp1ZF6(Gly)$_7$ and Sp1ZF6(Gly)$_{10}$ can induce DNA bending at the intervening region of the two distal binding sites and that the linker length between two 3-zinc finger motifs has a crucial effect on the entire DNA bending direction. To investigate the effect of linker type on DNA-binding properties, we have designed 6-zinc finger proteins with charged linkers, Sp1ZF6(GE)4 and Sp1ZF6(GR)4 (Imanishi et al., 2002). These proteins are also able to induce the DNA structural changes similar to the Sp1ZF6(Gly)$_{10}$ protein. However, the kinetic aspects of their DNA bindings were different in each case. In particular, the dissociation rate of Sp1ZF6(Gly)$_{10}$ was the fastest and that of Sp1ZF6(GR)4 was the slowest. These results indicate that the length of the linker region between two 3-zinc finger domain units affects on DNA structural changes and that the property of the linker regions gives different DNA binding kinetics.

2.3. 6-Zinc Finger Proteins Targeting Discontinuous 18 bp DNA

Though a lot of zinc finger motifs have been developed to contact with various DNA triplets, not all triplets can be recognized by any engineered zinc finger motifs. Therefore, binding to a discontinuous target with non-recognized DNA sequence is meaningful. Pabo et al. synthesized dimeric zinc finger peptides and succeeded in recognizing discontinuous target sequences (Pomerantz et al., 1998; Wolfe et al., 2003). We introduced a flexible long glycine linker between two 3-zinc finger units of transcription factor Sp1, Sp1ZF6(Gly)$_{10}$. Sp1ZF6(Gly)$_{10}$ can bind to discontinuous GC-box sequences separated by 10 bp. However, it binds to continuous GC-box sequences as well with almost same affinity (Imanishi et al., 2005). How to obtain selectivity to discontinuous recognition sequences with particular length of non-recognized gap is challenging.

A *Xenopus laevis* transcription factor TFIIIA contains nine contiguous zinc finger motifs. The NH_2-terminal six fingers of TFIIIA bind to 31 bp of the 5S rRNA gene but finger 4 does not make any specific base contacts with DNA. The crystal structure indicates that the finger 4 bridges over the DNA minor groove, allowing fingers 3 and 5 to bind to the separated DNA sequences in the major groove (Norte et al., 1998). Moore et al. used TFIIIA finger 4 to connect two 3-zinc finger units (Moore et al., 2001). The multi zinc finger peptide with inserted TFIIIA finger 4 showed DNA binding preference to discontinuous sequences with the expected gaps only when the linker is connected following to the 3-zinc finger motifs of

TFIIIA. Inserted between 3-zinc finger units derived from Zif268, TFIIIA finger 4 linker did not show the expected DNA binding preferences. In addition, they introduced a mutated Zif268 finger 2, in which the amino acid residues important for DNA recognition are changed not to bind to DNA. Because most of zinc fingers except for TFIIIA finger 4 would be predicted to span 3 or 4 bp, the multi zinc finger proteins with a linker of a mutated finger were expected to specifically span 3 or 4 bp gap. However, the peptides did not show obvious DNA binding preferences. Interestingly, the 6-zinc finger proteins with zinc finger structured linker bind to DNA with 0-10 bp non-bound gaps with higher affinity than that with flexible linker. The structural stability of the linker is maybe contributed to increase in binding affinity through the entropic factors.

We tried to create a multi zinc finger protein that binds to discontinuous target sequences but not to continuous ones. Eight arginine residues were introduced to connect 3-zinc finger units of Sp1 and Sp1ZF6(Arg)8 was created (Imanishi et al., 2005). Because the side chain of an arginine residue is bulky and positively charged, we expected that the steric hindrance, electrostatic repulsion, and/or nonspecific interaction with DNA might inhibit specific DNA binding to continuous GC-box sequences. Sp1ZF6(Arg)8 specifically bound to discontinuous two GC-box sites by a 10 bp gap. On the other hand, the binding affinity of Sp1ZF6(Arg)8 to continuous GC-box sites was extremely low compared with that to a discontinuous sequence. The bulky and cationic polyarginine linker seems to contribute to the selective binding to the discontinuous GC-box sites. Recently, we introduced the $(EAAAR)_4$ peptide sequence as a linker between two Sp1 zinc finger domains (Yan et al., 2007). The peptide is known to form α-helix and the length of the helix is estimated to be close to that of DNA one helical turn (10 bp). The α-helical content of the $(EAAAR)_4$ linker of the 6-zinc finger protein, Sp1ZF6$(EAAAR)_4$, was demonstrated to be higher in the complex with discontinuous two GC-boxes by a 10 bp gap than a 5 bp gap. In addition, Sp1ZF6$(EAAAR)_4$ showed DNA binding preference to 10 bp-gapped target than 5 bp-gapped target. The results indicate that a structured linker would also be a promissing approach to obtain DNA binding selectivity depending on the position of discontinuous recognition sites.

2.4. Sequence Specific Multiple Zinc Finger Type Endonuclease

A novel sequence-specific peptide can be created by adding a functional domain to a DNA binding domain. As mentioned later, artificial exonucleases have been created by fusing a zinc finger domain with a DNA cleavage domain of a restriction enzyme. We examined the possibility to add DNA cleavage function to the linker region (Nakatsukasa et al., 2005). An analogue of the metal binding loop was used to connect two zinc finger domains derived from Sp1 and GLI. The loop is known to exhibit DNA cleavage activity in the presence of cerium(IV) ions. The zinc finger peptide, Sp1(P1G)GLI, was demonstrated to cleave DNA at the region between two binding sites for Sp1 and GLI in the presence of Ce(IV). The result indicates that a novel sequence specific endonuclease can be created by introducing metal-binding loop as a linker of two zinc finger domains.

3. Artificial Zinc Finger Proteins with Sequence-Specific Functions

In addition to artificial zinc finger peptides with desired DNA binding specificities, many sequence-specific functional zinc finger proteins have been created taking advantage of the third property of DNA binding that a zinc finger domain binds to DNA as a monomer (figure 4).

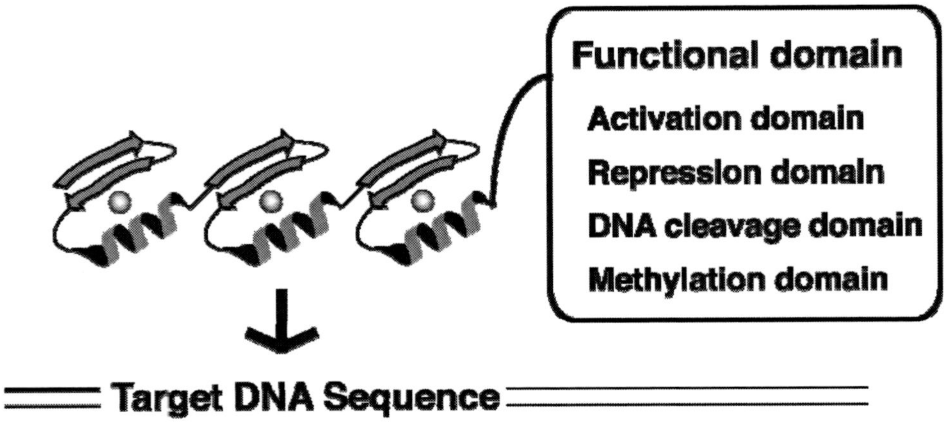

Figure 4. Artificial zinc finger proteins with sequence-specific various functions. A functional domain is attached to a DNA binding zinc finger domain.

By fusing with transcription effector domain, novel zinc finger transcription factors can upregulate and downregulate target genes (Beerli et al., 1998). The herpes simplex virus VP16 domain and the p65 domain of human transcription factor NFκB have been often used as an activator domain, and Kruppel associated box (KRAB) and SID as a repression domain. Even endogenous genes have been regulated by artificial zinc finger transcription factors (Beerli et al., 2000, Liu et al., 2001, Zhang et al., 2000).

A zinc finger nuclease is one of the most attractive tools for future gene therapy. Chandrasegaran's group fused a zinc finger domain with the DNA cleavage domain of the Type IIS restriction enzyme, *Fok*I, and showed that the fusion proteins cleave the DNA sequences at the position close to the zinc finger binding sequence (Kim et al., 1996, Smith et al., 1999). The double strand break induced by artificial zinc finger nucleases was shown to increase the rate of homologous recombination (Bibikova et al., 2003, Porteus and Baltimore, 2003). Urnov et al. succeeded in endogenous human gene correction using artificial zinc finger nucleases (Urnov et al., 2005). They used zinc finger nucleases targeting an X-linked severe combined immune deficiency (SCID) mutation in the IL2Rγ gene. Without any selection, about 20% of mutated chromosomes in a cell line was corrected by homologous recombination induced by double strand cleavage of zinc finger nucleases. Recently it was reported that zinc finger nucleases can drive the addition of not only short fragments but also an 8-kb sequences containing genes into an endogenous locus (Moehle et al., 2007).

4. Future Applications

Gene silencing by targeted DNA methylation is epigenetically very important. Recently trials of DNA methylation are developing by artificial zinc finger domain-methyltransferases. Several kinds of methyltransferases, e.g. M.SssI, M.HhaI, and their variants have been fused to zinc finger domains and tested in vitro, in bacteria, yeast, and mammalian cells (Xu and Bestor, 1997, McNanara et al., 2002, Carvin et al., 2003, Li et al., 2007, Smith and Ford, 2007). Li et al. demonstrated that targeted DNA methylation is effective in repressing Herpes Simplex Virus type 1 infection in cell culture (Li et al., 2007). Smith and Ford succeeded in specific methylation of a chromosomal locus with an integrated zinc finger targeting sequence and they showed that the methylation pattern is inherited through successive cell divisions (Smith and Ford, 2007). In the near future, inherited gene silencing of endogeneous genes would be possible by zinc finger methyltranscerases.

Not only chromosomal DNA, mitochondrial DNA could also be an important target of artificial zinc finger proteins. In general, natural zinc fingers tend to localize in the nucleus even without typical nuclear localization signals. In order to avoid the nuclear import of zinc finger proteins, Minczuk et al. fused a nuclear export signal (NES) in addition to a mitochondrial targeting sequence (MTS) (Minczuk et al., 2007). The fusion protein was revealed to localize in mitochondria. In addition, they demonstrated sequence-specific modification of mitochondrial DNA using a chimeric zinc finger methylase. The result provides variable information for designing zinc finger proteins targeting various organelles.

Conclusion

The first C_2H_2-type zinc finger protein, TFIIIA, was discovered in 1985 (Miller et al., 1985). During these two decades, zinc fingers have been amazingly developed, based on fundamental structural studies. Now, zinc fingers can be created targeting various DNA sequences. However, more information of design strategies would be necessary. Especially, towards desired endogeneous gene regulation, much more investigations are expected as to how to access chromatin targets. Previous researches assure a bright future for artificial zinc finger proteins as tools for both basic research and gene therapy.

Acknowledgments

I would like to thank Prof. Sugiura and the past and present members of Sugiura's laboratory for contributions to the research described herein and the Ministry of Education, Culture, Sports, Science, and Technology for financial support.

References

Beerli, R. R., Segar, D. J., Dreier, B. and Barbas, C. F., III. (1998) Toward controlling gene expression at will: specific regulation of the erbB-2/HER-2 promoter by using polydactyl zinc finger proteins constructed from modular building blocks, *Proc. Natl. Acad. Sci. U.S.A. 95*, 14628-14633.

Beerli, R. R., Dreier, B. and Barbas, C. F., III (2000) Positive and negative regulation of endogenous genes by designed transcription factors, *Proc. Natl. Acad. Sci. U.S.A. 97*, 1495-1500.

Bibikova, M., Beumer, K., Trautman, J. K. and Carroll, D. (2003) Enhancing gene targeting with designed zinc finger nucleases, *Science, 300*, 764.

Carvin, C. D., Parr, R. D. and Kladde, M. P. (2003) Site-selective in vivo targeting of cytosine-5 DNA methylation by zinc-finger proteins, *Nucleic Acids Res., 31*, 6493-6501.

Cheng, X., Boyer, J.L. and Juliano, R. L. (1997) Selection of peptides that functionally replace a zinc finger in the Sp1 transcription factor by using a yeast combinatorial library, *Proc. Natl. Acad. Sci. U.S.A. 94*, 14120-14125.

Choo, Y. and Klug, A. (1994) Toward a code for the interactions of zinc fingers with DNA: selection of randomized fingers displayed on phage, *Proc. Natl. Acad. Sci. U.S.A. 91*, 11163-11167.

Desjarlais, J. R. and Berg, J. M. (1992) Toward rules relating zinc finger protein sequences and DNA binding site preferences, *Proc. Natl. Acad. Sci. U.S.A. 89*, 7345-7349.

Desjarlais, J. R. and Berg, J. M. (1993) Use of a zinc-finger consensus sequence framework and specificity rules to design specific DNA binding proteins. *Proc. Natl. Acad. Sci. U.S.A. 90*, 2256-2260.

Dreier, B., Segal, D. J. and Barbas, C. F., III. (2000) Insights into the molecular recognition of the 5'-GNN-3' family of DNA sequences by zinc finger domains, *J. Mol. Biol. 303*, 489-502.

Dreier, B., Beerli, R. R., Segal, D. J., Flippin, J. D. and Barbas, C. F., III. (2001) Development of zinc finger domains for recognition of the 5'-ANN-3' family of DNA sequences and their use in the construction of artificial transcription factors, *J. Biol. Chem. 276*, 29466-29478.

Dreier, B., Fuller, R. P., Segal, D. J., Lund, C. V., Blancafort, P., Huber, A., Koksch, B. and Barbas, C. F., III. (2005) Development of zinc finger domains for recognition of the 5'-CNN-3' family DNA sequences and their use in the construction of artificial transcription factors, *J. Biol. Chem. 280*, 35588-35597.

Elrod-Erickson, M., Rould, M. A., Nekludova, L. and Pabo, C. O. (1996) Zif268 protein-DNA complex refined at 1.6 Å: a model system for understanding zinc finger-DNA interactions, *Structure 4*, 1171-1180.

Elrod-Erickson, M., Benson, T. E. and Pabo, C. O. (1998) High resolution structures of variant Zif268-DNA complexes: implications for understanding zinc finger-DNA recognition, *Structure 6*, 451-464.

Greisman H. A. and pabo, C. O. (1997) A general strategy for selecting high-affinity zinc finger proteins for diverse DNA target sites, *Science, 275*, 657-661.

Hirata, T., Nomura, W., Imanishi, M. and Sugiura, Y. (2006) Effects of linking 15-zinc finger domains on DNA binding specificity and multiple DNA binding modes, *Bioorg. and Med. Chem. Let., 15*, 2197-2201.

Hughes, M. D., Zhang, Z. R., Sutherland, A. J., Santons, A. F. and Hine, A. V. (2005) Discovery of active proteins directly from combinatorial randomized protein libraries without display, purification or sequencing: identification of novel zinc finger proteins, *Nucleic Acids Res., 33*, e32.

Ihara, H., Mie, M., Funabashi, H., Takahashi, F., Sawasaki, T., Endo, Y. and Kobatake, E. (2006) In vitro selection of zinc finger DNA-binding proteins through ribosome display, *Biochem. Biophys. Res. Commun., 345,* 1149-1154.

Imanishi, M., Hori, Y., Nagaoka, M. and Sugiura, Y. (2000) DNA bending finger: artificial design of 6-zinc finger peptides with polyglycine linker and induction of DNA bending, *Biochemistry 39*, 4383-4390

Imanishi, M. and Sugiura, Y. (2002) Artificial DNA-bending six-zinc finger peptides with different charged linkers: distinct kinetic properties of DNA bindings, *Biochemistry 41*, 1328-1334.

Imanishi, M., Yan, W., Morisaki, T. and Sugiura Y. (2005) An artificial six-zinc finger peptide with polyarginine linker: selective binding to the discontinuous DNA sequences, *Biochem. Biophys. Res. Commun. 333*, 167-173.

Isalan, M, Klug, A. and Choo, Y. (2001) A rapid, generally applicable method to engineer zinc fingers illustrated by targeting the HIV-1 promoter, *Nat. Biotechnol., 19*, 656-660.

Joung, J. K., Ramm, E. I. and Pabo, C. O. (2000) A bacterial two-hybrid selection system for studying protein-DNA and protein-protein interactions, *Proc. Natl. Acad. Sci. USA 97*, 7382-7387.

Kadonaga, J. T., Carner, K. R., Masiarz, F. R. and Tjian, R. (1987) Isolation of cDNA encoding transcription factor Sp1 and functional analysis of the DNA binding domain, *Cell, 51*, 1079-1090.

Kamiuchi, T., Abe, E., Imanishi, M., Kaji, T., Nabaoka, M. and Sugiura, Y. (1998) Artificial nine zinc-finger peptide with 30 base pair binding sites, *Biochemistry 37,* 13827-13834.

Kim, J.-S. and Pabo, C. O. (1998) Getting a handhold on DNA: design of poly-zinc finger proteins with femtomolar dissociation constants, *Proc. Natl. Acad. Sci. USA 97*, 1495-1500.

Kim, Y. G., Cha, J. and Chandrasegaran, S. (1996) Hybrid restriction enzymes: zinc finger fusions to Fok I cleavage domain, *Proc. Natl. Acad. Sci. USA 93*, 1156-1160.

Klevit, R. E., Herriot, J. R. and Horvath, S. J. (1990) Solution structures of a zinc finger domain of yeast ADR1, *Proteins, 7*, 215-226.

Laity, J. H., Dyson, H. J. and Wright, P. E. (2000) DNA-induced alpha-helix capping in conserved linker sequences is a determinant of binding affinity in Cys(2)-His(2) zinc fingers, *J. Mol. Biol. 295*, 719-727.

Lee, M. S., Gippert, G. P., Soman, K. V., Case, D. A. and Wright, P. E. (1989) Three dimensional solution structure of a single zinc finger DNA-binding domain, *Science 245*, 635-637.

Li, F., Papworth, M., Minczuk, M., Rohde, C., Zhang, Y., Ragozin, S. and Jeltsch, A. (2007) Chimeric DNA methyltransferases target DNA methylation to specific DNA sequences and repress expression of target genes, *Nucleic Acids Res, 35*, 100-112.

Liu, P. Q., Rebar, E. J., Zhang, L., Liu, Q., Jamieson, A. C., Liang, Y., Qi, H., Li, P. X., Chen, B., Mendel, M. C., Zhong, X., Lee, Y. L., Eisenberg, S.P., Spratt, S. K., Case, C. C. and Wolffe, A. P. (2001) Regulation of an endogenous locus using a panel of designed zinc finger proteins targeted to accessible chromatin regions. Activation of vascular endothelial growth factor A, *J. Biol. Chem., 276*, 11323-11334.

Liu, Q., Segal, D. J., Ghiara, J. B. and Barbas, C. F., III. (1997) Design of polydactyl zinc-finger proteins for unique addressing within complex genomes, *Proc. Natl. Acad. Sci. U.S.A. 94*, 5525-5530.

Mandell, J. G. and Barbas, C. F., III. (2006) Zinc Finger Tools: custom DNA-binding domains for transcription factors and nucleases, *Nucleic Acids Res., 34* (Web Server issue), W516-523.

Matthews, J. M. and Sunde, M. (2002) Zinc fingers-folds for many occasions, *IUBMB life, 54*, 351-355.

McNamara, A. R., Hurd, P. J., Smith, A. E. and Ford, K. G. (2002) Characterization of site-biased DNA methyltransferases: specificity, affinity and subsite relationships, *Nucleic Acids Res. 2002, 30*, 3818-3830.

Miller, J., McLachlan, A. D. and Klug, A. (1985) Repetitive zinc-binding domains in the protein transcription factor IIIA from Xenopus oocytes, *EMBO J. 4*, 1609-1614.

Minczuk, M., Papworth, M. A., Kolasinska, P., Murphy, M. P. and Klug, A. (2006) Sequence-specific modification of mitochondrial DNA using a chimeric zinc finger methylase, *Proc. Natl. Acad. Sci. U S A. 103*, 19689-19694.

Moehle, E. A., Rock, J. M., Lee, Y-L., Jouvenot, Y., DeKelver, R. C., Gregory, P. D., Urnov, F. D. and Holmes, M. C. (2007) Targeted gene addition into a specified location in the human genome using designed zinc finger nucleases, *Proc. Natl. Acad. Sci. U.S.A. 104*, 3055-3060.

Moore, M., Choo, Y. and Klug, A. (2001) Design of polyzinc finger peptides with structured linkers, *Proc. Natl. Acad. Sci. U.S.A. 98*, 1432-1436.

Nakatsukasa, T., Shiraishi, Y., Negi, S., Imanishi, M., Futaki, S. and Sugiura, Y. (2005) Site-specific DNA cleavage by artificial zinc finger-type nuclease with cerium-binding peptide. *Biochem. Biophys. Res. Commmun., 330*, 247-252.

Narayan, V. A., Kriwacki, R.W. and Caradonna, J. P. (1997) Structures of zinc finger domains from transcription factor Sp1. Insights into sequence-specific protein-DNA recognition, *J. Biol. Chem. 272*, 7801-7809.

Neuteboom, L. W., Lindhout, B. I., Saman, I. L., Hooykaas, P. J. J. and van der Zaal, B. J. (2006) Effects of different zinc finger transcription factors on genomic targets, *Biochem. Biophys. Res. Commmun., 339*, 263-270.

Nolte, R. T., Conlin, R. M., Harrison, S. C. and Brown, R. S. (1998) Differing roles for zinc fingers in DNA recognition: structure of a six-finger transcription factor IIIA complex, *Proc. Natl. Acad. Sci. U.S.A. 95*, 2938-2943.

Tupler, R. Perini, G. and Green, M. R. (2001) Expressing the human genome, *Nature*, *409*, 832-833.

Omichinski, J. G., Clore, G. M., Apella, E., Sakaguchi, K. and Gronenborn, A. M. (1990) High-resolution three-dimensional structure of a single zinc finger from a human enhancer binding protein in solution, *Biochemistry, 29*, 9324-9334.

Pomerantz, J. L., Wolfe, S. A. and Pabo, C. O. (1998) Structure-based design of a dimeric zinc finger protein, *Biochemistry 37*, 965-970.

Porteus, M. H. and Baltimore, D. (2003) Chimeric nucleases stimulate gene targeting in human cells, *Science. 300*, 763.

Rebar, E. J. and Pabo, C. O. (1994) Zinc finger phage: affinity selection of fingers with new DNA-binding specificities, *Science, 263*, 671-673.

Segal, D. J., Dreier, B., Beerli, R. R. and Barbas, C. F. III. (1999) Toward controlling gene expression at will: selection and design of zinc finger domains recognizing each of the 5'-GNN-3' DNA target sequences, *Proc. Natl. Acad. Sci. U.S.A. 96*, 2758-2763.

Smith, A. E. and Ford, K. G. (2007) Specific targeting of cytosine methylation to DNA sequences in vivo, *Nucleic Acids Res. 35*, 740-754.

Smith, J, Berg, J. M, and Chandrasegaran, S. (1999) A detailed study of the substrate specificity of a chimeric restriction enzyme, *Nucleic Acids Res., 27*, 674-681.

Urnov. F. D., Miller, J. C., Lee, Y-L., Beausejour, C. M., Rock, J. M., Augustus, S., Jamieson, A., Porteus, M. H., Gregory, P. D. and Holmes, M. C. (2005) Highly efficient endogenous human gene correction using designed zinc-finger nucleases, *Nature, 435*, 646-651.

Wolfe, S. A., Grant, R. A., Elrod-Erickson, M. and Pabo, C. O. (2001) Beyond the 'recognition code': structures of two Cys2-His2 zinc finger/TATA box complexes, *Structure, 9*, 717-723.

Wolfe, S. A., Grant, R. A. and Pabo, C. O. (2003) Structure of a designed dimeric zinc finger protein bound to DNA, *Biochemistry 42*, 13401-13409.

Xu, G. L. and Bestor, T. H. (1997) Cytosine methylation targetted to pre-determined sequences, *Nat. Genet., 17*, 376-378.

Yan, W., Imanishi, M., Futaki, S. and Sugiura, Y. (2007) α-Helical linker of an artificial 6-zinc finger peptide contributes to selective DNA binding to a discontinuous recognition sequence, *Biochemistry, 46,* 8517-8524.

Zhang, L., Spratt, S.K., Liu, Q., Johnstone, B., Qi, H., Raschke, E. E., Jamieson, A. C., Rebar, E. J., Wolffe, A. P. and Case, C. C. (2000) Synthetic zinc finger transcription factor action at an endogenous chromosomal site. Activation of the human erythropoietin gene, *J. Biol. Chem., 275*, 33850-33860.

In: Drug Design Research Perspectives
Editor: Stanley P. Kaplan, pp. 39-55
ISBN: 978-1-60021-823-1

Chapter III

P-Loop Pliability of Rho-Kinase for Inhibitor Binding

Keigo Gohda[1] and Toshio Hakoshima[2*]

[1] Computer-Aided Molecular Modeling Research Center Kansai,
1-3-12 Honjyo-cho, Higashinada-ku, Kobe, Hyogo 658-0012, JAPAN
[2] Nara Institute of Science and Technology, 8916-5, Takayama,
Ikoma, Nara 630-0192, JAPAN

Abstract

Rho-kinase is a leading player in the regulation of cytoskeletal events involving smooth muscle contraction and neurite growth-cone collapse and retraction, and is a promising drug target in the treatment of both vascular and neurological disorders. Recent crystal structures of Rho-kinase (ROCK 2) and ROCK 1 have revealed structural details of the ATP-binding site, which represents the target site for conventional small-molecule kinase inhibitors. Four series of kinase inhibitors are known which differ in scaffold chemical structures and includes isoquinolines, 4-aminopyridines, indazoles, and others derivatives containing amide and urea. The crystal structures of the fasudil- and Y27632-bound forms have provided valuable clues to compound development of the first two derivative series with improved selectivity. Especially, both kinase and drug conformational pliability and stability are underlined for conferring selectivity. In this study, we set out to predict the binding-modes of two series of the inhibitors, indazole and amide/urea series, in Rho-kinase by docking study in order to understand the structure-activity relationship of the Rho-kinase inhibitors from structural point of view. One isoquinoline inhibitor was also included in the study to examine our hypothesis that was previously addressed concerning the pliability of the P-loop. In the docking studies, three Rho-kinase structures with different P-loop conformations were served to examine

* To whom correspondence should be addressed: Professor Dr. Toshio Hakoshima; Nara Institute of Science and Technology, 8916-5, Takayama, Ikoma, Nara 630-0192, JAPAN. Tel: +81-743-72-5570; Fax: +81-743-72-5579; E-mail: hakosima@bs.naist.jp; Running title: Rho-kinase

the importance of the P-loop pliability in the binding mode and selectivity of all four derivatives including the indazoles and amide/urea derivatives. We show that the extended P-loop conformation is suitable to accommodate the indazoles and amide/urea derivatives that possess an aromatic pendant ring so as to bind a pocket beneath the phosphate-binding region which favors binding of a small aromatic ring, whereas the folded P-loop conformation accommodates the smaller compounds of the isoquinoline and 4-aminopyridine series. In the latter case, the conserved P-loop phenylalanine occupies the pocket, resulting in an increase of protein-ligand contacts. Thus, the P-loop pliability plays an important role in inhibitor binding affinity and specificity.

Introduction

Overview of Rho-Kinase

Rho-kinase (Rho-associated kinase, ROCK 2, ROKα, or ROCK II) (Kimura et al., 1996; Matsui et al., 1996; Leung et al., 1995, 1996; Ishizaki et al., 1996) is a serine/threonine protein kinase (1388 residues) that comprises an N-terminal extension region, a kinase core domain, a C-terminal extension region, a coiled-coil region with the Rho-binding region, and a C-terminal putative pleckstrin homology (PH) domain including a cysteine-rich subdomain. In the absence of Rho binding, the catalytic activity of the kinase is thought to be negatively regulated by an autoinhibitory interaction between the catalytic kinase domain and the C-terminal PH-like domain (Amano et al., 1999; Chen et al., 2002). For activation of the kinase, GTP-bound RhoA needs to bind the Rho-binding region of the coiled-coil domain, resulting in loss of interaction between the kinase domain and the C-terminal region (Ishizaki et al., 1996). Additionally, cleavage of the C-terminal domain by granzyme B during apoptosis was recently shown to activate the kinase (Sebbagh et al., 2005).

Rho-kinase has an isozyme (ROCK 1, ROKβ/ROCK I) that exhibits 92% sequence identity in the kinase domain (Nagawa et al., 1996; Amano et al., 2000; Riento and Ridley, 2003). The tissue distribution of the two enzymes differs where Rho-kinase is preferentially expressed in brain, while ROCK 1 shows the highest expression levels in nonneuronal tissues. Rho-kinase displays the greatest homology to myotonic dystrophy kinase (DMPK), DMPK-related cell division control protein 42 (Cdc42) binding kinase (MRCK) and citron kinase (CRIK).

Rho-kinase directly and indirectly regulates to facilitate myosin light chain phosphorylation, inducing interactions between actin and myosin, and thereby enhances myosin ATPase activity required for contraction (Amano et al., 1996). Since the first role of Rho-kinase was defined to mediate the formation of stress fibers and focal adhesions via RhoA activation (Leung et al., 1996; Amano et al., 1997; Ishizaki et al., 1997), many contractile processes in smooth muscle and nonmuscle cells including vascular smooth muscle contraction, stress-fiber formation, and cell migration have been linked to the Rho/Rho-kinase pathway (Kaibuchi et al., 1999; Riento and Ridley, 2003).

The pathophysiological implications of Rho/Rho-kinase signaling include hypercontraction (abnormal contraction of vascular smooth muscle), which is a major cause of disease such as hypertension and vasospasm of the coronary and cerebral arteries (Fukata et al., 2001). Moreover, abnormal activation of this pathway has been observed in various disorders of the central nervous system (Mueller et al., 2005). Thus, inhibition of this pathway represents a promising strategy for the treatment of these diseases.

Kinase Domain Structure and Its Inhibitors

The two series of Rho-kinase inhibitors widely used in a clinical setting or laboratory are the isoquinoline series and 4-aminopyridine series (figure 1). One of the most prominent isoquinoline derivatives is fasudil (HA-1077), which was one of the first small-molecule inhibitors of Rho-kinase identified (Uehata et al., 1997), and is clinically used to treat cerebral vasospasm caused by subarachnoid hemorrhage (Ono-Saito et al., 1999). Fasudil showed particular selectivity towards Rho-kinase compared its inhibitory effects on 27 serine/threonine kinases (Davies et al., 2000). In the 4-aminopyridine series, Y-27632 is representative of the series most widely utilized in biological and pharmacological experiments (Mueller et al., 2005).

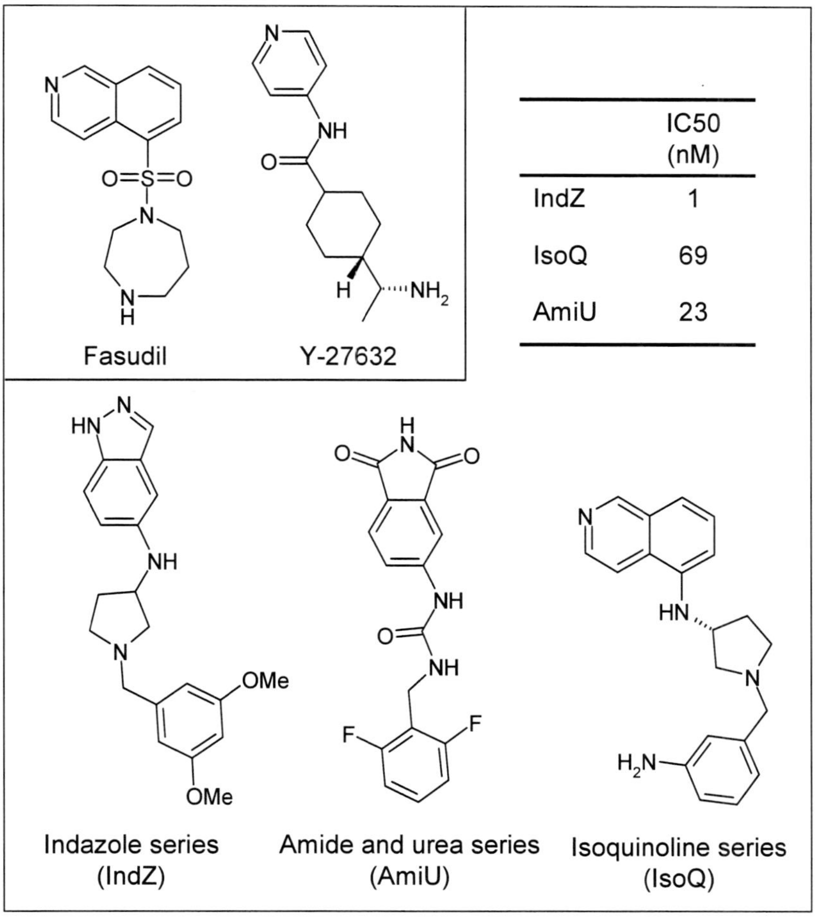

	IC50 (nM)
IndZ	1
IsoQ	69
AmiU	23

Figure 1. The chemical structure of the Rho-kinase inhibitors used in this study. The reported inhibitory activity is summarized.

Recently, the complex structures of the kinase domain of Rho-kinase with each of the two inhibitors and related compounds have been determined at the atomic level by the x-ray diffraction method (Yamaguchi et al., 2006a, b). These structures revealed a novel head-to-head dimer structure that is mediated by the dimerization domain formed by the N-terminal extension and the C-terminal extension (figure 2A). The overall fold of the kinase core domain shows the canonical protein kinase fold that forms a bilobal structure with a small N-terminal lobe (residues 85–174) and a larger C-terminal lobe (residues 175–350) (figure 2B). Rho-kinase belongs to the AGC family of eukaryotic protein kinases, which includes cAMP-dependent protein kinase (PKA), PKG, PKC, PKB/Akt, and PDK1. The N-terminal lobes of Rho-kinase and PKA consist of common secondary structure elements including a five stranded β sheet (strands β1–β5) with two α helices (αB and αC). The C-terminal lobe of Rho-kinase consists of eight helices (αD and αI) with seven short strands (β5D–β11). The relative disposition of the N- and C-terminal lobes represents a closed form, which is supposed to be an active conformation that orients active residues in proper positions and orientations.

Fasudil is identified in the interlobe cleft that corresponds to the conserved ATP binding site (figure 2B). Fasudil binds to the hydrophobic cleft between the N- and C-terminal lobes of the kinase domain. The planar isoquinoline ring of fasudil is inserted into the canonical adenine binding pocket (hereafter referred to as the "adenine binding site") of protein kinases, whereas the seven-membered homopiperazine ring is placed at the entrance of the cleft, where active residues are clustered. The isoquinoline ring is situated on a hydrophobic platform (Met172, Leu221, and Ala231) and is covered by β sheet residues of the N-terminal lobe (Ile98, Val106, Ala119, and Met169). The inserted isoquinoline ring is fixed in the pocket by conserved hydrogen-bonding between the ring nitrogen atom (N7) and the main chain of Met172 in the hinge region. Likewise, Y-27632 binds to the inter-subdomain cleft, and the pyridine ring of Y-27632 is placed on the adenine binding pocket in the cleft. Although contacts between Y-27632 and the kinase domain are mainly hydrophobic, two specific polar interactions hold the elongated Y-27632 molecule at both ends in the cleft; a hydrogen-bond is formed between the main-chain of Met172 and the pyridine nitrogen atom of the inhibitor, and at the other end the terminal amino group forms two hydrogen-bonds to two catalytic residues, Asn219 and Asp232.

Observed P-Loop Pliability

One of the most interesting aspects of the binding of fasudil and Y27632 to Rho-kinase compared to PKA is that we observed the plasticity or pliability of the phosphate-binding loop (P-loop) conformation on inhibitor binding (figure 2C). The P-loop is also known as a glycine-rich loop that is found in the ATP/GTP binding site in many nucleotide-binding proteins, possessing a commonly conserved (the GxGxFG motif) sequence. In the fasudil complex, two crystallographically independent molecules with similar structures (molecule A and molecule B) are identified in the unit cell; the root-mean square deviation between the two molecules is small (0.65 Å for 321 Cα-carbon atoms). In molecule A, the P-loop adopts a similar conformation to that of PKA, and fasudil is likely to adopt a natural conformation. In contrast to molecule A, the loop of molecule B folds down to increase surface complementarity with the inhibitor. The aromatic ring of Phe103 flips inside the loop to form a number of contacts with the homopiperazine ring of fasudil, capping the hydrophobic pocket to which the inhibitor binds. In molecule B, the inhibitor appears to significantly alter the conformation of the homopiperazine ring to accommodate the Phe103 side-chain. Similarly, the Rho-kinase P-loop in the Y-27632-bound form displays an induced-fit conformational change upon binding Y-27632. The P-loop in Y-27632-bound Rho-kinase is folded towards the inside with a flip of the Phe103 side-chain to increase surface complementarity with the inhibitor. The overall structures of the Y-27632- and fasudil-bound forms, except for the P-loop region, align well.

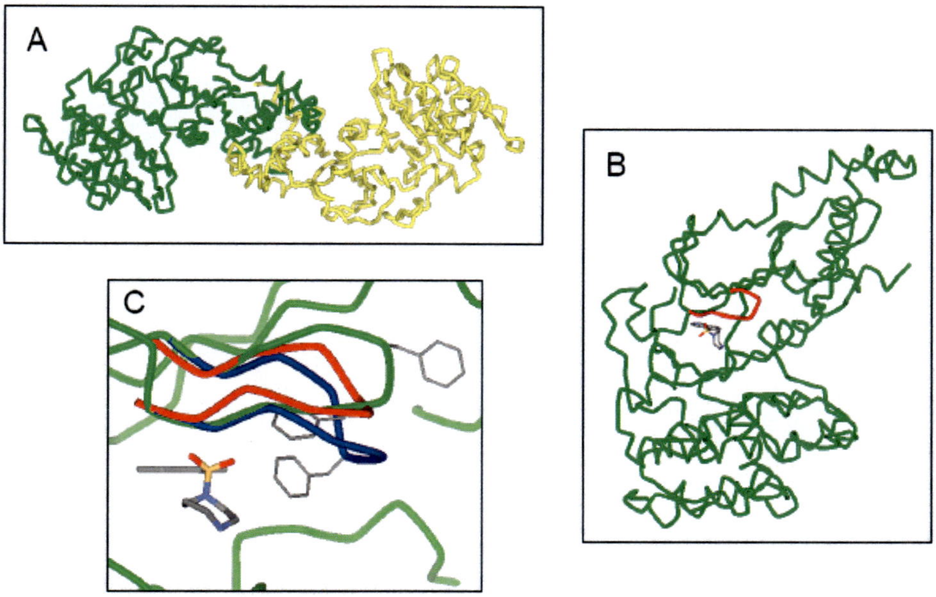

Figure 2. The crystallographic structure of the Rho-kinase domain (PDB entry ID: 2F2U). (A) Overall dimer structure of the kinase domain (B) Carbon α trace of the kinase domain (molecule A of the fasudil-Rho-kinase complex) with fasudil. The kinase domain is shown in green and the P-loop in the domain is represented in red. The structure of fasudil is shown as a stick model. (C) Superimposing the P-loop structures of molecule A (green) and molecule B (blue) of the fasudil-bound form, and the molecule of the Y-27632-bound form (red, PDB entry ID: 2H9V). The side-chain atom of Phe103 in the P-loop is shown as a stick model. The binding conformation of fasudil in the ATP binding site of molecule B in the fasudil complex is also represented as a stick model.

In general, the P-loop is known to be mobile in protein kinases, and can readily accommodate induced-fit changes upon ligand binding. The very first case of such an induced-fit mechanism of inhibitor binding in the P-loop region was observed in the structure of the FGF receptor cytoplasmic tyrosine kinase domain in complex with the specific inhibitor SU5402 (Mohammadi et al., 1997). This conformational change of the P-loop reflects its flexible nature to function as a flap on the bound substrate ATP. P-loop pliability in Rho-kinase contributes to increase contacts when fasudil or Y-27632 binds, something which is not observed in the PKA-fasudil complex (Breitenlechner et al., 2003). Y-27632 possesses more contacts than fasudil in the Rho-kinase catalytic domain, which might explain the stronger binding of Y-27632 compared with fasudil. Thus, the orientation of Phe103 in the Rho-kinase structure might represent an important part of the inhibitor binding mechanism. The P-loop pliability was not observed in the crystal structures of ROCK 1 which were recently determined (Jacob et al., 2006). The failure to detect the pliability in these crystals probably is attributed to differences in crystallization conditions or crystal packing effects rather than differences between ROCK1 and ROCK 2.

In addition to isoquinoline and 4-aminopyridine inhibitors, two other series of inhibitors that possess different chemical scaffolds include the indazole (hereafter referred to as "IndZ"), and the amide and urea (hereafter referred to as "AmiU") series (figure 1) (Mueller et al., 2005). In an effort to glean some insight into the structure-activity relationship (SAR) of the Rho-kinase inhibitors, we computationally modeled the binding conformation of representative derivatives from each of the two series by employing a docking study.

Methods

Starting Structures

For the structures of Rho-kinase, three protein structures of Rho-kinase with different P-loop conformations were subjected to the targets of the computational docking study: molecule A of the fasudil complex (molA), molecule B of the fasudil complex (molB) (Yamaguchi et al., 2006a), and molecule of the Y-27632 complex (molY27) (Yamaguchi et al., 2006b), since the P-loop pliability should influence binding of the inhibitors. In the previous report of the Y-27632-bound form, the position of Phe103 in the Y-27632 complex overlapped well with the position of aromatic rings of inhibitors complexed with PKA as determined by superimposing analysis of the Rho-kinase complex with various inhibitor-PKA complexes. Then, we made a hypothesis for an inhibitor-design that the introduction of hydrophobic functionality reachable to that position into isoquinoline or 4-aminopyridine derivatives could enhance the inhibitory activity. To examine our hypothesis, one of the isoquinoline derivatives (hereafter referred to as "IsoQ"), a derivative of fasudil with longer substituents following the isoquinoline ring, was included for this docking study.

Computation

Docking of Rho-kinase inhibitors to Rho-kinase structures was performed with AutoDock, version 3.0 (Morris et al., 1998). This program allows ligand structures to dock in a conformationally flexible manner to a protein, and adopts a rigid-body protein approximation to speed up the calculation of docking energy. The initial structures of the ligands were constructed using the Maestro molecular modeling package (Schrodinger Inc., Portland, OR). AMBER (Weiner et al., 1984) and Gasteiger (Gasteiger and Marsili, 1980) atomic charges were loaded on the protein and inhibitor atoms, respectively, as partial charges. A docking space of the ligands was defined with a rectangular box (30.0 × 26.2 ×

22.5 Å^3) which emcompassed the ATP binding site of the kinase domain. Grids of probe-atom interaction energies were computed on the boxes. The number of grid points in the x, y, z-axis was 80 × 70 × 60 with grid points separated by 0.375 Å. The inhibitors were then docked using the Lamarckian genetic algorithm (LGA). Flexibility was allowed for all exocyclic rotational bonds of the inhibitors by freely rotating the rotational bonds. One LGA calculation generated 27000 conformers by varying the molecular position in the docking space and rotational bonds of the inhibitor, and ranked the conformers by their docking energy. Each docking experiment consisted of a series of 200 LGA simulations. A hydrophobic contact was counted in this study if two intermolecular carbon-atoms between a protein side-chain and an inhibitor were located within a 4.0 Å distance.

Results

Indazole Series (IndZ)

molA

The indazole ring of IndZ was placed at the adenine binding site as in the case of fasudil or Y-27632 in their respective complexes (figure 3A). The key hydrogen-bonds which recognize the adenine ring of a substrate ATP were formed between the main-chain carbonyl group of Glu170 and the ring N1 hydrogen atom of the inhibitor, and between the main-chain amino group of Met172 and the ring N2 nitrogen atom of the inhibitor. The pyrrolidine moiety following the indazole ring was bound at the place where the ribose moiety of ATP was situated (hereafter referred to as the "ribose binding site"), and had no hydrogen-bonds between the amino group of the pyrrolidine and the protein, while one hydophobic contact was observed with the side-chain atom of Val106 (figure 3A). The terminal dimethoxyphenyl ring of the inhibitor protruded into the area under the P-loop (hereafter referred to as the "lipophilic region under the P-loop"), and one of the ring methoxy groups made a hydrogen-bond with the side-chain amino group of Lys121 which is a catalytically relevant residue that forms the Glu-Lys ion pair with Glu140 in the active form. The hydrophobic contacts were found between the terminal ring and the side-chain atoms of Lys121, Leu123, and Phe136 (figure 3A). The side-chain atoms of Lys121 were aligned along the face of the dimethoxyphenyl ring, making multiple hydrophobic contacts (figure 3B). In addition, more noteworthy is that the positively charged side-chain amino group of Lys121 was positioned on the face of the dimethoxyphenyl ring of the inhibitor at a distance of 3.6 Å from the centroid of the ring, which very likely makes a cation-π interaction. The orientation and relative position of the side-chain atom of Lys121 is fixed by an ionic interaction with the negatively charged side-chain atom of Glu140.

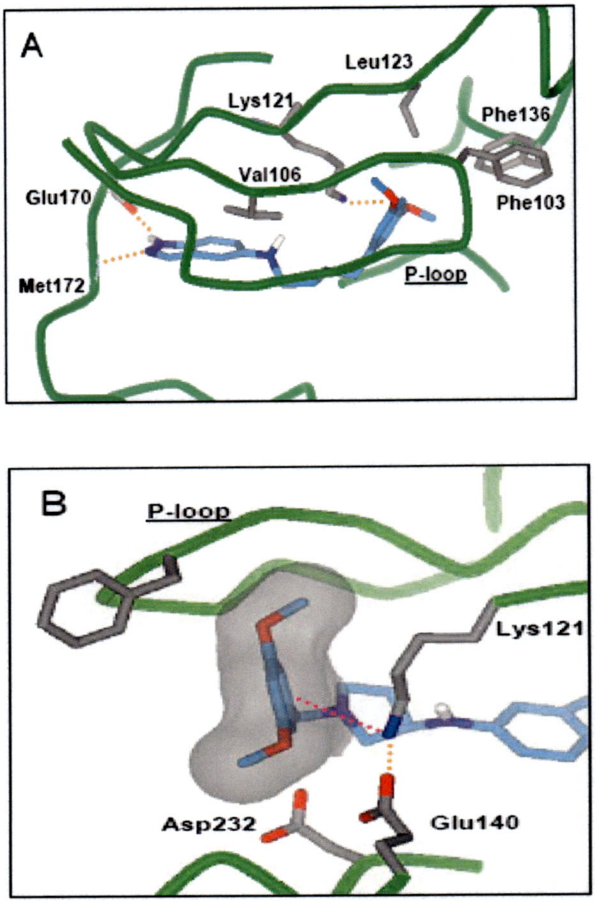

Figure 3. Indazole-inhibitor (IndZ) binding to molecule A. Hydrogen-bonds are indicated as orange dotted lines. (A) The inhibitor is shown as a stick model and residues surrounding the inhibitor are shown. (B) Close-up view of the IndZ-molA complex around the P-loop. Solvent accessible surface area model of the dimethoxyphenyl ring of the inhibitor is also depicted. Possible cation-π interaction between the centroid of the ring and the side-chain amino group of Lys121 is shown as red dotted lines.

molY27

The structure of molY27 possesses a bent conformation of the P-loop. This conformational bending changes the binding conformation of IndZ compared with that in molA (figure 4A). The indazole ring of the inhibitor was bound to the hinge region at the adenine binding site with two hydrogen-bonds between the main-chain atom of Glu170 and the ring N1 hydrogen atom and between the main-chain atom of Met172 and the ring N2 atom, similar to the complex model with molA. Contrarily, the pyrrolidine moiety was pushed away from the ribose binding site, and resulting in the dimethoxyphenyl ring of the inhibitor being located out of the ATP binding region. The dimethoxyphenyl ring made hydrophobic contacts with surrounding residues including as Val106 and Phe384 (data not shown), but no hydrogen-bonds were formed with the protein.

molB

The P-loop of molB takes a more bent conformation than that of molY27 (figure 4B). In molB, the indazole ring of the inhibitor stayed in adenine binding site with one hydrogen-bond between the main-chain carbonyl group of Glu170 and the ring N1 hydrogen atom of the inhibitor. However, the face of the ring was turned over compared with that found in molA and molY27. This flipping-over altered the direction of the N2 atom of the ring towards the inside of the kinase domain, leading to no hydrogen-bonding with the main-chain atom of Met172. The pyrrolidine moiety and the dimethoxyphenyl ring of the inhibitor were located outside of the ATP binding site since the spatial space of the binding site became narrower by the side-chain atom of Phe103 in the P-loop falling down deeply towards the ATP binding site. One methoxy group of the dimethoxyphenyl ring formed a hydrogen-bond with the main-chain amino group of Asp385.

The Most Probable Model of the Complex

The most probable model of the IndZ complex with Rho-kinase was selected from the three complexed models with molA, molB, and molY27 according to the geometric and physicochemical properties of the inhibitor in the complexes. For the physicochemical properties, the docking energy derived from the AutoDock calculation and the number of hydrogen-bonds formed in the complex were used. The lowest docking energy and the largest number of hydrogen-bonds were observed in the molA complex, being -11.30 kcal/mol and 3, respectively (figure 4). Concerning the geometric properties of the complexes, IndZ adopted an extended conformation only in molA, suggesting the most natural geometry. Therefore, the most probable model is considered to be the molA complex.

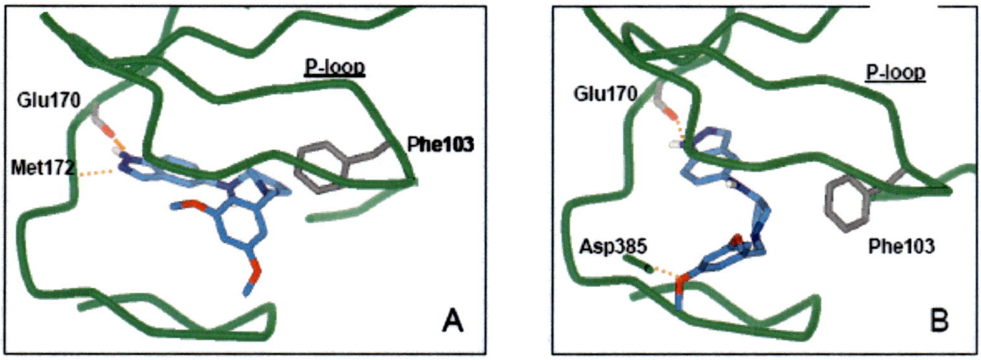

	Docking energy (kcal/mol)	Number of HBonds
molA	-11.30	3
molY27	-10.87	2
molB	-11.14	2

Figure 4. Indazole-inhibitor (IndZ) binding to (A) the molecule of the Y-27632-Rho-kinase complex and (B) molecule B of the fasudil-Rho-kinase complex. Hydrogen-bonds are indicated as orange dotted lines. Summary of docking energy and number of hydrogen-bonds for the three models of the IndZ complex with molA, molY27, and molB is shown in the Table.

Amide and Urea Series (AmiU)

In molA, the overall conformation of AmiU was similar to that of IndZ in the IndZ-molA complex (figure 5A). The phthalimide ring of AmiU was placed at the adenine binding site with hydrogen-bonding to the main-chain atoms of Glu170 and Met172, but no hydrogen-bonds were observed between the urea moiety of the inhibitor and the protein. The terminal difluorophenyl ring of the inhibitor was placed at the lipophilic region under the P-loop, and made hydrophobic contacts with the side-chain atoms of Val106 and Lys121. In molY27 and molB, the phthalimide ring of AmiU was partially positioned on the adenine binding site with hydrogen-bonds to the hinge region (figures 5B and C), however, the urea and difluorophenyl moieties of the inhibitor were pushed away from the ATP binding site as a result of steric hindrance derived from the side-chain atom of Phe103 which is turned down towards the ATP binding site in both protein structures. The urea moiety of the inhibitor was not involved in hydrogen-bonding to the protein in either complex models.

From a comparison of the docking energy, the number of hydrogen-bonds, and the binding conformation of the inhibitor among the three complexed models (figure 5), the most probable conformation of AmiU was likely to be the conformation in molA.

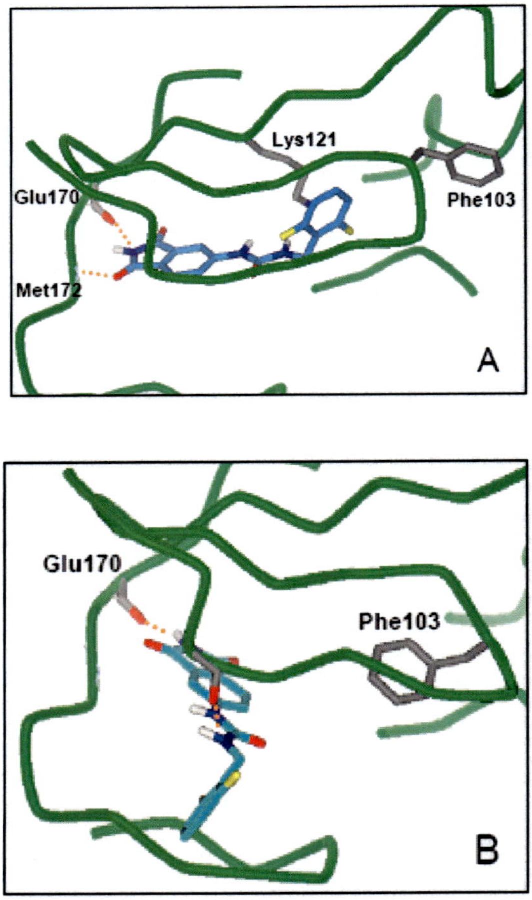

Figure 5 Continued on the next page

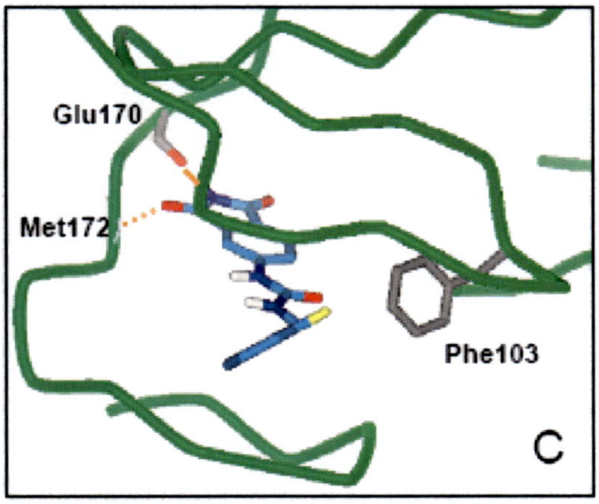

	Docking energy (kcal/mol)	Number of HBonds
molA	-10.12	2
molY27	-9.55	1
molB	-9.48	2

Figure 5. Amide and urea-inhibitor (AmiU) binding to (A) the molecule A of the fasudil-Rho-kinase complex, (B) the molecule of the Y-27632-Rho-kinase complex (C) the molecule B of the fasudil-Rho-kinase complex. Hydrogen-bonds are indicated as orange dotted lines. Summary of docking energy and number of hydrogen-bonds for the three models of the AmiU complex with molA, molY27, and molB is shown in the Table.

Isoquinoline Series (IsoQ)

IsoQ is a derivative of fasudil and possesses a pyrrolidine and aniline moiety following the isoquinoline ring. In molA, the isoquinoline ring of the inhibitor was located in the adenine binding site with a hydrogen-bond between the main-chain atom of Met172 and the ring nitrogen atom of the inhibitor (figure 6A). Compared with the binding conformation of fasudil in molA, the location of the ring of the inhibitor overlaps well with that of fasudil; the ring nitrogen atom occupies the same position in both (figure 6B). Two hydrogen-bonds with the amino group of the terminal aniline ring situated at the lipophilic region under the P-loop were formed to the main-chain atoms of Glu105 and Leu122, and hydrophobic contacts with the side-chain atom of Lys121 were observed with the aniline ring. In molY27 and molB, IsoQ took bent conformations (figures 7A and B). However, both conformations bound differently to the ATP binding site. In molY27, the isoquinoline ring of the inhibitor was not placed on the adenine binding site, but more towards the inside of the kinase domain. The pyrrolidine moiety was positioned away from the ribose binding site, while the terminal aniline ring of the inhibitor was directed back to the position of Phe103. There were no hydrogen-bonds between the inhibitor and the protein. Contrarily, the isoquinoline ring in molB was positioned on the adenine binding site with a hydrogen-bond to the main-chain atom of Met172 (figure 7B). The moieties following the isoquinoline ring were stuck out of the ATP binding site due to the bent conformation of the P-loop. The physicochemical and geometric properties of the inhibitor in molA indicate that the most probable model is the molA complex.

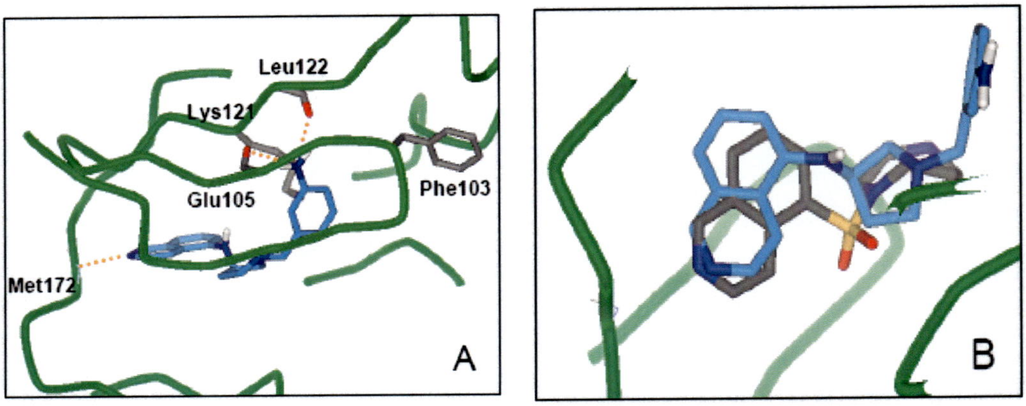

Figure 6. Isoquinoline-inhibitor (IsoQ) binding to molecule A. (A) The inhibitor is shown as a stick model and residues surrounding the inhibitor are shown. Hydrogen-bonds are indicated as orange dotted lines. (B) Superimposing the binding conformations of IsoQ and fasudil in molecule A. Blue and gray lines denote the structures of IsoQ and fasudil, respectively.

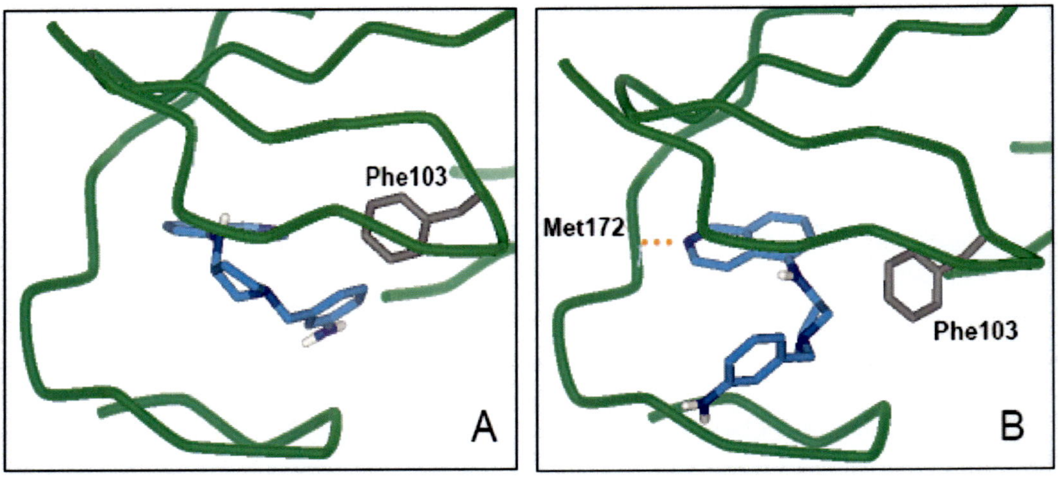

	Docking energy (kcal/mol)	Number of HBonds
molA	-11.49	3
molY27	-10.79	0
molB	-11.39	1

Figure 7. Isoquinoline-inhibitor (IsoQ) binding to (A) the molecule of the Y-27632-Rho-kinase complex and (B) the molecule B of the fasudil-Rho-kinase complex. Hydrogen-bonds are indicated as orange dotted lines. Summary of docking energy and number of hydrogen-bonds for the three models of the IsoQ complex with molA, molY27, and molB is shown in the Table.

Discussion

We have so far determined the binding-mode of fasudil and Y-27632 with Rho-kinase by the diffraction method (Yamaguchi et al., 2006a, b). These inhibitors are representative of the isoquinoline and aminopyridine series of Rho-kinase inhibitors, which are categorized to be four groups (Mueller et al., 2005). Indazole, and amino and urea series are the other two groups of the inhibitors that are unknown their binding-mode in Rho-kinase. Some groups have reported on a model of Rho-kinase complexed with several inhibitors, although the protein structures of Rho-kinase used were models built from related kinase structures (Kimura et al., 2004; Iwakubo et al., 2007a, b; Stavenger et al., 2007; Goodman et al., 2007). Here, using an experimentally determined Rho-kinase structure, we predicted the binding-modes of two series of the inhibitors, the indazole and amide and urea series, by the docking study to understand the SAR of the Rho-kinase inhibitors from structural point of view. In addition, one isoquinoline inhibitor was included for the study to examine our previously addressed hypothesis (Yamaguchi et al., 2006b).

The inhibitors modeled to bind Rho-kinase adopted extended conformations and overlapped well each other (figure 8). The fused aromatic rings of all inhibitors were bound at the adenine binding site and made hydrogen-bonds with the hinge region of the kinase, likewise seen in the recognition of the adenine ring in the ATP binding site. All terminal hydrophobic rings of the inhibitors were commonly located at the lipophilic region under the P-loop, and sandwiched by the side-chain atoms of Lys121 with hydrophobic contacts. To understand the SAR of the inhibitors, the inhibitory activity and physicochemical property of the inhibitors estimated in the model complexes were compared (table 1). The docking energies of IndZ and IsoQ are very similar, although the inhibitory activity of the two inhibitors differed by more than 50-fold. The number of rotational bonds, which indicates the degree of entropic penalty on binding to a protein, was the same for both. Only in the IndZ complex, the cation-π interaction between the terminal ring and Lys121 was foreseen to be made. This additional interaction could contribute to the high activity of IndZ. On the other hand, the inhibitory activity of AmiU was relatively high notwithstanding its weaker docking energy. This could be explained by the presence of a small number of rotational bonds compared with the others. The presence of many rotational bonds in a molecule impacts negatively on the binding free energy since fixation of one rotational bond in a molecule is estimated to lose 0.3113 kcal/mol entropic energy (Bohm, 1994). In addition, the rigid structure of the inhibitor can allow for straightforward placement of the terminal ring into the lipophilic region under the P-loop.

In a previous report, we assumed that the introduction of hydrophobic functionality into isoquinoline or 4-aminopyridine derivatives could enhance the inhibitory activity, since the position of Phe103 in the Y27632-Rho-kinase complex overlapped well to the position of the terminal aromatic ring of inhibitors complexed with PKA (Yamaguchi et al., 2006b). IsoQ possessing a terminal hydrophobic ring was predicted to have an extended conformation in Rho-kinase with the terminal ring being located at the lipophilic region under the P-loop as observed in PKA complexes, and forming hydrophobic contacts with the side-chain atoms of Lys121. These features of the interaction may increase the inhibitory activity 2-fold compared with fasudil (IC50 = 180 nM, Iwakubo et al., 2007a), supporting our hypothesis.

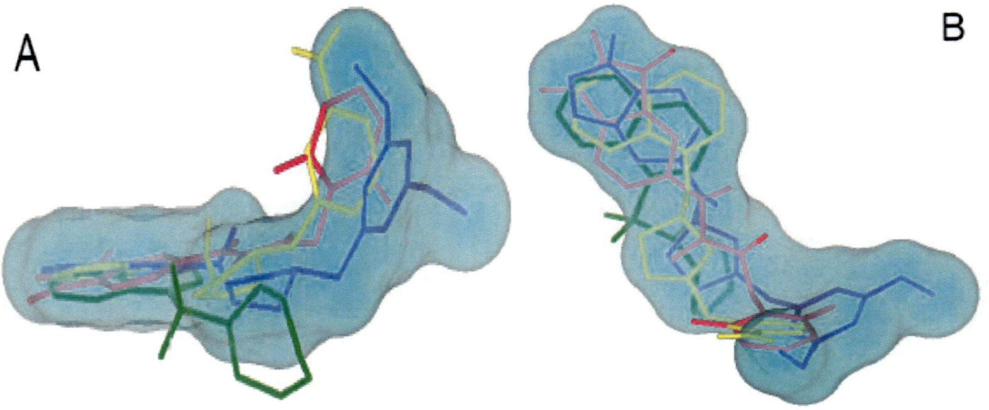

Figure 8. Superimposing of inhibitors bound to molecule A. Fasudil, IndZ, AmiU, and IsoQ are colored green, blue, red, and yellow, respectively. Solvent accessible surface area model for IndZ is also depicted. Side and top views of the superimposed inhibitors are shown in (A) and (B) , respectively.

Table 1. Inhibitory activity and physicochemical property of the inhibitors

	IC50 (nM)[a]	Docking energy (kcal/mol)	Number of HBonds	Number of rot. bonds[b]
IndZ	1	-11.30	3	4
AmiU	23	-10.12	3	2
IsoQ	69	-11.49	2	4

[a] See references cited in Mueller et al., 2005;
[b] The rotational bonds in the terminal functionality and cyclic moiety were not counted.

We employed the three Rho-kinase structures, molA, molB, and molY27, for the prediction of complexed structures with inhibitors to consider the pliability of the P-loop. Consequently, the kinase structure with the extended conformation of the loop, i.e, molA, was considered as the most probable model for the complex following assessment of all three inhibitors based on structural geometry and physicochemical features. The results suggest that the conformation of the P-loop of Rho-kinase changes according to the size of the inhibitor molecules in an effort to maximize the steric complementarity in the binding site. Although the P-loop sequences of other related kinases are not very varied, the pliability of the loop might differ among the kinases since the surrounding environment of the loop may be altered. In fact, the P-loop of PKA complexed with various types of inhibitors for the most part display no significant changes in conformation, while the sequence of the loop in PKA and Rho-kinase are identical except for one amino acid: Ala102 in Rho-kinase and Ser at the corresponding position in PKA. This pliability difference among the kinases could be utilized in the design of selective and/or potent inhibitors of Rho-kinase.

Unlike other protein kinases, Rho-kinase requires both N- and C-terminal extension segments outside the kinase domain for activity. These extensions have been found to form a novel dimerization domain that regulates kinase activation by directly contacting with the N-lobe of the kinase domain. The interface may be a novel target site for synthetic peptides or peptide-mimic compounds and might spawn a new development scheme of newly-designed compounds that are distinct from current protein kinase inhibitors.

Conclusion

Rho-kinase is a promising drug target in the treatment of both vascular and neurological disorders. Four series of Rho-kinase inhibitors target the ATP-binding site, isoquinolines, 4-aminopyridines, indazoles, and others derivatives containing amide and urea. We discussed the importance of the P-loop pliability in the binding mode and selectivity of all four derivative types including indazoles and amide/urea derivatives. We show that the extended P-loop conformation is suitable to accommodate the indazoles and amide/urea derivatives, that possess an aromatic pendant ring so as to bind a pocket beneath the phosphate-binding region which favors binding of a small aromatic ring, while a folded P-loop conformation accommodates the smaller compounds of the isoquinolines and 4-aminopyridines series. In the latter case, the conserved P-loop phenylalanine occupies the pocket, resulting in an increase of protein-ligand contacts. Thus, the P-loop pliability plays an important role in inhibitor binding affinity and specificity.

References

Amano, M., Ito, M., Kimura, K., Fukata, Y., Chihara, K., Nakano, T., Matsuura, Y., and Kaibuchi, K. (1996) Phosphorylation and activation of myosin by Rho-associated kinase (Rho-kinase). *J. Biol. Chem.* 271, 20246–20249.

Amano, M., Chihara, K., Kimura, K., Fukata, Y., Nakamura, N., Matsuura, Y., and Kaibuchi, K. (1997) Formation of actin stress fibers and focal adhesions enhanced by Rho-kinase. *Science.* 275, 1308–1311.

Amano, M., Chihara, K., Nakamura, N., Kaneko, T., Matsuura, Y., and Kaibuchi, K. (1999) The COOH terminus of Rho-kinase negatively regulates rho-kinase activity. *J. Biol. Chem.* 274, 32418–32424.

Amano, M., Fukata, Y., and Kaibuchi, K. (2000) Regulation and function of Rho-associated kinase. *Exp. Cell Res.* 261, 44–51.

Bohm H.J. (1994) The development of a simple empirical scoring function to estimate the binding constant for a protein-ligand complex of known three-dimensional structure. *J. Comput.-Aided Mol. Des. 8*, 243–256.

Breitenlechner, C., Gassel, M., Hidaka, H., Kinzel, V., Huber, R., Engh, R.A., and Bossemeyer, D. (2003) Protein kinase A in complex with Rho-kinase inhibitors Y-27632, Fasudil, and H-1152P: structural basis of selectivity. *Structure.* 11, 1595–1607

Chen, X.Q., Tan, I., Ng, C.H., Hall, C., Lim, L., and Leung, T. (2002) Characterization of RhoA-binding kinase ROKα implication of the pleckstrin homology domain in ROKα function using region-specific antibodies. *J. Biol. Chem.* 277, 12680–12688.

Davies, S.P., Reddy, H., Caivano,M., and Cohen, P. (2000) Specificity and mechanism of action of some commonly used protein kinase inhibitors. *Biochem. J.* 351, 95–105.

Fukata, Y., Amano, M., and Kaibuchi, K. (2001) Rho-Rho-kinase pathway in smooth muscle contraction and cytoskeletal reorganization of non-muscle cells. *Trends Pharmacol. Sci.* 22, 32–39.

Gasteiger, J., and Marsili, M. (1980) *Tetrahedron.* 36, 3219-3228.

Goodman, K.B., Cui, H., Dowdell, S.E., Gaitanopoulos, D.E., Ivy, R.L., Sehon, C.A., Stavenger, R.A., Wang, G.Z., Viet, A.Q., Xu, W., Ye, G., Semus, S.F., Evans, C., Fries, H.E., Jolivette, L.J., Kirkpatrick, R.B., Dul, E., Khandekar, S.S., Yi, T., Jung, D.K., Wright, L.L., Smith, G.K., Behm, D.J., Bentley, R., Doe, C.P., Hu, E., and Lee, D. (2007) Development of dihydropyridone indazole amides as selective Rho-kinase inhibitors. *J. Med. Chem.* 50, 6-9.

Ishizaki, T., Maekawa, M., Fujisawa, K., Okawa, K., Iwamatsu, A., Fujita, A., Watanabe, N., Saito, Y., Kakizuka, A., Morii, N., and Narumiya, S. (1996) The small GTP-binding protein Rho binds to and activates a 160 kDa Ser/Thr protein kinase homologous to myotonic dystrophy kinase. *EMBOJ.* 15, 1885–1893.

Ishizaki, T., Naito, M., Fujisawa, K., Maekawa, M., Watanabe, N., Saito, Y., and Narumiya, S. (1997) p160^{ROCK}, a Rho-associated coiled-coil forming protein kinase, works downstream of Rho and induces focal adhesions. *FEBS Lett.* 404, 118–124.

Iwakubo, M., Takami, A., Okada, Y., Kawata, T., Tagami, Y., Ohashi, H., Sato, M., Sugiyama, T., Fukushima, K., and Iijimaa, H. (2007a) Design and synthesis of Rho kinase inhibitors (II). *Bioorg. Med. Chem.* 15, 350–364.

Iwakubo, M., Takami, A., Okada, Y., Kawata, T., Tagami, Y., Sato, M., Sugiyama, T., Fukushima, K., Taya, S., Amano, M., Kaibuchi, K., and Iijimaa, H. (2007b) Design and synthesis of Rho kinase inhibitors (III). *Bioorg. Med. Chem.* 15, 1022–1033.

Jacobs, M., Hayakawa, K., Swenson, L., Bellon, S., Fleming, M., Taslimi, P., Doran, J. (2006) The structure of dimeric ROCK I reveals the mechanism for ligand selectivity. *J. Biol. Chem.* 281, 260-268.

Kaibuchi, K., Kuroda, S., and Amano, M. (1999) Regulation of the cytoskeleton and cell adhesion by the Rho family GTPases in mammalian cells. *Annu. Rev. Biochem.* 68, 459–486.

Kimura, K., Ito, M., Amano, M., Chihara, K., Fukata, Y., Nakafuku, M., Yamamori, B., Feng, J., Nakano, T., Okawa, K., Iwamatsu, A., and Kaibuchi, K. (1996) Regulation of myosin phosphatase by Rho and Rho-associated kinase (Rho-kinase). *Science.* 273, 245–248.

Leung, T., Manser, E., Tan, L., and Lim, L. (1995) A novel serine/threonine kinase binding the Ras-related RhoA GTPase which translocates the kinase to peripheral membranes. *J. Biol. Chem.* 270, 29051–29054.

Leung, T., Chen, X.Q., Manser, E., and Lim, L. (1996) The p160 RhoA-binding kinase ROKα is a member of a kinase family and is involved in the reorganization of the cytoskeleton. *Mol. Cell. Biol.* 16, 5313–5327.

Matsui, T., Amano, M., Yamamoto, T., Chihara, K., Nakafuku, M., Ito, M., Nakano, T., Okawa, K., Iwamatsu, A., and Kaibuchi, K. (1996) Rho-associated kinase, a novel serine/threonine kinase, as a putative target for small GTP binding protein Rho. *EMBO J.* 15, 2208–2216.

Ono-Saito, N., Niki, I., and Hidaka, H. (1999) H-series protein kinase inhibitors and potential clinical applications. *Pharmacol. Ther.* 82, 123–131.

Mohammadi, M., McMahon, G., Sun, L., Tang, C., Hirth, P., Yeh, B.K., Hubbard, S.R., and Schlessinger, J. (1997) Structures of the tyrosine kinase domain of fibroblast growth factor receptor in complex with inhibitors. *Science.* 276, 955–960.

Morris, G.M., Goodsell, D.S., Halliday, R.S., Huey, R., Hart, W.E., Belew, R.K., and Olson, A.J. (1998) Automated docking using a Lamarckian genetic algorithm and an empirical binding free energy function. *J. Comput. Chem.* 19, 1639-1662.

Mueller, B.K., Mack, H., and Teusch, N. (2005) Rho kinase, a promising drug target for neurological disorders. *Nat. Rev. Drug Discov.* 4, 387–398.

Nakagawa, O., Fujisawa, K., Ishizaki, T., Saito, Y., Nakao, K., and Narumiya, S. (1996) ROCK-I and ROCK-II, two isoforms of Rho-associated coiled-coil forming protein serine/ threonine kinase in mice. *FEBS Lett.* 392, 189–193.

Sebbagh, M., Hamelin, J., Bertoglio, J., Solary, E., and Bréard, J. (2005) Direct cleavage of ROCK II by granzyme B induces target cell membrane blebbing in a caspase-independent manner. *J. Exp. Med.* 201, 465–471.

Riento, K., and Ridley, A. (2003) ROCKs: multifunctional kinases in cell behaviour. *Nature Rev. Mol. Cell. Bio.* 4, 446–456.

Stavenger, R.A., Cui, H., Dowdell, S.E., Franz, R.G., Gaitanopoulos, D.E., Goodman, K.B., Hilfiker, M.A., Ivy, R.L., Leber, J.D., Marino, J.P. Jr., Oh, H.-J., Viet, A.Q., Xu, W., Ye, G., Zhang, D., Zhao, Y., Jolivette, L.J., Head, M.S., Semus, S.F., Elkins, P.A., Kirkpatrick, R.B., Dul, E., Khandekar, S.S., Yi, T., Jung, D.K., Wright, L.L., Smith, G.K., Behm, D.J., Doe, C.P., Bentley, R., Chen, Z.X., Hu, E., and Lee, D. (2007) Discovery of aminofurazan- azabenzimidazoles as inhibitors of Rho-kinase with high kinase selectivity and antihypertensive activity. *J. Med. Chem.* 50, 2-5.

Takami, A., Iwakubo, M., Okada, Y., Kawata, T., Odai, H., Takahashi, N., Shindo, K., Kimura, K., Tagami, Y., Miyake, M., Fukushima, K., Inagaki, M., Amano, M., Kaibuchi, K., and Iijimaa, H. (2004) Design and synthesis of Rho kinase inhibitors (I). *Bioorg. Med. Chem.* 12, 2115–2137.

Uehata, M., Ishizaki, T., Satoh, H., Ono, T., Kawahara, T., Morishita, T., Tamakawa, H., Yamagami, K., Inui, J., Midori Maekawa, M., and Narumiya, S. (1997) Calcium sensitization of smooth muscle mediated by a Rho-associated protein kinase in hypertension. *Nature.* 389, 990–994.

Weiner, S.J., Kollman, P.A., Case, D.A., Singh, U.C., Chio, C., Alagona, G., Profeta, S., and Weiner, P. (1984) *J. Am. Chem. Soc.* 106, 765-784.

Yamaguchi, H., Kasa, M., Amano, M., Kaibuchi, K., and Hakoshima, T. (2006a) Molecular mechanism for the regulation of Rho-kinase by dimerization and its inhibition by fasudil. *Structure.* 14, 589–600.

Yamaguchi, H., Miwa, Y., Kasa, M., Kitano, K., Amano, M., Kaibuchi, K., and Hakoshima, T. (2006b) Structural basis for induced-fit binding of Rho-kinase to the inhibitor Y-27632. *J. Biochem.* 140, 305–311.

In: Drug Design Research Perspectives
Editor: Stanley P. Kaplan, pp. 57-102

ISBN: 978-1-60021-823-1

Chapter IV

Click Chemistry in Carbohydrate Based Drug Development and Glycobiology

Brendan L. Wilkinson,[1] Laurent F. Bornaghi,[1] Todd A. Houston[2] and Sally-Ann Poulsen[1]
[1] Eskitis Institute for Cell and Molecular Therapies,
Griffith University, 170 Kessels Road, Nathan, Queensland 4111, Australia
[2] Institute for Glycomics, Gold Coast Campus,
Griffith University, Queensland 4222, Australia

Abstract

The 1,3-dipolar cycloaddition reaction (1,3-DCR) of a 1,3-dipole to a dipolarophile for the synthesis of heterocycles is a ubiquitous transformation in synthetic organic chemistry. Recently, the Sharpless and Meldal groups have reported the dramatic rate enhancement (up to 10^7 times) and improved regioselectivity of the Huisgen 1,3-DCR of an organic azide to a terminal acetylene to afford, specifically a 1,4-disubstituted-1,2,3-triazole in the presence of a Cu^I catalyst. This Cu^I catalysed 1,3-DCR has successfully fulfilled the requirements of "click chemistry" as prescribed by Sharpless and within the past few years has become a premier component of the click chemistry paradigm. Click chemistry has proven to be of remarkable utility and broad scope, not only in organic synthesis, but in chemical biology and drug discovery. Click chemistry is highly modular and simplifies difficult syntheses. The biocompatibility of the reaction, tolerance towards a broad range of pH and relative inertness of acetylenes and azides within highly functionalised biological milieus has allowed click chemistry to become a viable bioconjugation strategy both for labelling biomolecules and for *in situ* lead discovery applications. More recently, click chemistry has emerged as a powerful conjugation strategy for the preparation of structurally diverse neoglycoconjugates of biomedical interest. This chapter aims to highlight recent developments appearing within the literature concerning the use of click chemistry in carbohydrate based drug discovery and glycobiology. Topics range from small molecule probes and drug leads, multivalent neoglycoconjugates acting as lectin inhibitors and potential vaccines, to bioconjugation strategies for labelling of engineered cell surface glycans. The chapter aims to be comprehensive with commentary on future perspective.

1. Introduction

1.1. Background and Significance

Protein and lipid glycosylation is an omnipresent and life governing process. Oligosaccharides and polysaccharides, either free or anchored to proteins and lipids to form glycoconjugates, exert a multitude of biological effects ranging from nascent protein folding and stabilisation, energy storage, structural support and protection, and as molecular recognition motifs for cell-cell communication events underlying physiological processes including cellular differentiation and development, hormone trafficking and immune surveillance [1]. Overwhelming evidence is now available to implicate both glycan binding and defective metabolic pathways in pathological processes including inflammation and microbial virulence, neurodegenerative conditions and tumour metastasis [2]. The clinical intervention of a defective metabolic pathway, the inhibition of a glycan recognition or pathogen binding event and the elucidation of aberrant glycosylation patterns expressed on certain cancer cell surfaces, are each primary avenues for rational carbohydrate based drug design and vaccine development [3].

The acquisition of a well defined, single glycoform is a prerequisite for correlating structure and function and ultimately developing carbohydrate based therapeutics. Glycan biosynthesis occurs within the secretory pathway of the ER-golgi complex and is not template driven nor under direct transcriptional control [1a]. Naturally occurring glycoproteins are therefore expressed as collections of glycoforms which contain the same underlying peptide sequence but display variable patterns of glycosylation [4]. Due to this considerable microheterogeneity, the isolation of a single glycoform for structure-activity purposes from a biological extract and/or from genetic engineering methods alone is often exceedingly difficult. Whilst genetic engineering and "knock-out/knock-in" methods have contributed enormously to understanding the implications of protein glycosylation, arguably the most reliable means of procuring homogeneous glycoforms is via their *de novo* chemical synthesis [5].

Chemical methods have emerged as powerful tools for elucidating cellular mechanisms of health and pathology associated with glycosylation [4-6]. Improvements in our understanding of glycan metabolism and structure-function relationships using molecular tools allow for the development novel drug leads for the treatment of carbohydrate related disease processes. Two possible pathways can be envisaged, either through interrogation of cell surface glycoprotein/lipid binding events with artificial oligosaccharides, glycoconjugates and their analogues; or the perturbation of glycan metabolic pathways with small molecule enzyme inhibitors, leading to downstream changes in glycosylation patterns [5].

1.2. Glycomimetics as Probes and Drug Leads

Although the knowledge base concerning the structure and function of glycans has increased dramatically in recent years, our ability to routinely apply this knowledge in a drug development context is still far from complete [3b]. The sheer numbers of potential carbohydrate related drug targets and the inherent difficulties associated with the synthesis of natural glycans have fuelled efforts to prepare carbohydrate mimics from natural or synthetic origins as novel drug leads or as biochemical tools [7]. Natural and synthetic carbohydrates are often considered "privileged" combinatorial scaffolds because of the unrivalled structural diversity arising from a high chiral density, variable branching patterns and polyfunctionality [8]. The synthetic modification and inherent reactivity of these powerful building blocks can in turn lead to a tremendous explosion in structural and stereochemical diversity.

Many of the carbohydrate based drugs currently in clinical use are indeed mimics of endogenous carbohydrates and act as inhibitors of carbohydrate-protein binding events or modulators of glycan metabolism (Table 1) [7d,9]. Carbohydrates linked to many natural products are often a prerequisite for biological activity and can thus heavily influence the pharmacokinetics, drug targeting and mechanism of action. Indeed, the literature is replete with examples of carbohydrate containing natural products possessing antimicrobial and anticancer properties, such as the aminoglycosides, saponins and anthracyclines [10]. Despite the plethora of disease states that involve carbohydrate recognition phenomena, comparatively few carbohydrate based drugs are in clinical use. The failure of carbohydrate based therapeutics during the 1980's and 1990's can however be attributed to a poor understanding of their mechanism and pharmacokinetic profiles, rather than inherent flaws in carbohydrate based drugs [11]. When a detailed knowledge is applied in a drug design setting, excellent results have been achieved [12]. Heparin, for example, a polysaccharide displaying potent anticoagulant properties is amongst the worlds most widely used drug [3b,13]. The aminoglycoside and glycopeptide antibiotics remain the frontline defence against a wide range of drug resistant bacterial infections [3c]. Owing to their unique physicochemical and biochemical properties, glycoconjugates have also been employed as vehicles and prodrugs for cell selective drug targeting [3c,14].

Despite the widespread physiological and pathophysiological effects of carbohydrates, there are relatively few clinically used carbohydrate based drugs. Carbohydrates possess less than ideal therapeutic profiles concerning solubility, *in vivo* stability, target binding affinity and cell permeability (although notable exceptions do exist) [9] and as such have received little attention compared to other small molecule drug classes. The synthetic modification of natural polyfunctional carbohydrate scaffolds can however lead to analogues with improved bioactivity and pharmacokinetic profiles. Unnatural functionalities can be introduced into the carbohydrate ring to exploit a specific structural element within the target receptor that is otherwise neglected in the native ligand-receptor binding event [7,8]. The substitution or elimination of ring hydroxyl groups often leads to improved solubility and *in vivo* clearance. In addition, the bioavailability of carbohydrate based therapeutics can be improved by the substitution of a natural, labile glycosidic linkage with a robust, artificial linkage that is impervious to digestive degradation and enzymatic catabolism. Importantly, the installation of artificial linkages can potentially simplify complex multistep syntheses by exploiting chemistry which is facile, high yielding, orthogonal and stereoselective, whilst retaining as much as possible the geometry and electronics of the native linkage [16].

Table 1. Clinically used carbohydrate based drugs and glycomimetic drugs [3,6b,8]

Compound	Target	Indication
Acarbose, Voglibiose, Miglitol (eg. Glyset®)	Intestinal α-glucoamylase, α-glucosidase	Non-insulin dependent Diabetes mellitus (NIDDM)
Heparin analogues (eg. Lovenox®)	Antithrombin III (factor Xa)	Thrombosis
2-Deoxy-2,3-deanhydro-*N*-acetylneuraminic acid derivatives (eg. Relenza®, Tamiflu®)	Influenza sialidase	Antiinfluenza
N-butyl deoxynojirimycin (Zavesca®)	Glucosylceramide synthase, glycan processing α-glucosidase	Lysosomal storage diseases Type 1 Gauchers disease
Vancomycin (Vancocyn®, Lyphocin®)	Bacterial transpeptidase	Methicillin-resistant *Staphylococcus aureus* (MRSA)
Topirimate (Topamax®)	Brain carbonic anhydrase (CA) isozymes*	Antiseizure, migraine

* Postulated anticonvulsant mechanism is inhibition of brain CA isozymes [15].

1.3. Click Chemistry: Background and Perspective

The 1,3-dipolar cycloaddition reaction (1,3-DCR) of a 1,3-dipole to a dipolarophile (i.e. an acetylene or alkene) for the synthesis of five membered heterocycles are ubiquitous transformations in synthetic organic chemistry [17]. Recently, the Sharpless [18a] and Meldal [18b] groups have reported the dramatic rate enhancement (up to 10^7 times) and improved regioselectivity of the Huisgen 1,3-DCR of an organic azide to a terminal acetylene to afford, specifically, the 1,4-disubstituted-1,2,3-triazole in the presence of a Cu^I catalyst (Scheme 1). The Cu^I catalysed 1,3-DCR has successfully fulfilled the requirements of "click chemistry" as prescribed by Sharpless, and within the past few years has become a premier component of the click chemistry paradigm [19].

The classical non-catalysed process proceeds by a concerted mechanism under thermal conditions to afford a mixture of 1,4- and 1,5-disubstituted 1,2,3-triazole regioisomers. The relative proportion of regioisomers and the rate can be predicted from electronic and steric effects [20]. Increased 1,4-regioselectivity is increasingly observed with electron deficient acetylene dipolarophiles (increased dipolarophilicity) and electron rich azides. The Cu^I catalysed ("click") process has been postulated to occur by a stepwise mechanism on the basis of recent thermal and kinetic studies [21]. The substantial rate increase of the Cu^I catalysed process in aqueous solvents is rationalised in terms of a stepwise process which lowers the activation barrier relative to that of the non-catalysed process by as much as 11.8 kcal mol^{-1} [21]. The proposed catalytic cycle involves several postulated and transient Cu^I-

acetylide complexes, starting with the complexation of the alkyne to the Cu^{I} metal centre to form a Cu^{I}-alkyne π-complex (A) (Scheme 2) [21]. The enhanced reaction rate in water relative to organic solvents can be rationalised in terms of the endothermic ligand disassociation in organic media, for example, acetonitrile (endothermic by 0.6 kcal mol^{-1}) relative to water (exothermic by 11.8 kcal mol^{-1}) [21]. The formation of the Cu^{I}-acetylide complexes is also water assisted, since water lowers the pKa of the acetylene C-H by 9.8 pKa units [20]. Formation of the Cu^{I}-acetylide species allows for subsequent ligand displacement with azide and results in a dimeric copper species (B). Complexation with azide activates it toward nucleophilic attack at the N-3 with the acetylide C-4 (the numbering is presented according to triazole nomenclature). The resulting metallocycle (C) undergoes facile ring contraction via a transannular association of the N-1 lone pair with the C5-Cu π*-orbital to give the copper-triazole complex (D). Protonation of the triazole species, possibly with water and dissociation of the labile copper complex affords the 1,4-disubsituted-1,2,3-triazole (E), thus regenerating catalyst and ending the cycle.

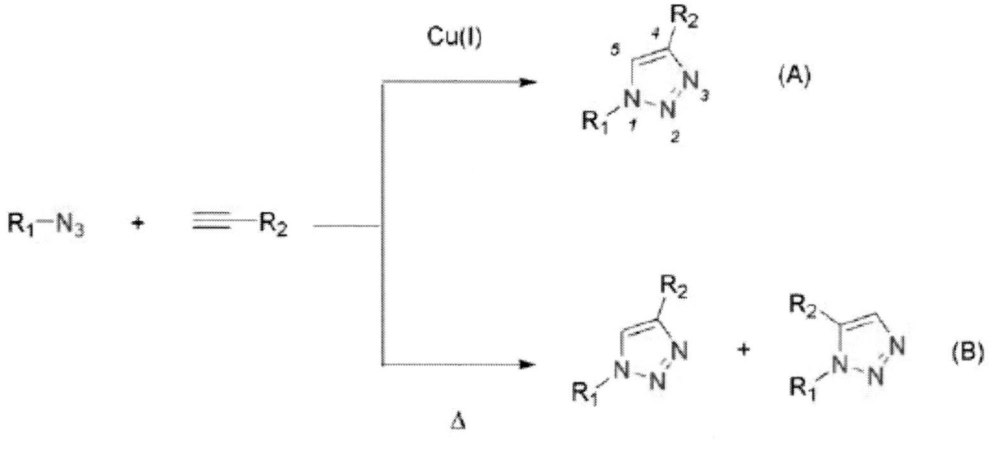

Scheme 1. (A) The Cu^{I} catalysed 1,3-dipolar cycloaddtion reaction of an organic azide and terminal acetylene affords, exclusively the 1,4-disubstituted-1,2,3-triazole regioisomer. (B) The non-catalysed cycloaddition reaction affords a regioisomeric mixture of 1,4- and 1,5-disubstituted 1,2,3-triazoles.

Within a short time frame, click chemistry has proven to be of remarkable utility and broad scope, not only in organic synthesis, but in chemical biology and drug discovery [22]. Azides and acetylenes are by definition kinetically stable entities possessing high built in energy and are tolerant to a wide range of synthetic conditions [20]. Click chemistry is highly modular and simplifies difficult syntheses, thus enabling a more cost effective and efficient surveillance of structural space. The biocompatibility of the reaction, tolerance towards a broad range of pH and relative inertness of acetylenes and azides within highly functionalised biological milieus has allowed click chemistry to become a viable bioconjugation strategy for labelling biomolecules and for *in situ* lead discovery applications [23]. The 1,2,3-triazole

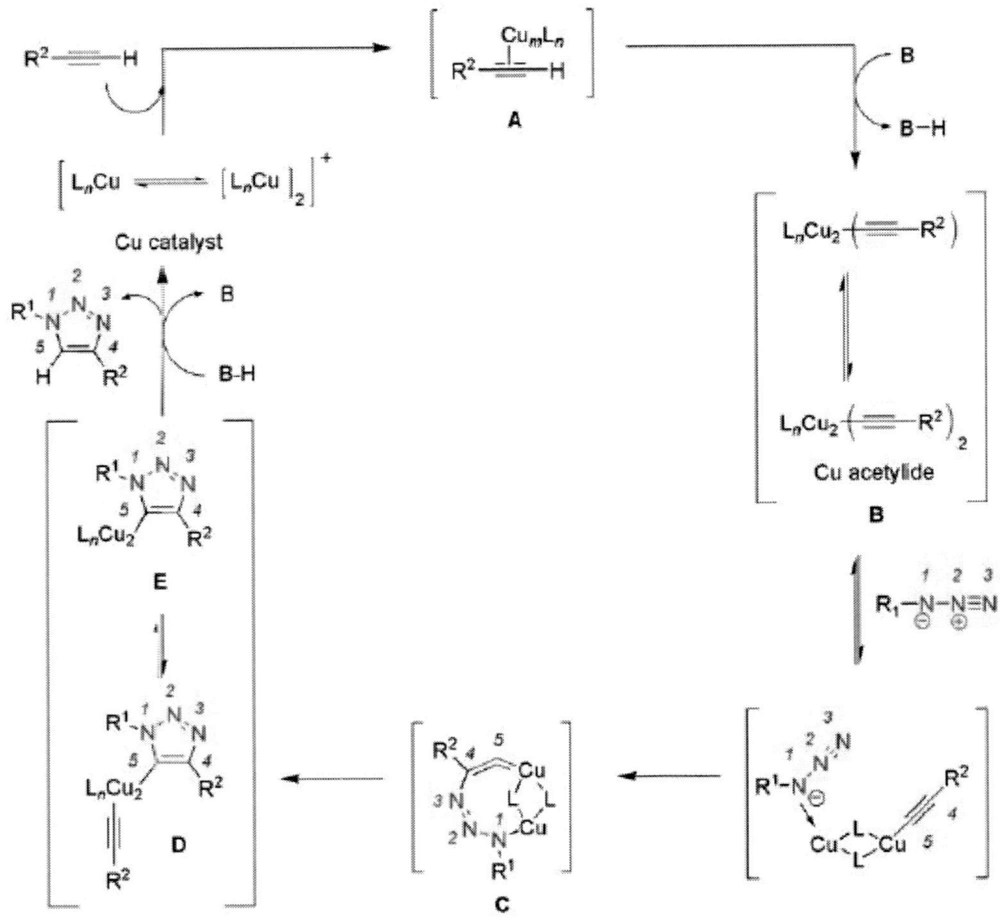

Scheme 2. Proposed stepwise mechanism for the stepwise, Cu^{I} catalysed 1,3-DCR of an azide and acetylene [21].

moiety is a potential pharmacophore owing to its moderate dipole character and rigidity and can therefore be readily incorporated into a design strategy, rather than used as a passive linker between two respective fragments of structural space [24]. Indeed, several examples exist within the literature which describe the biological activity of 1,2,3-triazoles, including antiHIV-1 [25], antibacterial [26], selective β3 adrenergic receptor inhibition [27], antiplatelet activity [28], and antiinflammatory [29] agents to name a few.

More recently, click chemistry has emerged as a powerful conjugation strategy for the preparation of structurally diverse neoglycoconjugates of biomedical interest. This chapter aims to highlight recent developments appearing within the literature concerning the use of click chemistry in carbohydrate based drug discovery and glycobiology. Topics include small molecule probes and drug leads, multivalent neoglycoconjugates acting as lectin inhibitors and potential vaccines, and bioconjugation strategies for labelling of engineered cell surface glycans. This chapter aims to be comprehensive with commentary on future perspective.

2. Small Molecule Probes and Drug Leads

2.1. General

Heterocyclic glycoconjugates are widespread in nature. Obvious examples are the nucleosides and nucleotides, the primary building blocks of the nucleic acids. The design of inhibitors of DNA replication based on the mimicry of nucleoside scaffolds has, for decades, been the cornerstone of rational antiinfectives and anticancer drug research [30]. The 1,3-DCR's on carbohydrates, in this case, have proven to be of tremendous utility in such research programmes and have been the subject of previous reviews [31]. Heterocyclic carbohydrates from natural product origin which contain exocyclic nitrogens act as strong inhibitors of certain sugar processing enzymes and are thus valuable antibiotic, antifungal and antiparasitic lead compounds [32]. Examples include Nagstatin, a natural product consisting of a fused tetrahydroimidazo[1,2-*a*]pyridine ring system and a potent inhibitor of β-hexosamidases [33], the Nikkomycin nucleoside-peptide antibiotics isolated from *Streptomyces sp.* which are potent chitin synthase inhibitors [34], the trehazolins [35] and allosamidines containing exocyclic nitrogen heterocycles acts as potent trehalase and chitinase inhibitors, respectively [36]. Intramolecular 1,3-DCR's between azides and alkynes have also been utilised to generate novel fused bicyclic polyhydroxylated alkaloids as potential analogues of bioactive analogues of the indole and indolizodine type [37].

2.2. Glycosidase and Glycosyltransferase Inhibitors

Heterocyclic glycoconjugates are potential mechanistic probes or drug leads owing to their hydrolytic stability towards glycosidases and glycosyltransferases. The strong dipole character and rigidity of the triazole ring can potentially mimic the charge build up and flattening of the carbohydrate ring that is normally associated with the oxocarbenium like ion transition state [32]. In addition, the modular synthesis of carbohydrate triazole analogues by click chemistry lends itself particularly well to the parallel synthesis of small molecule carbohydrate based libraries as potential drug leads.

Recently, Basu and Rossi have demonstrated the β-glycosidase inhibition of 1-β-D-glucosyl-4-phenyl triazole **1** and 1-β-D-galactosyl-4-phenyl triazole **2** as a new class of aglycone modified β-glycosidase inhibitor (Figure 1) [38]. Compounds were screened at 0.24 mM against sweet almond β-glucosidase, *Escherichia coli* β-galactosidase (ECG), and bovine liver β-galactosidase in a preliminary fashion using known glycosidase inhibitors deoxynojirimycin (DNJ) and deoxygalactonojirimycin (DGJ) as comparative reference compounds. The phenyl triazoles were shown to be weak inhibitors of glycosidases and importantly were not substrates for enzyme hydrolysis. In a similar fashion, Périon et al. have employed a mild non-catalysed 1,3-DCR to generate a series of saccharidyl-1,4,5-trisubstituted-1,2,3-triazoles as heterocyclic analogues of the well known glucoamylase inhibitor and antidiabetic drug, acarbose [39].

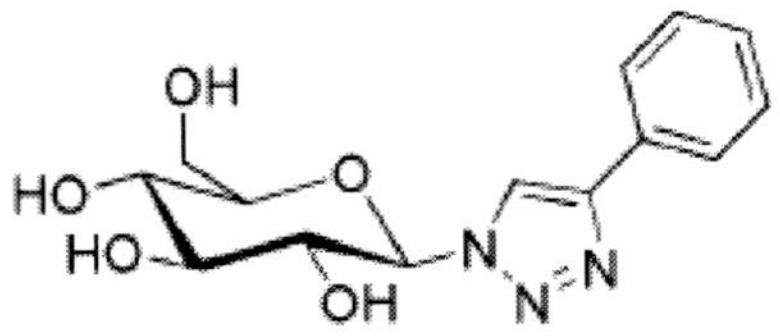

1

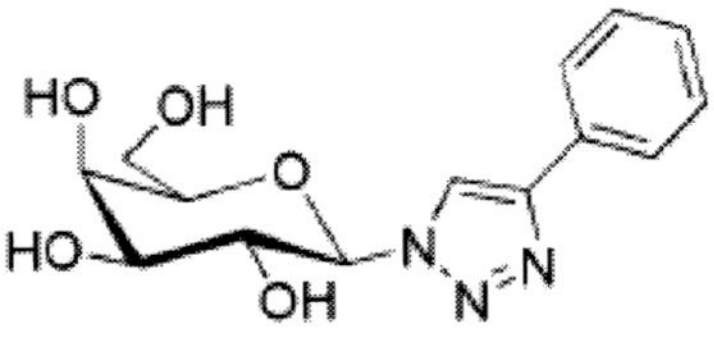

2

Figure 1. 1-Glycosyl-4-phenyl triazoles 1 and 2 as a new class of aglycone modified β-glycosidase inhibitor [38].

Influenza virus A and B Sialidase (EC 3.2.1.18) is strongly inhibited by 2,3-dianhydro-*N*-acetylneuraminic acid derivatives such as the clinically used drug, Relenza **3** (Zanamivir®) [40]. Motivated by the emergence of avian flu strains (e.g. H5N1) and the need for new frontline antiviral agents, Jiang and colleagues [41] reported the antiinfluenza activity of a library of triazole analogues of Relenza as avian influenza inhibitors. Systemic modification of C-4 ring substituents with amino or guanidino has led to potent and selective inhibition of viral neuraminidase [33, 42]. Thus, the C-4 azido neuraminic acid derivative **4**, which has previously served as a synthetic precursor to **3**, was used as the scaffold for library design (Figure 2). A library of 16 triazole analogues modified in the C-4 position were prepared by click chemistry and were assayed using a neutral red (NR) assay with the data reported as percent protection of MCDK infected cells at 50 μM inhibitor concentration. The protective rate of the triazole analogue **5** against the H5N1 strain was found to be comparable to that of Zanamivir (>64% *cf.* 86%).

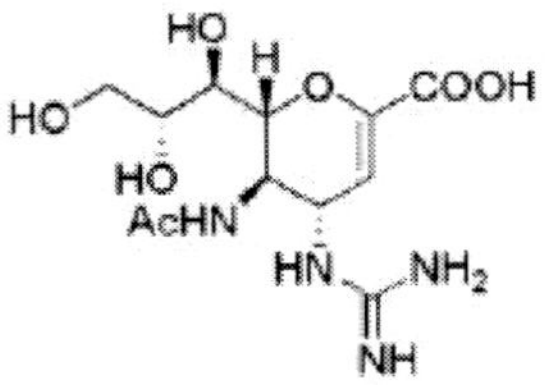

3

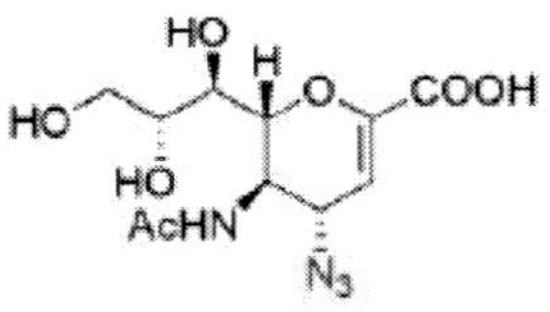

4

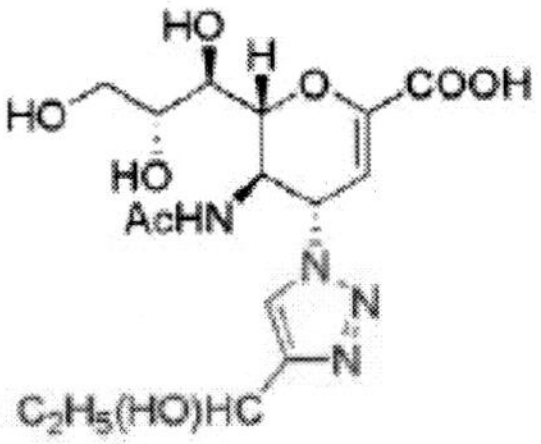

5

Figure 2. Relenza 3 and analogues 4 (azide) and 5 (the C-4 modified 1,2,3-triazole generated by click chemistry) [41].

There are at least five distinct human α-1,3-fucosyltransferases (α-1,3-FucT III to VIII, EC 2.4.1.152) known which catalyse the transfer of L-fucose from the β-configured GDP-fucose donor to a 3-OH group of an *N*-acetyl lactosamine acceptor to give the Lewis X trisaccharide antigen (Le^X) [43]. Alternatively, some fucosyl transferases (eg αFucT V) utilise the 3'-*O*-sialylated lactosamine acceptor to give the sialyl Lewis X antigen (sialyl Le^X). Both antigenic oligosaccharides are key components implicated in oncogenesis and tumour metastasis, as well as inflammation. Inhibitors of this enzyme are thus important leads in the development of antiinflammatory and antitumour agents [43, 44].

Owing to the complex four partner enzymatic transition state and the low catalytic proficiency of α-1,3-FucT, the successful design of potent inhibitors has been limited. Strongest inhibition lies in the design of substrate donor analogues since most of the binding energy lies in the retention of the GDP moiety [43], yet the strongest inhibitors to date have mainly been in the μM range and too weak to constitute effective leads. Recently, Wong et al. [45] identified a nanomolar inhibitor of α-1,3-FucT from a library of hydrophobic aglycone donor substrate analogues using a novel *in situ* click chemistry screening approach. The library of 85 donor substrate analogues were prepared by click chemistry in microtitre plate format and screened *in situ* within the same well in which they were synthesised (Scheme 3). The GPD donor analogue **6** was identified as the most potent FucT VI inhibitor known to date (K_i = 62 nM).

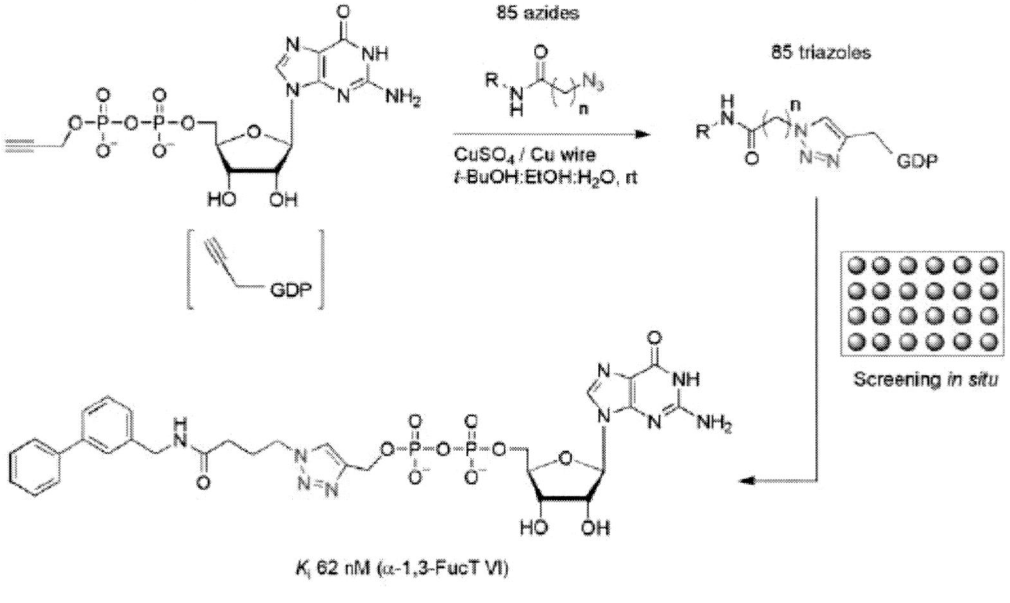

6

Scheme 3. Hydrophobic donor substrate analogue 6 identified by click chemistry and *in situ* screening as a potent inhibitor α-1,3-FucT VI (K_i = 62 nM) [45].

Leishmania protozoa are the causative agent of Leishmaniasis infection which currently affects 12 million people worldwide [46]. Leishmaniasis is implicated in a host of disease states ranging from subcutaneous lesions to lethal visceral infections. Current therapy of leishmaniasis is far from adequate owing to the emergence of drug resistant strains, as well as the low efficacy, severe side effects, and high toxicity and cost of currently used drugs. The primary carbohydrate reserve of *Leishmania sp.* is a β-1,2 linked mannan (with an average degree of polymerisation of 4–40) which is biosynthesised by a specific β-1,2 mannosyltransferase [44]. It has been shown from previous genetic studies that the β-1,2 linked mannan oligosaccharides are important for parasite survival during intracellular infectivity stages within host macrophages [44, 47]. A detailed investigation into the substrate specificity and active site structure of the β-1,2-mannnosyl transferase is necessary for the development of potent inhibitors of this enzyme and novel Leshmaniasis therapies. Williams and co-workers [48] have recently reported the synthesis of a library of α-mannosides using click chemistry and were assessed as potential substrates for β-1,2-mannosyl transferase (*L. Mexicana*). Thus, two libraries of 14 compounds were prepared by conjugating seven aliphatic and aromatic azides and acetylenes to the structurally conserved α-propynyl mannoside and α-methyl 6-azido-6-deoxy mannoside scaffolds, respectively. Based on HPTLC and fluorography analysis of digest fractions, it was concluded that library members derived from the α-propynyl mannoside scaffold were tolerated as substrates, but not the library derived from the 6-azido mannoside. Also, the enzyme was found to tolerate a broad range of aromatic substituents, an important step in the development of β-1,2-mannosyl transferase inhibitors and new Leishmania therapies.

2.3. Carbonic Anhydrase Inhibitors

The carbonic anhydrase (CA, EC 4.2.1.1) Zn(II)-metalloenzymes are ubiquitous to all kingdoms. These ancient and evolutionary conserved enzymes are efficient catalysts for the reversible hydration of carbon dioxide (CO_2) to give bicarbonate (HCO_3^-) and a proton (H^+), a regulatory and life sustaining reaction underlying physiological processes such as respiration and photosynthesis, the provision of HCO_3^- for crucial biosynthetic pathways (gluconeogenesis, lipogenesis, ureagenesis), the regulation of CO_2 and pH, and electrolyte and fluid secretion [49]. To date, 15 isoforms belonging to α-CA gene family have been discovered in humans (designated hCAs). Their expression dysregulation is implicated in a host of disease states including oncogenesis, elevated intraocular pressure, obesity, memory loss, depression, arteriosclerosis and renal pathology [49]. The modulation of CA activity through inhibition or activation is a potential avenue for therapeutic intervention [50]. The primary pharmacophore for CA recognition and inhibition is the primary aryl sulfonamide (or groups bioisoteric to this functional group, such as sulfamates and sulfamides) [49]. The variable tissue expression and subcellular locations of CA isozymes provides a pharmacological footprint wherein broad specificity CA inhibitors such as acetazolamide exhibit unwanted side effects. This has prompted the development of isozyme selective CA inhibitors as well tissue specific delivery systems. Owing to the strong conservation of amino acid sequence and CA active site architecture, the rational design of isozyme selective inhibitors is a challenging process.

Our group has recently prepared a library of benzene sulfonamides conjugated to monosaccharide and disaccharide "tails" as a dual isozyme-differentiating and solubilising strategy [51, 52, 53]. A total of 50 glycoconjugate benzene sulfonamides were efficiently prepared in parallel fashion using a "click-tailing" strategy developed in our laboratory (Scheme 4). The triazole-linked glycoconjugates were screened using a CO_2 hydration assay against the cytosolic hCA I and II isozymes, as well as the transmembrane hCA IX, XII and XIV isozymes possessing extracellular catalytic domains. Inhibition of these isozymes was in most cases non-clustered indicating that the carbohydrate tails were able to impart selectivity for the isozymes relative to the nonselective parent sulfonamide scaffolds. The methyl β-D-glucuronate triazole **7** was found to be a potent inhibitor of the tumour associated isozyme, hCA IX (K_i = 23 nM), comparable to the nonselective bis-sulfonamide indisulam **8**, which is in phase II clinical trials for the treatment of solid tumours. Compound **7** was 16-fold selective for hCA IX over the physiologically dominant hCA II isozyme, an important finding in the quest for new cancer chemotherapeutics.

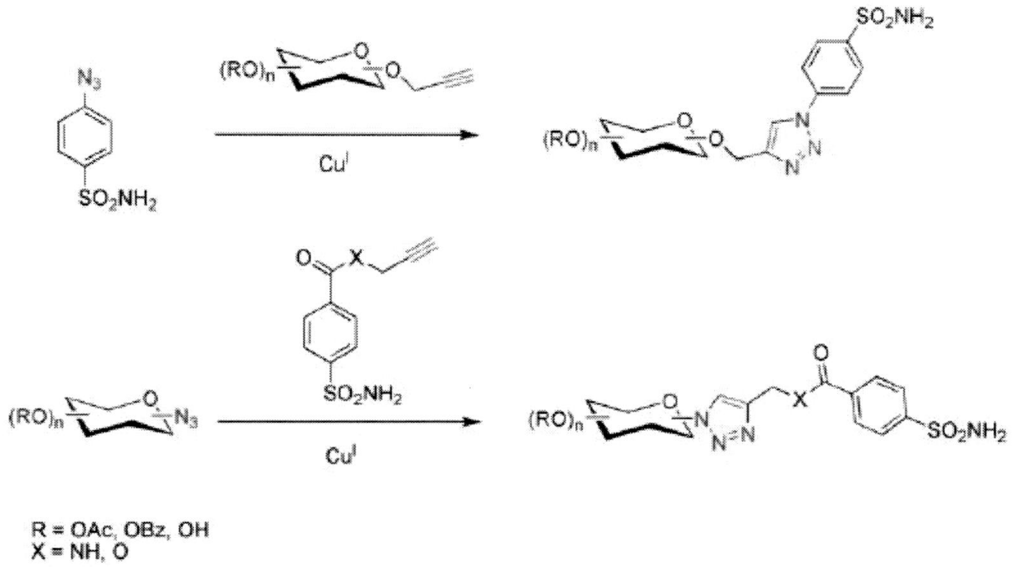

Scheme 4. "Click-tailing" of sugar azides and acetylenes onto benzene sulfonamide scaffolds to generate isozyme selective human carbonic anhydrase inhibitors [51-53].

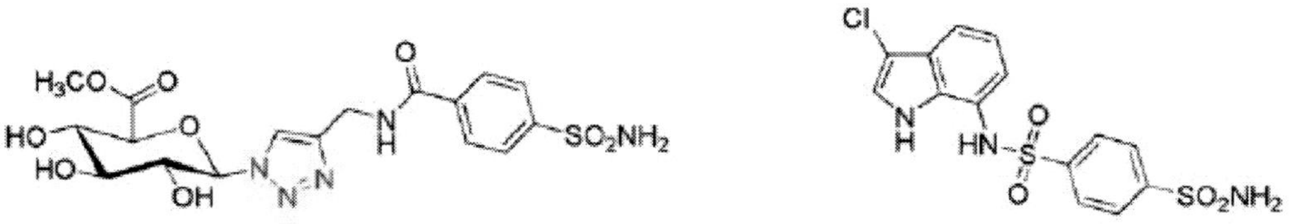

7

K_i 23 nM, K_i (hCA II/hCA IX) = 16.4

8

K_i 24 nM, K_i (hCA II/hCA IX) = 0.63

Figure 3. Methyl β-D-glucorunate derivative 7 was a potent and selective inhibitor of tumour-associated carbonic anhydrase isozyme IX [51].

2.4. Glycoamino Acid and Glycopeptide Mimics

Glycoprotein based therapeutics constitute more than one third of approved biopharmaceuticals [3]. Glycoproteins and peptides that contain glycans commonly *N*-linked to an asparagine (belonging to the Asn-X-Ser/Thr consensus sequence) or *O*-linked to Ser or Thr residues play pivotal roles in biology. The branched glycan component (as an poly-, oligo-, or monosaccharide) are not only important for protein folding, stability and specificity, but can determine the primary function of the mature glycoprotein as a modulator of intercellular communication [1]. The *de novo* synthesis method is a viable approach for accessing native glycoforms and artificial glycoproteins (neoglycoproteins) of biomedical interest [4, 54]. The conjugation of a saccharide to a protein or peptide often results in the enhancement of an immune response to a carbohydrate antigen or the *in vivo* stability and biological activity of a peptide drug [55].

The theme strongly reverberated in the mimicry of endogenous glycopeptides is the substitution of labile *N*- and *O*-glycosidic bonds with robust linkages that also mimic the geometry and electronics of the native glycopeptide bond [56]. Current impetus is now directed toward the ligation of preformed oligosaccharide and peptide fragments containing mutually reactive functional groups in aqueous solution, a term coined native chemoselective ligation [16]. Click chemistry of preformed carbohydrate and peptide fragments bearing azide and acetylene functionalities has numerous advantages over traditional glycosylation and peptide bond formation chemistry and has emerged as a promising ligation strategy. It allows for the mild, chemoselective and stereospecific conjugation of preformed building blocks/fragments under aqueous conditions and is insensitive to global protecting group effects or synthetic conditions. Azide and acetylenes are also powerful functional groups since they are relatively inert to a wide range of synthetic transformations occurring on the peptide or oligosaccharide component (e.g. glycosylation, protecting group chemistry, peptide coupling etc). Triazole linked glycomimetics are also chemically robust entities towards a wide range of synthetic transformations commonly encountered in carbohydrate chemistry, as has recently been demonstrated by our group [57].

Click chemistry can therefore greatly simplify synthesis and can be introduced at virtually any stage of the glycoprotein/peptide synthetic pathway (convergent or linear strategies). From a mimetic point-of-view, 1,2,3-triazole linked glycoamino acid mimics share similar structural and electronic features to that of the native amide counterparts and are relatively impervious to enzyme or digestive hydrolysis [22a]. 1,2,3-Triazoles have a larger dipole moment compared with amide bonds (~5.0 D by *ab initio* calculation cf. 3.8 D of *N*-methylacetamide) and can potentially enhance amide mimicry due to their weak hydrogen bond acceptor properties [22]. The dipole character can also be fine-tuned depending on the electronic effects of the neighbouring groups. For example, the C-5 bond can function as a potential hydrogen bond donor, much like an amide proton given suitable polarisation exists between the N-2 and N-3 bonds of the triazole ring. Distances between the atoms in the 1 and 4 position of the triazole ring are similar compared to the distance between the α carbon (R) and the carbon attached to the amide (R'), 5.0 Å cf. 3.8 Å (Figure 4) [22a].

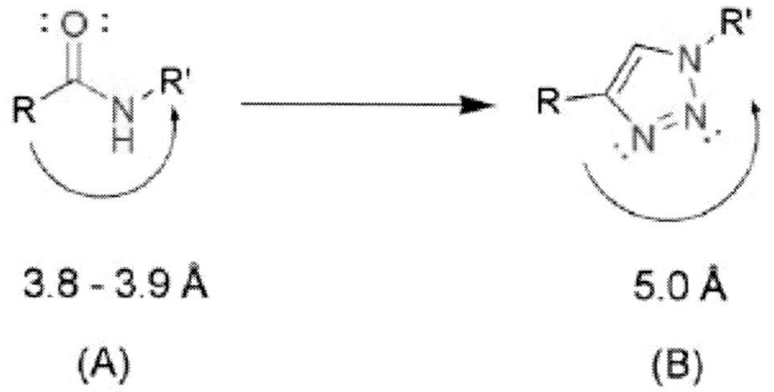

Figure 4. 1,4-Disubstituted-1,2,3-triazoles (B) can function as a potential mimics of glycopeptide bonds (A) due to similar geometric features [22a].

Since the pioneering work of Meldal and colleagues [18b] who described the first use of click chemistry to generate glycopeptides on solid support, there has been a flurry of activity within the literature concerning the development of glycoamino acid and glycopeptides using this synthetic strategy. In an a comprehensive study, Rutjes et al. [58] have recently reported the expedient synthesis of *N*-glycosyl and *C*-glycosyl triazoles as glycoamino acid mimics from glycosyl azide and ethynyl glycosides, respectively (Scheme 5). Click chemistry between acetylene or azide functionalised amino acids and the respective saccharide building blocks in solution resulted in the formation of triazole- glycoamino acids under mild conditions in moderate to good yields. The bifunctional glycoamino acids could potentially then undergo subsequent peptide couplings at either the carboxy or amino terminus. In a preliminary study, the triazole moiety was also shown to be stable to mineral acid and base treatment. Likewise, ethynyl glycosides have also been exploited by Massi and co-workers [59] for the efficient generation of *C*-glycosyl isoxazole and *C*-glycosyl triazaole α-amino acids. In the latter case, the Cu^I catalysed cycloaddition of the various ethynyl glycosides with a suite of azide functionalised amino acids proceeded smoothly at room temperature in toluene in good yields.

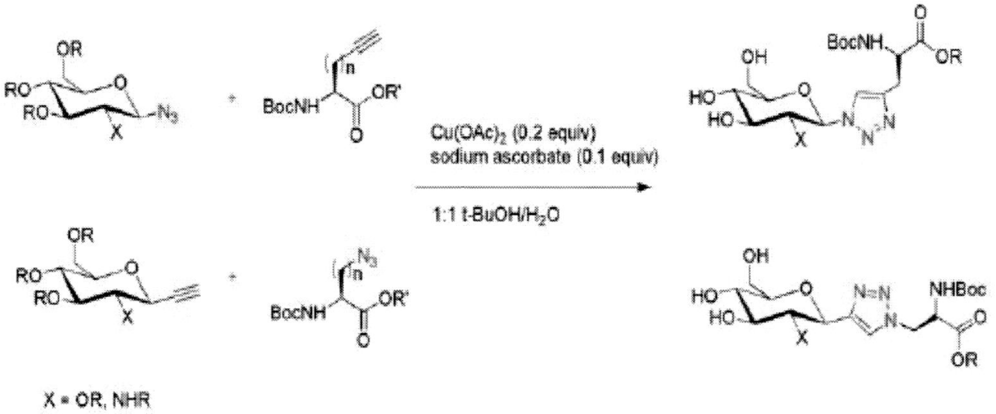

Scheme 5. Synthesis of glycosyl triazole linked glycoamino acid and glycopeptide mimics [58].

In a simple, yet elegant study Macmillan and Blanc [60] have demonstrated the compatibility of click chemistry to a native chemical ligation strategy employing a chemoselective nucleophilic displacement of alkyl bromides with side chain cysteine thiols. The glycoamino acid mimics were based on *N*-acetyl glucosamine, *N*-acetyl lactosamine and chitobiose, all major constituents in endogenous *N*-linked glycoproteins. An important finding revealed the unimportance of the order in which the chemoselective ligation and click chemistry events took place in constructing the triazole modified neoglycopeptides. In this respect, the click reaction itself could potentially function as the chemoselective ligation event, as demonstrated by the model reaction of benzyl mercaptan with 2-bromoacetyl propynyl amide or with the preformed glycosyl triazole glycoamino acid mimic containing the bromoacetamide moiety (Scheme 6).

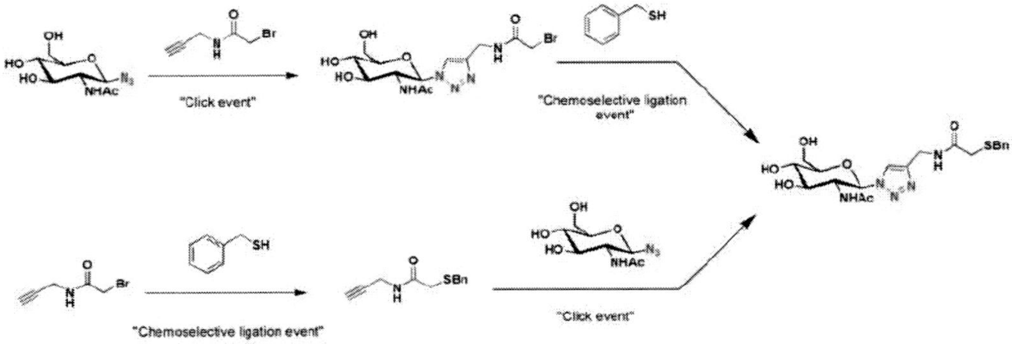

Scheme 6. Synthesis of glycopeptide mimics using a dual chemoselective ligation/click chemistry approach [60].

Click chemistry has recently been used as a glycosylation tool to modify the physicochemical properties of an azabicycloalkane amino acid peptidomimetic of a homoSer-Pro dipeptide [61]. The click reaction between acetylated and free propynyl β-D-glucopyranoside with various azide functionalised peptidomimetic scaffolds afforded the triazole conjugate in good yields (72-84%) under aqueous conditions. In a similar fashion, 1,2,3-triazole analogues of pyrazinone based peptidomimetics have been generated by click chemistry of various glycosyl azides with substituted ethynyl pyrazinone scaffolds [62].

Many carbohydrate macrocycles of natural product origin, including glycosylated cyclic peptides, possess remarkable biological activities and diverse structural characteristics. Indeed, several are used clinically as antibiotics and antifungal drugs [9]. The macrocyclisation and glycosylation are two key structural modifications that occur late in the biosynthesis and confer much of the unique physicochemical properties and biological activity *per se*. Westermann and colleagues [63] have recently utilised click chemistry in combination with ring closing metathesis in the preparation of a series of macrocyclic glycolipid mimics. The synthetic strategy involved a consecutive click ligation and ring closing metathesis using dual functionalised carbohydrate building blocks incorporating azide and olefin groups.

Motivated by the emergence of vancomycin resistant strains of *Enterococcus faecium* (VRE), Thorson and co-workers [64] have reported the synthesis and biological activity of a library of 1,2,3-triazole modified vancomycin analogues. From a library of 15 compounds, they identified a triazole derivative that was twice as effective as vancomycin against antibiotic resistant strains *E. faecium* and *S. aureus*. Walsh and Lin [65] have also exploited click chemistry as part of a chemoenzymatic approach to prepare carbohydrate modified cyclic peptide tyrocidine (tyc) antibiotics with lowered toxicity. A library of 247 glycopeptides was prepared by employing a thioesterase mediated macrocyclisation followed by Cu^{I} catalysed triazole formation in 96-well microtitre plate format. Following antibacterial and haemolytic assays, two glycopeptides (Tyc4PG-14 and Tyc4PG-15) were identified as retaining the antibacterial potency of the native peptide, whist increasing its therapeutic index by a factor of 6 (Figure 5).

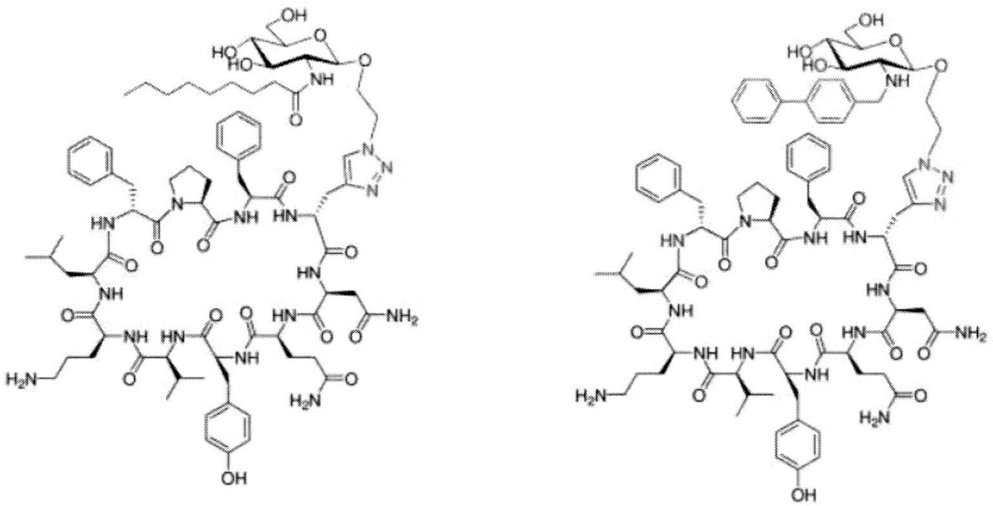

Tyc4PG-14 **Tyc4PG-15**

Figure 5. Glycopeptide antibiotic analogues of tyrocidine (tyc), Tyc4PG-14 and Tyc4PG-15. [65]. Reprinted with permission from Lin, H., Walsh, C. T. *J. Am. Chem. Soc.* 2004, *126* (43), 13998-14003. Copyright ©2004 American Chemical Society.

2.5. Bio-Orthogonal Chemical Reporter Strategies

The development of bioorthogonal chemical methods for profiling disease related glycoproteins expressed on living cell surfaces is an important and recent development in glycomics research and carbohydrate based drug development [66]. In particular, the biosynthetic incorporation of unnatural saccharides into glycan epitopes expressed on living cell surfaces is a recent strategy described by the Bertozzi group and has provided considerable insight into glycosylation structure and function *in vivo* [47, 67]. The principle is called metabolic oligosaccharide engineering and involves the incorporation of synthetically modified saccharides containing an inert but mutually reactive functional group into a host glycoprotein by exploiting a particular metabolic pathway. The favourable size and inertness of azide and acetylene functionalised substrates within living cellular environments as well as the promiscuity of certain glycosyl transferases enables these abiotic substrates it to "slip through" the cellular catalytic machinery and become incorporated into cell surface glycans. The residues then can serve as chemoselective handles for bioconjugation to a fluorescent marker for visualisation, or an affinity tag for proteomic enrichment. In this respect, click chemistry has been instrumental as a bio-orthogonal tagging strategy, owing largely to the biocompatibility of the reaction and the inertness of azides and acetylenes within complex biological environments.

In an early example, Bertozzi and co-workers [68] utilised a strain promoted cycloaddition of a cyclooctane functionalised biotin probe with recombinant glycoproteins and cell surface glycoproteins displaying *N*-azidoacetyl sialic acid (Sia-NAz) residues. The 1,2,3-triazole products were detected in both recombinant systems and on the cell surface of human T-lymphoma Jurkat cells by Western blot analysis and flow cytometry, respectively. In the latter case, fluorescence measurements were shown to be dose dependent and proportional to the density of the Sia-NAz residues on the cell surface glycoproteins as well as the incubation time with the cyclooctane biotin probe.

L-Fucose is a ubiquitous component in sialyl Lewis^y (sialyl Le^y) and sialyl Lewis^x (sialyl Le^x) antigenic selectins found unregulated in certain cancers [8b, 69]. The Bertozzi group recently has recently utilised click reaction as a key bioconjugation step in a fluorescent reporter strategy for the chemical profiling of fucosylated glycoproteins as new cancer biomarkers (Scheme 7) [70]. Utilising the fucose salvage pathway and the promiscuity of fucosyl transferases, three synthetic azido fucose analogues, functionalised with azide at the 2, 4 and 6 positions, respectively, were incorporated into cell surface glycans of human T-lymphoma cell line Jurkat. The azide markers were incorporated into the glycan and presented on the cell surface by culturing the cell culture with 125 μM of each azide analogue. The cells were lysed with an NP-40 detergent which is known to be compatible with click chemistry. The click bioconjugation of the cell lysates with acetylene biotin marker, followed by binding of triazole tagged glycoprotein to α-biotin-horse-radish peroxidise conjugate (HRP) allowed for Western blot analysis. Significant protein labelling (up to 40% of all available fucosylation sites) of only the 6-azido analogue was observed by Western blot analysis of the hydrolysates, although this analogue was deemed cytotoxic at elevated culture concentrations (100 μM). It was postulated that azide analogues substituted at C-2 and C-4 were not acceptable substrates for the fucose salvage pathway.

Very recently, Wong and co-workers [71] have developed a click activated fluorogenic labelling technique to analyse cell surface glycoprotein fucosylation *in vivo* (Scheme 8). In a similar manner to that reported by Bertozzi and colleagues, the artificial 6-azido and 6-ethynyl fucose substrates were incorporated into cell surface glycoproteins of human lymphoma Jurkat cell lines via the fucose salvage pathway. The extent of cell surface glycoprotein fucosylation was analysed *in vivo* by flow cytometry and fluorescent microscopy following the click bioconjugation of the fluorogenic probe to the functionalised L-fucose analogues. Fluorescence was only detected for the 6-azido analogue, following fluorogenic bioconjugation with 4-ethynyl-*N*-ethyl-1,8-naphthalimide, indicating selective uptake and integration of this artificial substrate into cell surface glycans.

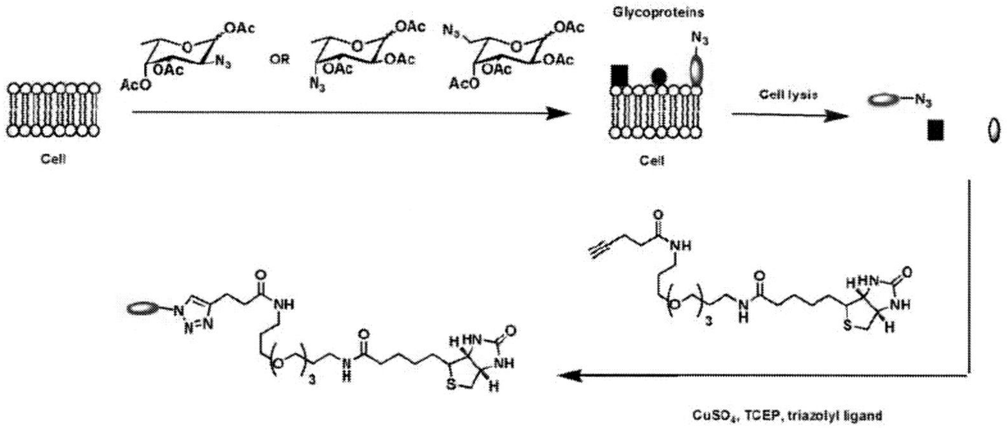

Scheme 7. Metabolic engineering of human T-lymphoma cell line Jurkat and labelling of azido L-fucose derivatives with a biotin reporter probe using click chemistry [70].

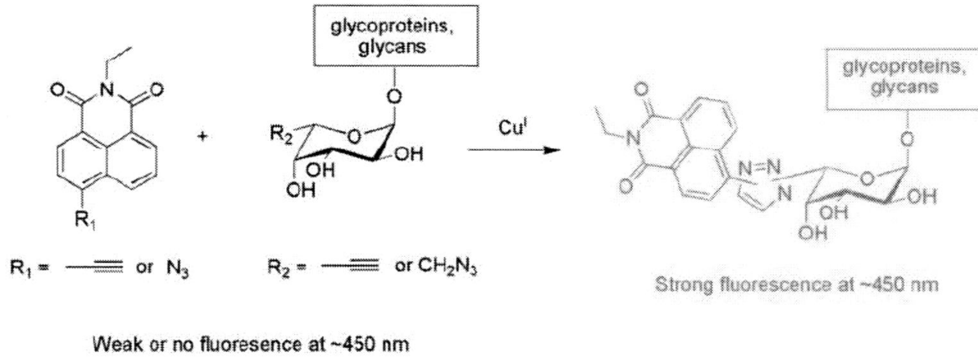

Scheme 8. The bioconjugation of the fluorogenic naphthalene probe to a functionalised L-fucose derivative by click chemistry [71].

3. Multivalent Neoglycoconjugates

3.1. General

Nature compensates for the weak binding affinity between a monovalent carbohydrate ligand and cognate receptor protein (K_a in the ranges of $10^{3\text{-}4}$ M^{-1}, although exceptions are known) by expressing multiple copies of the carbohydrate recognition epitope, or multiple receptor binding sites on a cell surface receptor - a concept known as the "multivalent" or "glycoside cluster effect" [72]. The overall effect is the increase in binding affinity to an extent which exceeds the sum of the constituent binding events in sufficient strength to effect or inhibit a cellular or physiological response. By exploiting one or more of these mechanisms, chemists have assembled a wide range of structurally diverse polymeric artificial carbohydrate polymers (glycopolymers) as effectors or inhibitors of cellular processes and analytical tools for investigating carbohydrate-protein binding events [73]. Artificial glycopolymers hold great promise as inhibitors of bacterial adhesion [74] and anticancer vaccines [75], but also sustainable and economic biomaterials with a wide range of biomedical and drug delivery applications, including biocompatible matrices for tissue engineering [76] and sustained drug release and as polymeric vehicles for targeted drug and gene delivery [77]. Apart from their widespread biomedical applications, glycopolymers can also function as mimics of naturally occurring polysaccharides and display antiinflammatory, anticoagulant and antitumour properties and are thus viable leads for drug development and vaccine development [78,79].

3.2. Glycopolymers and Polysaccharide Mimics

Novel synthetic methods are critical for the development of innovative glycopolymers of biomedical and pharmaceutical importance. Yet, devising novel synthetic pathways for developing biocompatible glycopolymers that are well defined and tailored for a specific purpose is a continuing challenge faced by carbohydrate chemists [79]. Subtle changes in polymer structure and pendant attachments can have profound effects on the mechanical properties and biological activity of the material. Motivated by the requirement for more expedient synthesis of multivalent glycopolymers, Haddleton and co-workers [80] have recently reported the use of a novel, dual click chemistry/transition-metal-mediated-living-radical-polymerisation (TMM-LRP) strategy for the facile preparation of a linear glycopolymer library containing pendantly attached carbohydrates. The acetylene functionalised poly(methylacrylate) polymer was assembled through TMM-LRP and was used as a subsequent scaffold for combinatorial design. A series of glycosyl azides and azido ethyl glycosides were then grafted by click chemistry in a highly efficient manner to the acetylene functionalised polymer backbone providing rapid access to glycopolymer libraries with different bulk properties and potential protein binding affinity. The work demonstrates the powerful utility of click chemistry in preparing pendant-functionalised glycopolymers in a combinatorial fashion.

Reineke et al. [81] have recently utilised click chemistry as a pivotal step in the preparation of glycopolymer-nucleic acid complexes (polyplexes) as gene delivery agents. The synthetic poly(glycoamidoamine) co-polymers were chemically tailored to be non-toxic and biocompatible, whilst improving genetic transfection efficiency by stabilising plasmid DNA (pDNA) complexation, and preventing cellular disassociation, aggregation and biological degradation of the polyplex system prior to delivery at the therapeutic site. On the basis of previous studies implicating heterocyclic-amide co-functionalised polymers as strong DNA binding agents, the triazole-amide moiety (two per repeat unit) was postulated as a strong DNA binding motif via hydrogen bonding and van der Waals interactions. Diazido trehalose units were efficiently polymerised by click chemistry to a variety of dialkyne-amide-terminated oligoamine monomers containing 1, 2 and 3 secondary amine(s) (Scheme 9). *N*-Boc deprotection of the ethyleneamine groups resulted a library of biocompatible and polycationic carbohydrate co-polymers with similar degrees of polymerisation, dispersity, molecular weights and polymer stiffness in aqueous solution. The highly cationic nature rendered the polymers capable of phosphate charge neutralisation and polyplex formation. To elucidate the effect of the 1,2,3-triazole ring and the relative number of ethyleneamine groups in the polymer backbone on pDNA complexation, a series of gel electrophoresis and fluorescent ethidium bromide assays were conducted. Cellular delivery efficacy into HeLa cells was also determined using fluorescent flow cytometry analysis.

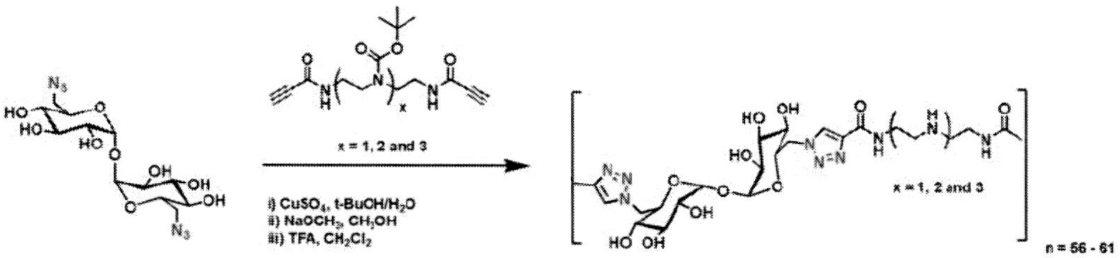

Scheme 9. Click chemistry polymerisation of triazole functionalised, trehalose poly(glycoamidoamine) co-polymers as stabilised polyplex delivery agents [81].

Nature's ability to perform myriad chemical transformations under exquisite regio- and stereocontrol within complex, polyfunctional chemical environments has captured the imagination of synthetic organic chemists for many years. The ability to perform multiple transformations on a single substrate *in vitro* using orthogonal chemistries is a key challenge facing chemists in the quest for novel materials and drugs [82]. Two possible routes have emerged; either conducting multiple transformations simultaneously on separate functional groups within the same molecule, or multiple transformations on the same functional group in a "cascade-type" fashion. Impressive examples have emerged thus far and these are primarily concerned with total synthesis of natural products and peptides [83]. However its adaptation to polymer science has been limited owing to the difficult nature of these polyfunctional macromolecules. The orthogonal functionalisation of polymeric macromolecules requires methods which operate under high fidelity, mild conditions and with strong regio- and stereocontrol [77]. Click chemistry provides appropriate scope in glycopolymer chemistry, as it can be used as a polymerisation tool or to functionalise preformed polymers with equal fidelity under orthogonal reaction conditions.

Hawker and colleagues [84] have recently exploited click chemistry as a mild, regio- and stereoselective strategy for performing cascade functionalisations on glycopolymers. In the simultaneous method, a pendant acetylene functionalised polystyrene terpolymer was reacted with 2,3,4,6-tetra-*O*-β-D-glucopyranosyl azide using click chemistry whilst a hydroxyl pendant side chain was reacted simultaneously with an *N*-hydroxysuccinaimde (NHS) activated ester, affording the functionalised co-polymer containing a triazole and ester pendant sequence (Scheme 10A). On the other hand, the cascade mechanism proceeded via the sequential reaction of a dual ester-functionalised polyacrylate with propargyl amine affording the propynyl amide which was subsequently grafted to the glucosyl azide in a one-pot manner (Scheme 10B).

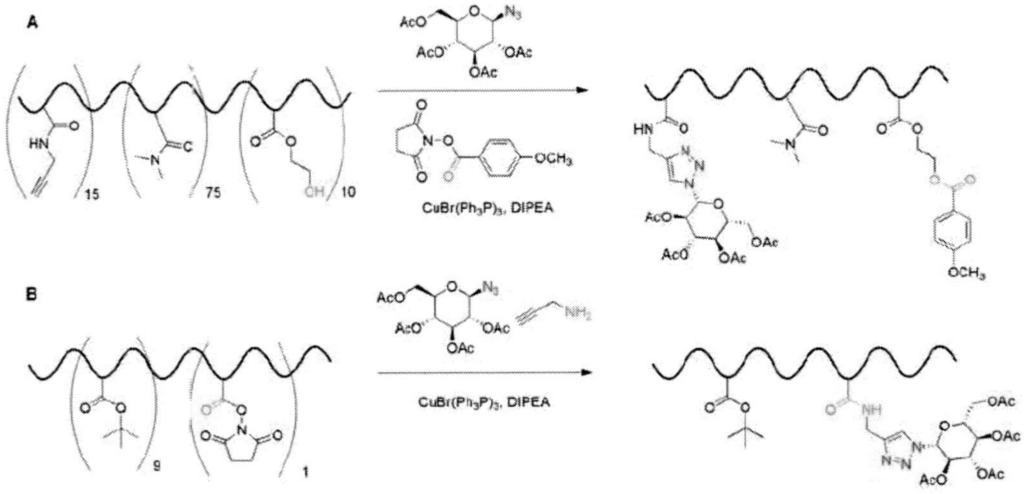

Scheme 10. Preparation of glycopolymers with pendant functionality through (A) the simultaneous functionalisation of a water soluble terpolymer and (B) the cascade functionalisation of an ester functionalised polyacrylamide [84].

Cellulose is an abundant natural biomaterial and its synthetic modification can lead to materials possessing bulk properties that are suitable for a wide range of industrial and biomedical applications [76]. In the pharmaceutical industry, functionalised cellulose homopolymers (cellulosics) have been extensively used as biodegradable enteric coatings for controlled drug release formulations and more recently for bone, cartilage and brain tissue engineering owing to theri excellent biocompatibility [76]. The first example of the chemical modification of cellulose using click chemistry was recently reported by Heinze and co-workers [85]. Click chemistry of the 6-azido-6-deoxy cellulose scaffold afforded triazole spaced carboxyester, thiophene and aniline functionalised cellulosics displaying degree of substitution (DS) values up to 0.9 and good solubility in organic solvents (DMF, DMSO). The click methodology offers hydrolytically stable bonds between the polymer backbone and functional groups and overcomes the formation of polymer crosslinking byproducts that are typically observed with conventional cellulose esterification.

The conjugation of multivalent antigenic oligosaccharides to preformed biomacromolecules, in particular polypeptides, often results in an increase in an immune response and has become a primary strategy in the development of antitumour vaccines [75]. Research efforts in this area are now focused towards the preparation of multiantigenic constructs with the aim of stimulating the immune system towards a broad range of malignant transformations rather than one (i.e. using monomeric antigenic constructs). In this respect, more sophisticated and complex synthetic strategies will be required and are being sought [86]. In a proof-of-concept manner, the Danishefsky group has recently demonstrated the viability of click chemistry as a strategy for linking antigenic oligosaccharides to keyhole limpet hemocyanin (KLH) polypeptide carriers [87]. The authors concluded click chemistry to be a practical and orthogonal linker strategy owing to its high degree of chemoselectivity and tolerance towards other synthetic transformations. In addition, the compatibility of the chemistry under aqueous conditions permits the use of a broad range of carrier proteins/peptides proteins that are otherwise unstable or insoluble within organic solvents.

Many (1→3)-β-D-glucopyranan polymers, such as Lentinan, possess immunostimulating and anticancer properties [88]. Natural and functionalised (1→3)-β-D-glucopyranan polymers have been utilised as novel biomaterials possessing lectin and RNA binding affinity owing to their unique triple stranded helical superstructure [89]. In an effort to generate more regioselective methods for the preparation of functionalised glucan polymers, Shinkai and co-workers [90], have utilised click chemistry to chemically modify the (1→3)-β-D-glucan, 6-azido-6-deoxycurdlan. In a pilot study, the click reactions were performed with a variety of acetylene terminated modules with potential lectin binding, redox active, phytoactive, and fluorescent properties in order to develop novel polysaccharide based materials.

Dondoni and co-workers [91] have recently prepared a series of triazole linked (1,6)-oligomannose analogues as potential mimics of lipoglycan structures and high density oligomannose-type antimicrobial vaccines [92]. The orthogonally substituted ethynyl α-*C*-mannoside building block was conjugated to the 6-azido-α-*C*-mannoside platform to afford the triazole linked dimannoside coupling product. Using an iterative click strategy, involving the necessary installation of an azide group at C-6 and coupling to the ethynyl building block resulted in the facile construction of a structurally rigid triazole linked oligomannose over a total of four consecutive cycles. Whilst the click coupling reactions required up to 30 h for completion at room temperature, microwave irradiation at 100 °C dramatically shortened this (to five minutes), whilst retaining good yields and regioselectivity.

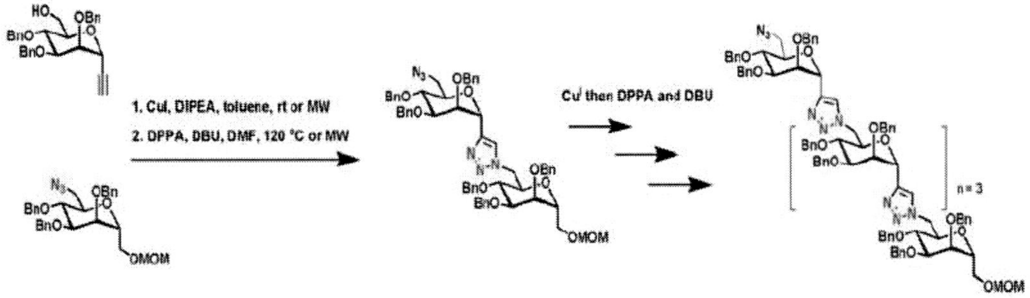

Scheme 11. Iterative assembly by of a triazole linked α-1,6-oligomannose analogue using click chemistry [91].

The first reported example of the use of click chemistry in the modular synthesis of branched oligosaccharide mimics was demonstrated by the work of Field and colleagues [93]. The facile synthesis of amylopectin fragments occurred in a single step between dipropynylated maltoside core fragments and glycosyl azide building blocks based on β-D-glucose, maltose, maltotriose and maltoheptose, respectively. Employing propynyl functionalised maltose templates substituted at C-4/C-6' and C-6/C-6', respectively, allowed for the template assisted click synthesis of the respective regioisomeric amylopectin analogues.

3.3. Cyclodextrins and Artificial Receptors

Cyclodextrins (CDs) are water soluble, cyclic malto-oligosaccharides derived from starch which possess a hydrophilic outer surface and a relatively lipophilic central cavity [94]. For quite some time, CDs have been used in drug and cosmetic formulations as solubilising carriers of otherwise hydrophobic and water insoluble molecules through the reversible formation of water soluble inclusion complexes but recently have been gaining considerable attention as topical delivery agents to improve the bioavailability of drugs administered by this route [95]. There has also been much interest in CDs as vehicles for the cell specific delivery of hydrophobic drugs to target sites through multivalent interactions with carbohydrate receptors [71]. CDs are also non-immunogenic and offer enhanced biocompatibility and water solubility compared to other multivalent carbohydrate constructs based on rigid macrocyclic scaffolds, such as azacrown ethers, cyclic peptides, calixarenes and cyclophanes [94, 96].

The first example of an artificial CD analogue incorporating a heterocyclic motif within the binding cavity was demonstrated by Vasella and co-workers [97] during ongoing structural and binding investigations of artificial CD analogues. The intramolecular, thermal Huisgen 1,3-DCR between a dual azide/acetylene functionalised amylose fragment proceeded under thermal conditions to afford the regioisomeric mixture of a heterocyclic β-CD mimic.

Click chemistry offers numerous advantages over existing cycloglycosylation and cyclooligomerisation methods, which are often impeded by long synthetic sequences, low cyclisation yields, as well as difficulties in resolving anomeric mixtures and/or macrocycles of different sizes. Click chemistry was originally proposed by Gin et al [98] as a high yielding, stereo- and regiospecific cyclodimerisation strategy for preparing well defined, modified β-CD analogues. The work reported the convergent click cyclodimerisation of a dual azide/acetylene functionalised trimannoside using a CuI/DBU catalytic system, affording the cyclic dimer in good yields (80%) with the concomitant formation of cyclic trimer (15%) (Scheme 12). The mannose based macrocycle also showed similar propensity to form inclusion complexes with hydrophobic fluorogenic molecules compared to native β-CD, as determined by fluorescence emission spectroscopy. In a recent extension from this work, a series of C_2 and C_3-symmetric oligosaccharide macrocycles were prepared using click chemistry in a convergent manner from a suitably difunctionalised monosaccharide and disaccharide, respectively [99]. The investigation of macrocycle cavity shape and the potential for host-guest interactions was conducted using computational and NMR shift titration methods, revealing a functionalised and sterically congested cavity which may prove useful as novel molecular pores or receptors for studying host-guest interactions.

Click chemistry has been used to prepare C_2-symmetric carbohydrate/amino acid hybrid macrocycles as potential water soluble inclusion complexes [100]. The cyclodimerisation of bifunctional azido-acetylene glycopeptide building blocks proceeded in acetonitrile using a CuI/DIPEA catalyst system in moderate yields. Interestingly, the analogous reaction was attempted using aqueous conditions prescribed by Sharpless and showed minor product conversion at elevated temperatures (70 °C). The novel carbohydrate/amino acid hybrid macrocycles were investigated as potential water soluble artificial receptors using

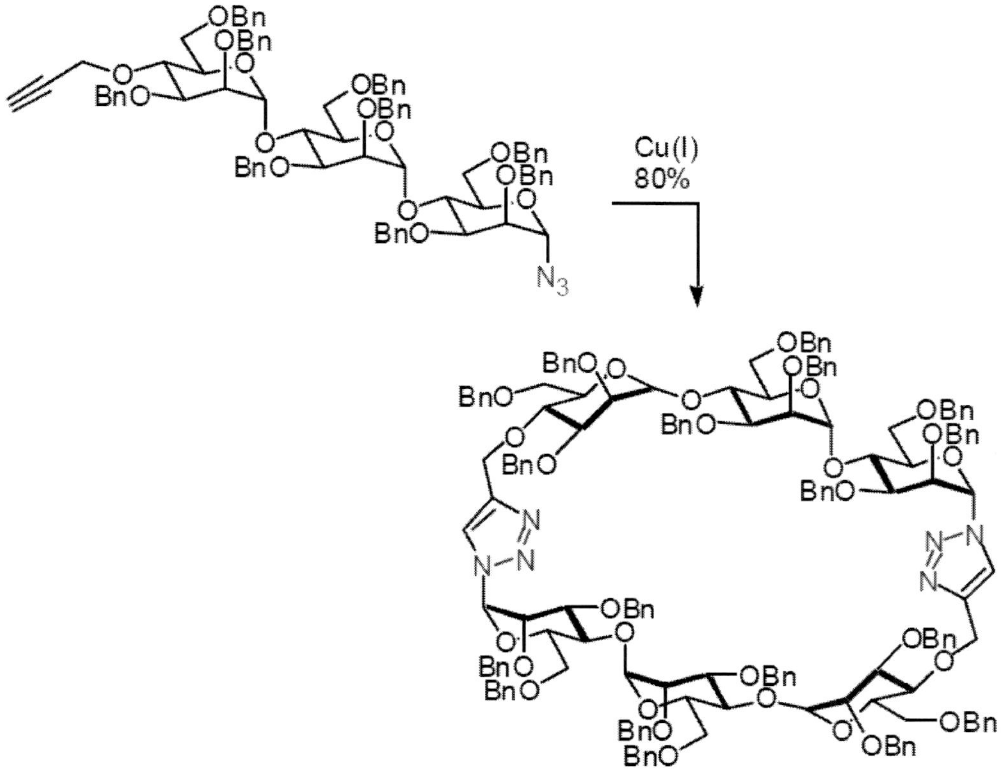

Scheme 12. Convergent synthesis of a manno-β-cyclodextrin using click chemistry [98].

computational methods. It was demonstrated that the 1,2,3-triazole motif imparted increased macrocycle rigidity and cavity size, relative to previously reported nontriazole analogues.

The synthetic modification of preformed CDs provides an alternative route to artificial CDs with particular properties. Exploiting differences in reactivity between primary and secondary hydroxyl groups orientated at the outer and inner faces of CDs, respectively, has allowed for the selective substitution of these groups. Schubert et al. [101] have recently reported the use of click chemistry in the preparation of seven armed star shaped polymer, heptakis-pεCL-β-cyclodextrin. Star shaped polymers have added advantages of over linear polymers since they have lower hydrodynamic volume and a larger number of functionalised end groups for further modification. The acetylene functionalised poly(ε-caprolactone) (pεCL) was prepared via a ring opening polymerisation of ε-caprolactone utilising 5-hexyn-1-ol initiator and tin(II)octanoate catalyst, which was in turn appended to heptakis-azido-β-cyclodextrin via click chemistry and microwave irradiation, affording poly(ε-caprolactone)-β-cyclodextrin in a single step.

3.4. Glycodendrimers, Glycoclusters and Lectin Inhibitors

Glycodendrimers are highly globular, monodisperse, oligo- or polymeric carbohydrate containing nanostructures. Carbohydrates are typically presented at the periphery of a dendritic core in high-densities with well defined, symmetrical branching patterns, and are prepared through a series of iterative steps via convergent or divergent synthetic strategies [102]. The innate structural control, monodispersity and relative size of glycodendrimers - in between that of glycoclusters and polyvalent glycopolymers, endow numerous advantages which have been successfully exploited for a variety of biological and biomedical applications, including multivalent tools for investigating carbohydrate-protein interactions, microbial antiadhesives, cancer vaccines, and as biocompatible scaffolds for drug and gene delivery purposes [78, 103].

The successful synthesis of structurally well defined glycodenrimers and glycoclusters must employ chemistries that maximise atom efficiency. Such chemistries should display a high degree of tolerance towards polyfunctionality and provide access to structurally diverse architectures in the smallest number of steps. Click chemistry has allowed for flexible and rapid synthetic design of a wide range of sophisticated carbohydrate nanostructures of biomedical importance. The first example of the use of Cu^I catalysis in generating triazole linked glycodenrimers was demonstrated Santoyo-González and colleagues [104]. The employment of organic soluble copper catalysts ($(Ph_3P)_3$•CuBr and $(EtO)_3$•CuI) enabled the facile construction of a diverse range of 1,4-disubstituted-1,2,3-triazole linked neoglycoconjugates and glycodendrimers.

Synthetic modifications of the original click chemistry conditions have recently appeared within the literature with the aim of improving the efficiency and scope of click chemistry in multivalent neoglycoconjugate design and synthesis. For example, Van der Eycken and colleagues [105] have reported a highly efficient, one-pot, microwave assisted click chemistry method based on a three component system employing sodium azide, alkyl halide and acetylene. Pieters et al. [106] have reported the facile synthesis of glycodendrimers containing up to nine peripheral carbohydrate units in high yields and shortened reaction times using microwave-assisted click chemistry. Another recent example is provided by Moran and co-workers who have described microwave-assisted click chemistry to rapid label oligonucleotides on solid support with pendant carbohydrate attachments [107] Testament to the impressive scope and compatibility with other synthetic tools, Chang and co-workers have recently explored the utility of sonication (sonochemistry) as a facile means for generating triazole linked neoglycoconjugates under mild, ambient conditions [108]. In addition, a one-pot, orthogonal transformation of glycosyl azides from glycosyl halides followed by click chemistry has also been reported for the efficient preparation of multivalent neoglycoconjugates in solution [109].

Perhaps the greatest advantage of the click methodology in multivalent neoglycoconjugate synthesis is the ability to conduct convergent or divergent syntheses in aqueous solution. This effectively enables the anchoring of deprotected saccharides to the core in a single step, furnishing the target glycodendrimer in high yields and purity, whilst bypassing otherwise necessary activation and/or protection/deprotection steps. Click chemistry has been utilised by Sharpless, Hawker and colleagues [110] as a general and facile strategy for the preparation of a series of bifunctionalised and unsymmetrical dendritic co-polymers incorporating dual purpose binding/detection chain terminals (Scheme 13). Multi-functionalised dendrimers have potential recognition/detection/delivery applications and can be utilised as multivalent ligands for *in situ* monitoring or as novel drug delivery agents. The facile nature of the reaction between the appropriately functionalised dendrons with azide or acetylene substituents allowed for the highly modular, flexible and divergent assembly of the difunctionalised glycodendrimer. The binding element was a polymannose functionalised periphery capable of binding to Concanavalin A (Con A). The fluorescent coumarin functionalised periphery was located at the opposite chain terminal and was employed for potential diagnostic purposes. Multi-functionalised glycodendrimers allow for the introduction of binding elements, reporter elements and/or drug molecules at defined locations within the same nanostructure. The bifunctionalised glycodendrimer was assayed for binding with Con A in a standard hemagglutination assay. From this assay, a relative binding affinity of 240 was observed relative to free mannose, which equated to a per unit affinity of 15.

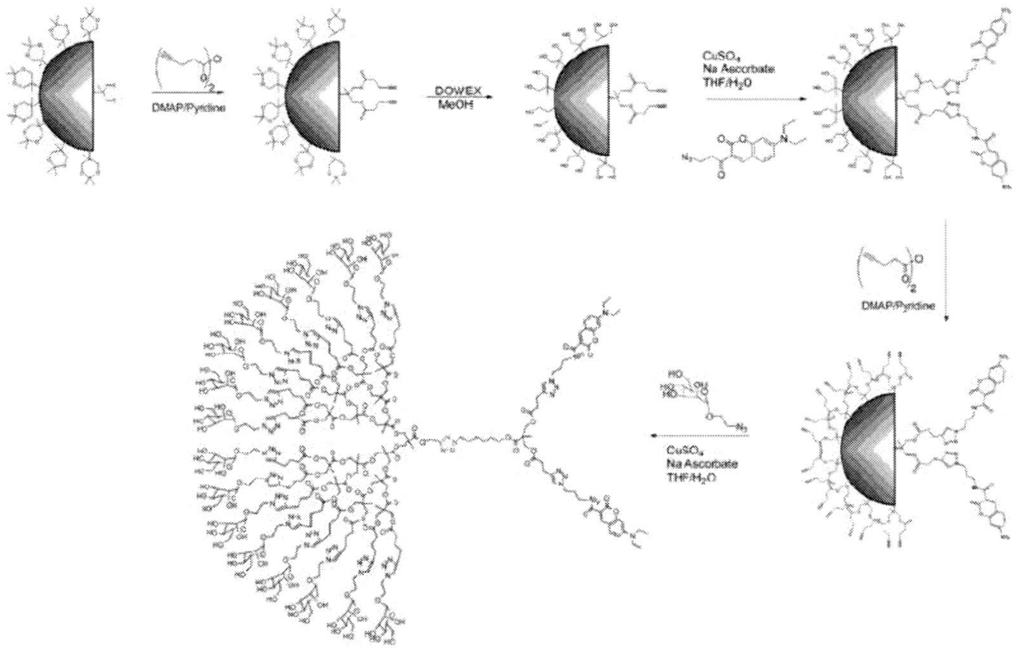

Scheme 13. The synthesis of an unsymmetrical and bifunctionalised dendrimer using click chemistry [110]. From: P. Wu, et al. *Chem. Commun.* 2005, *46*, 5775–5777. Reproduced by permission of The Royal Society of Chemistry ©2005.

Riguera and colleagues [111] have recently reported the facile preparation of unprotected glycodendrimers based on click chemistry. Several *O*-propynyl glycosides based on fucose, mannose, and lactose were appended to three generations of azido terminated gallic acid-triethylene glycol dendrimers in a highly efficient and divergent manner under aqueous conditions and in the presence of catalytic quantities of Cu^I catalyst (1 mol%). Impressive atom efficiency and reported yields (up to 92%) were observed, with the incorporation of up to 27 free propynyl glycosides in a single step. In a separate study by the same group, the applicability of the click methodology was further investigated on other co-polymeric core materials, such as azido terminated PEG-dendritic block co-polymers [112]. Similar atom efficiency and fidelity was also achieved using this versatile and novel co-polymer.

Galectins are a ubiquitous family of cytosolic β-D-galactoside binding proteins, of which 14 are known in mammalian systems [113]. They display broad and diffuse biological activities and their expression during cellular development and oncogenesis is highly variable. In particular, Galectin-1 (Gal-1) and Galectin-3 (Gal-3) are implicated in a wide range of biological and pathological processes including cell apoptosis, innate immune function, oncogenesis as well as HIV-1 infectivity. The chemical modulation of these ambiguous proteins using multivalent inhibitors and effectors is therefore of great interest to glycobiology and clinical research [113b]. Roy and colleagues [114] have prepared a suite of β-D-galactose containing neoglycoconjugates as potential galectin inhibitors from a series of sugar and non-sugar building blocks containing azide or acetylene functionalities. The resulting triazole neoglycoconjugates were rapidly prepared using click chemistry and screened individually against three Galectin proteins; Gal-1, Gal-2 and Gal-3. A C_3-symmetrical triazolo lactose analogue was prepared by a Cu^I catalysed cycloaddition in good yield (83%) and was identified as a strong inhibitor of Gal-1 (20 μM) and a moderate inhibitor of Gal-3 (250 μM). In particular, a 13.3-fold increase in potency for each lactose unit in the trivalent array was observed relative to free lactose against Gal-1, leading to a total relative potency of ~ 40.

In a similar study, Nilsson et al. [115] have examined the binding affinity of a wide range of multivalent lactose-bearing neoglycoconjugates against cancer- and tumour-related galectins, Gal-1, -3, -4N, -4C, -4, -7, -8N and -9N. The multivalent neoglycoconjugates were prepared by a Cu^I catalysed 1,3-DCR of the readily available 2-azidoethyl-β-D-galactopyranosyl-(1→4)-β-D-glucopyranoside and various acetylene functionalised phenyl-bis-alanine and phenyl-tris-alanine building blocks, affording the di- and trivalent neoglycoconjugates, respectively. To investigate the potential effect of the aglycone on binding, the corresponding monovalent homologues were prepared and used as a reference for relative binding affinity along with methyl β-D-lactoside. From a total of eight potential galectin ligands, the divalent carbamate species was found to have a binding affinity for Gal-1 of one order of magnitude stronger than all other ligands studied, with a K_d of 3.2 μM and a cluster effect of 7.7 and 30 relative to the corresponding monomer and methyl-β-D-lactoside, respectively. In contrast, the corresponding monomer displayed markedly weaker affinity for Gal-1, with a K_d of 49 μM.

Deprotected 2-azidoethyl glycosides of lactose and *N*-acetyl glucosamine have been employed by Gao and co-workers [116] as water soluble building blocks in multivalent ligand design. The facile click reaction between the deprotected ω-azido glycosides and the uniformly propynyl alkylated methyl-β-D-glucoside scaffold resulted in the rapid, modular generation of glycoclusters under aqueous conditions (Scheme 14). The two glycoclusters were assayed by a competitive ligand displacement assay (AGP-asialoglycan) using capillary electrophoresis. For both glycoclusters, an impressive 400-fold increase in binding affinity for plant lectin protein RCA120 was observed relative to free lactose, which is indicative of a strong multivalent effect.

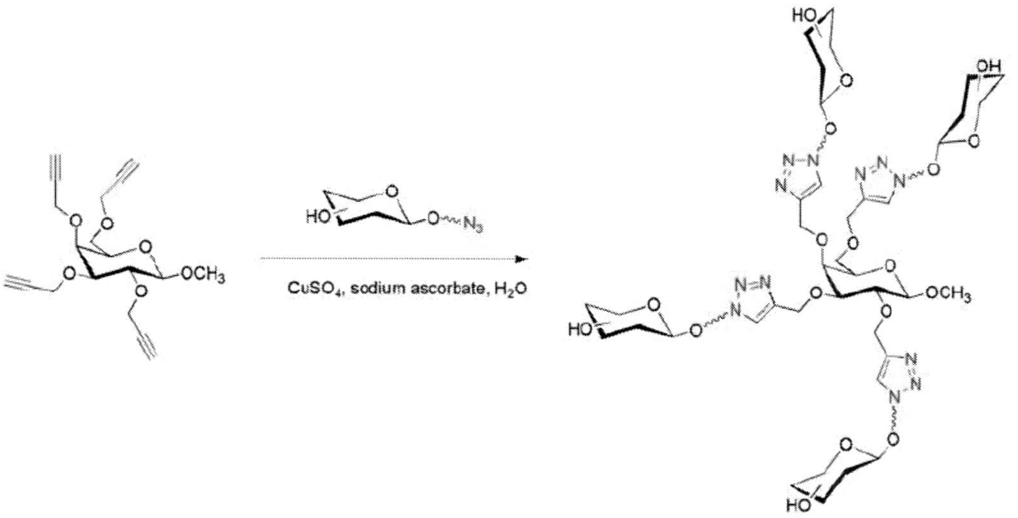

Scheme 14. The preparation of glycoclusters using click chemistry [116].

Du and co-workers [117] have previously reported the mild antitumour and increased white blood cell proliferation activity of linear β-D-(1→6)-*N*-acetyl-β-D-glucosamine hexa- and nonasaccharides. Based on these findings, synthetic efforts were directed towards multivalent models in order to enhance *in vivo* potency. The facile convergent synthesis of a C_3-symmetric (1→6)-*N*-acetyl-β-D-glucosamine octadecasaccharide and the corresponding monovalent hexasaccharide was achieved in good yields using click chemistry. In a preliminary *in vivo* assay, the C_3-symmetic, triantennary octadecasaccharide showed moderate tumour growth inhibition relative to the linear oligosaccharides and the corresponding monovalent hexasaccharide.

4. Carbohydrate Microarrays and Self-Assembled Monolayers

4.1. General

Progress in glycomics research and carbohydrate drug discovery has arisen not only from an impressive array of synthetic and separation technologies, but also the development of robust, high throughput screening tools for the identification of protein-carbohydrate receptor interactions [118]. Carbohydrate microarrays (glycoarrays) have emerged as versatile, high throughput tools for the screening of potential lectin proteins and more recently for investigating the binding of pathogenic bacteria to carbohydrates [119]. Carbohydrate microarrays consist of several hundred or thousands of carbohydrate ligands immobilised on a solid surface. The screening of test proteins for carbohydrate affinity occurs simultaneously in chip format using a variety of design strategies, whilst consuming only minute quantities of test compound and protein, an important prerequisite when using complex oligosaccharides and test compounds which have been painstakingly acquired. Self-assembled monolayers (SAMs) are also gaining attention and consist of homogeneous saccharides packed onto surfaces in a well defined and controllable manner. SAMs technology allows for exquisite control over carbohydrate density and concentration, which allows for real time monitoring and quantitation of binding events [113c]. Owing to the near quantitative product conversion and high fidelity, click chemistry is well poised to be the forerunner biomolecular immobilisation strategy for potential *in situ* screening of carbohydrate-protein interactions.

4.2. Glycoarrays

The earliest example of the construction of glycoarrays using click chemistry was demonstrated by Wong and co-workers [120] in an attempt to develop an accessible, high throughput screening method in microtitre plate format for the identification lectin-like proteins selective for galactose. An azidoethyl galactoside was used as the model carbohydrate for the study and was tethered via a Cu^I/DIPEA catalysed cycloaddition to the hydrophobic propionamide which, inturn was attached noncovalently to the polystyrene support of the well surface (Scheme 15).

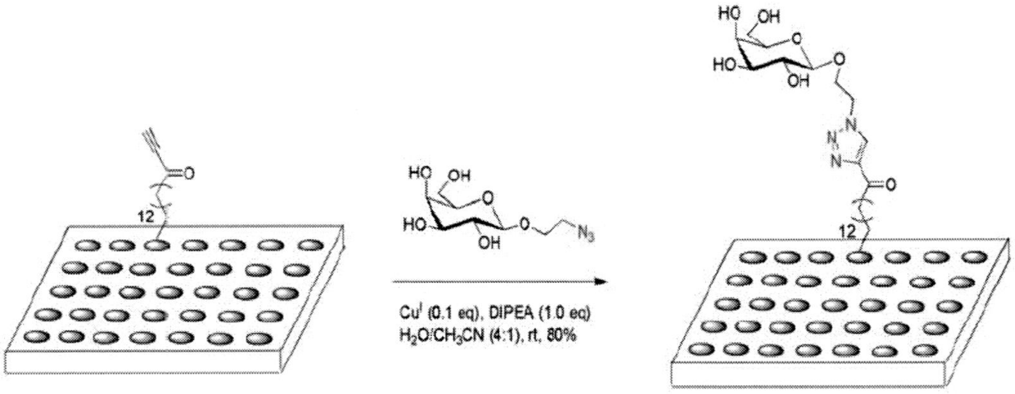

Scheme 15. The immobilisation of an azido ethyl galactoside to a noncovalent, polystyrene supported acetylene linker using click chemistry [120].

The click reaction optimised to allow conjugation on micromolar scales (4-80 μM), and was found complete under all conditions by ESI-MS in excellent yields (up to 92%), according to a previously reported sulfuric acid-phenol (SAP) assay. The Ricin B lectin protein isolated from *Ricinus communus* (castor bean), which selectively binds β-linked D-galactosides, was screened against the galactolipid array. Using the same click chemistry immobilisation strategy, a suite of 11 oligosaccharides displaying variable terminal branching patterns and degrees of terminal sialylation and/or fucosylation were then screened against potential lectin proteins. The enzyme catalysed fucosylation of the immobilised trisaccharide allowed for the *in situ* generation of immobilised sialyl Le^X ligands. The *in situ* generated sialyl Le^X ligand bound selectively to a fucose specific lectin from *Tetragonolobus purpureas* which did not show any affinity for the nonfucosylated trisaccharide precursor.

Wong and colleagues have also reported a similar noncovalent glycoarray utilising click chemistry as a key immobilisation strategy for the high throughput screening of Fuc-T inhibitors [121]. *N*-Acetyl lactosamine was immobilised to the solid support by click chemistry and was incubated with the triazole inhibitor, Fuc-T and the substrate, GDP-Fuc. The *in situ* fucosylation of the displayed *N*-acetyl lactosamine array was quantified by binding to *T. purpureas* lectin conjugated to a peroxidase. In this manner, four compounds were detected with nanomolar inhibition constants, three of which contained a biphenyl substituent which is consistent with earlier findings [45]. Mei et al. [122] have also employed click chemistry as an immobilisation strategy in conjunction with quantum dots (QDs) as a fluorescent label for the detection and quantitation of carbohydrate-protein binding.

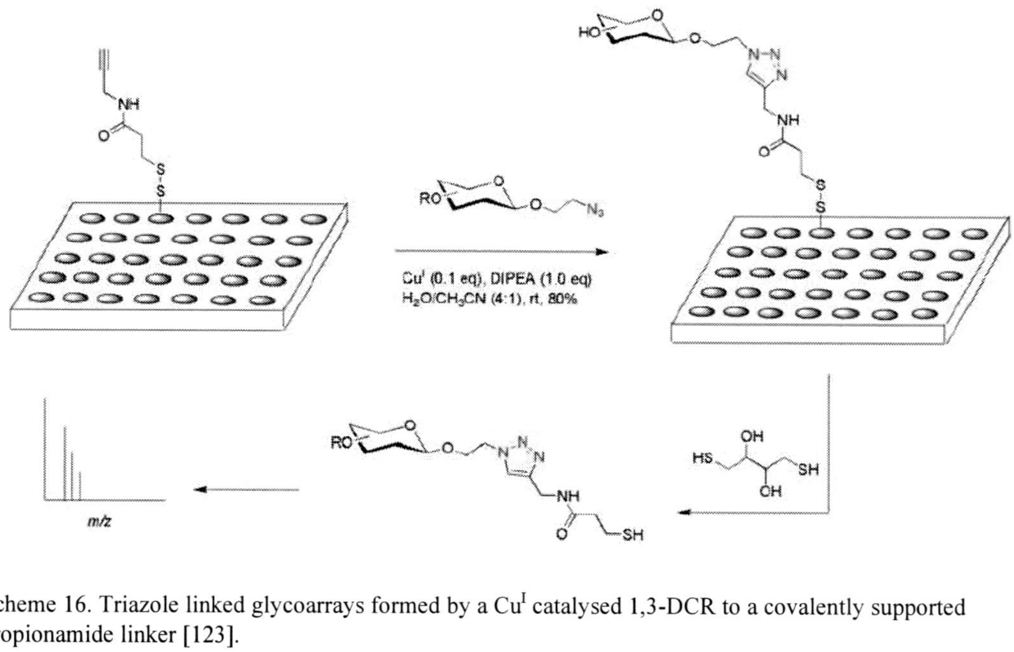

Scheme 16. Triazole linked glycoarrays formed by a Cu^I catalysed 1,3-DCR to a covalently supported propionamide linker [123].

In an effort to generate glycoarrays with enhanced stability and robustness towards high throughput screening protocols, Wong and co-workers [123] have exploited click chemistry to immobilise carbohydrate arrays to *covalently* attached solid supported linkers via thiourea and amide bonds. The immobilisation of azido ethyl glycosides to an alkyne terminated propionamide linker proceeded with high fidelity and efficiency using click chemistry (Scheme 16). Analysis of the immobilised ligand by ESI-MS could be achieved by the mild, reductive cleavage of the disulfide linker from the covalent anchor and subsequent dissolution. In a similar approach to the previous study, two glycoarrays were screened against lectins selective for α-fucose (*Lotus tetragonolobus*, LTL) and galactose (*Erythrina cristagalli,* EC) residues. The tetrasaccharide breast cancer antigenic determinant, Globo-H was also screened against LTL and was found to bind to this lectin in a concentration dependent manner (K_d = 22.8 μM), this was comparable with previous findings. In the same study, the immobilisation of carbohydrate ligands using click chemistry was also shown to be applicable to ELISA type formats. The covalently attached glycoarrays were also assessed for their ability to bind a GP-120-antibody (2G12), which recognises dense oligomannose clusters presented on the surface of HIV-1 gp120 ($Man_9GlcNAc_2$). Several triazole tethered oligomannose arrays were reported to display very strong affinity for the antibody (K_d = 0.1-1.0 μM).

4.3. Carbohydrate Self-Assembled Monolayers

Click chemistry between azide functionalised sugar linkers with solid-supported alkynes has recently been exploited by Wang and co-workers [124] for the fabrication of carbohydrate containing SAMs, using an approach similar to that developed by Chidsey and co-workers [125] for the fabrication of oligonucleotide SAMs. In this case, traditional alkane thiol linkers were employed, owing to their clean and spontaneous adsorption to the solid gold surface. Further functionalisation at the linker terminus with an alkyne group allowed for the fabrication of the carbohydrate SAM by the chemoselective delivery of various azido(triethylene)glycol-*O*-glycosides, based on α-D-mannose, α-D-galactose and β-D-lactose via click chemistry (Scheme 17). The carbohydrate ligands were successively interspersed with azido(triethylene)glycol forming a stable, rigid and well dispersed SAM. The relative binding affinities (reported as affinity constants, K_A) of various lectin proteins selective for these carbohydrate epitopes were determined by real time analytical techniques – surface plasmon resonance (SPR) and quartz crystal microbalance (QCM). Relative binding affinities to the lectin proteins, in particular that of the triazole-mannose SAM with Con A lectin ($K_A = 3.9 \pm 0.2 \times 10^6$ M^{-1}), were found to be in good agreement with previously reported values.

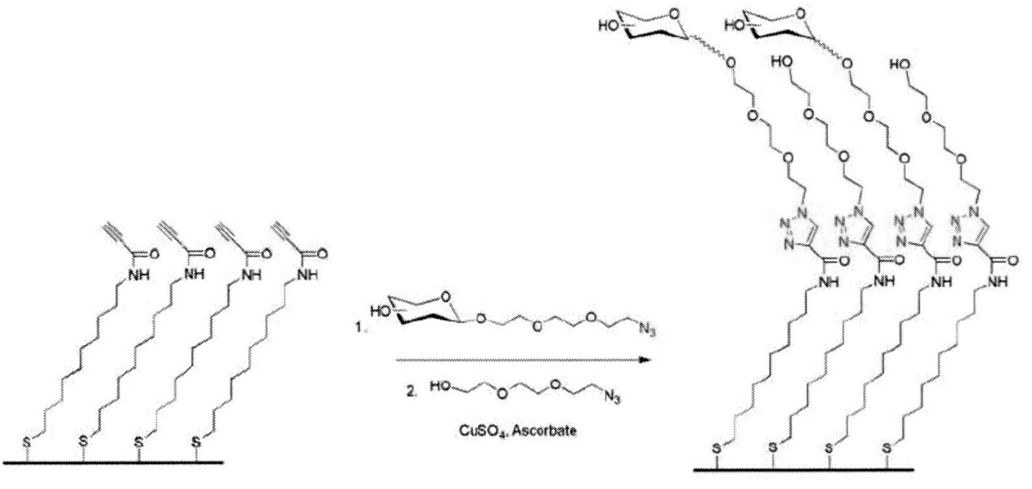

Scheme 17. Fabrication of carbohydrate SAM using Cu^I catalysed 1,3-DCR [124].

Chaikof and co-workers [126] have recently demonstrated the potential of click chemistry to immobilise carbohydrates to glass surfaces using flexible, biocompatible polyethylene glycol (PEG) linkers. The attachment of the PEG linker to the maleimide terminated solid support was achieved using a PEG chain functionalised at both termini with cyclopentadiene and an alkyne. Successive immobilisation of the azide functionalised carbohydrate was then achieved using click chemistry under buffered aqueous conditions (PBS, pH 7.0) with impressive fidelity and homogeneity.

5. Conclusions

Click chemistry has had a considerable impact on drug discovery and biotechnology since its conception some 7 years ago. However, the application to glycomics research and carbohydrate based drug development is more recent, yet the sheer volume of literature which has appeared the past 3 to 4 years is testament to its broad scope and versatility. Click chemistry holds great promise as a premier bioconjugation strategy for a wide variety of preparative and analytical applications. It has become a powerful synthetic tool for preparing a wide range of structurally diverse neoglycoconjugates, either as probes for elucidation of carbohydrate-protein binding or as drug leads. It has also shown potential as a bioorthogonal strategy for the labelling of chemically engineered cell surface glycans. These tools serve to correlate glycan structure and function in physiological states of health and disease, leading to enhanced understanding of the nature of protein and lipid glycosylation and the identification of novel targets for clinical intervention and vaccine development. Click chemistry has also become a robust immobilisation strategy for the fabrication of surface based carbohydrate arrays and monolayers that facilitate high throughput lectin carbohydrate investigations for the identification of lectin proteins of particular relevance in pathology. Owing to its impressive scope and utility in many aspects of carbohydrate based drug design processes, it is reasonable to envisage that click chemistry within carbohydrate based drug research will flourish.

6. References

[1] (a) Varki, A., Cummings, R., Esko, J., Freeze, H., Hart, G., Marth, J. Eds. In *Essentials of Glycobiology*. Cold Spring Harbor Laboratory Press, Cold Spring Harbor, NY, 1999; (b) Haltiwanger, R. S., Lowe, J. B. Role of glycosylation in development. *Annu. Rev. Biochem.* 2004, *73*, 491-537; (c) Rudd, P. M., Elliott, T., Cresswell, P., Wilson, I. A., Dwek, R. A. Glycosylation and the Immune System. *Science* 2001, *291* (5512), 2370-2375.

[2] (a) Ohtsubo, K., Marth, J. D., Glycosylation in Cellular Mechanisms of Health and Disease. *Cell* 2006, *126*, 855-867; (b) Ley, K. The role of selectins in inflammation and disease. *Trends Mol. Med.* 2003, *9* (6), 263-268; (c) Hölemann, A., Seeberger, P. H., Carbohydrate diversity: synthesis of glycoconjugates and complex carbohydrates. *Chem. Biotech.* 2004, *15*, 615-622.

[3] (a) Walsh, G., Roy Jefferis, R. Post-translational modifications in the context of therapeutic proteins. *Nature Biotechnol.* 2006, *24* (10), 1241-1252; (b) Shriver,

Z., Raguram, S., and Sasisekharan, R. Glycomics: A pathway to a class if new and improved therapeutics. *Nat. Rev. Drug Discov.* 2004, *3*, 863–873; (c) Doores, K. J., Gamblin, D. P., Davis, B. G. Exploring and exploiting the therapeutic potential of glycoconjugates. *Chem. Eur. J.* 2006, *12* (3), 656-665; (d) Dove, A. The bittersweet promise of glycobiology. *Nat. Biotechnol.* 2001, *19*, 913-917.

[4] (a) Davis, B. G., Synthesis of Glycoproteins. *Chem. Rev.* 2002, *102*, 579-601. (b) Spiro, R. G. Protein glycosylation: nature, distribution, enzymatic formation, and disease implications of glycopeptide bonds. *Glycobiology* 2002, *12*, 43R-50R.

[5] (a) Bertozzi, C., R., Kiessling, L. L. Chemical Glycobiology. *Science* 2001, *291* (5512), *2357-2364*; (b) Buskas, T., Ingale, S., Booms, G.–J., Glycopeptides as versatile tools for glycobiology. *Glycobiology* 2006, *16* (8), 113R-136R; (c) Seitz, O. Glycopeptides and Glycoproteins: Synthetic Chemistry and Biology. In *Carbohydrate-based Drug Discovery*. Wong, C.–H. ed., Wiley VCH, Weinheim, Germany, 2003. pp 169-209.

[6] (a) Saotome, C., Kanie, O. Synthetic Methodologies. In *Carbohydrate-based Drug Discovery*. Wong, C.–H. ed., Wiley VCH, Weinheim, Germany, 2003. pp 1-33; (b) Seeberger, P. H., Solid-Phase Oligosaccharide Synthesis. In *Carbohydrate-based Drug Discovery*. Wong, C.–H. ed., Wiley VCH, Weinheim, Germany, 2003. pp 103-107; (c) Dudkin, V. Y., Miller, J. S., Danishefsky, S. J. Chemical Synthesis of Normal and Transformed PSA Glycopeptides. *J. Am. Chem. Soc.* 2004, *126* (3), 736-738; (d) Zhang, J., Shao, J., Kowal, P., Wang, P. G. Enzymatic Synthesis of Oligosaccharides. In *Carbohydrate-based Drug Discovery*. Wong, C.–H. ed., Wiley VCH, Weinheim, Germany, 2003. pp 129-162.

[7] (a) Sears, P., Wong, C.–H., Carbohydrate Mimetics: A New Strategy for Tackling the Problem of Carbohydrate-Mediated Biological Recognition. *Angew. Chem. Int. Ed.* 1999, *38* (16), 2300-2324; (b) Ernst, B., Kolb, H. C., Schwardt, O. Carbohydrate mimetics in drug discovery. In *Organic Chemistry of Sugars*. Levy, D. E. Ed., CRC Press LLC, Boca Raton, FL, USA, 2006, pp. 803-861.

[8] (a) Marcaurelle, L. A., Seeberger, P. H. Combinatorial carbohydrate chemistry. *Curr. Opin. Chem. Biol.* 2002, *6* (3), 289-296; (b) Velter, I., La Ferla, B., Nicotra, F., Carbohydrate-Based Molecular Scaffolding. *J. Carbohydr. Chem.* 2006, *25* (2-3), 97-138; (c) Gabius, H.–J., Siebert, H.–C., Andry, S., Jimenez-Barbero, J., Rüdiger, H. Chemical Biology of the Sugar Code. *ChemBiochem.* 2004, *5* (6), 741-764.

[9] (a) Macmillan, D., Daines, A. M. Recent Developments in the Synthesis and Discovery of Oligosaccharides and Glycoconjugates for the Treatment of Disease. *Curr. Med. Chem.*, 2003, *10* (24), 2733-2773; b) Kannagi, R. Carbohydrate-Based Treatment of Cancer Metastasis. In *Carbohydrate-based Drug Discovery*. Wong, C.–H. ed., Wiley VCH, Weinheim, Germany, 2003. pp 803-829; (c) Drinnan, N. Vari, F. Aspects of the Stability and Bioavailability of

carbohydrates and Carbohydrate Derivatives. *Mini-Rev. Med. Chem.* 2003, *3* (7), 633-649; (d) Dube, D. H., Bertozzi, C. R., Glycans in cancer and inflammation - potential for therapeutics and diagnostics. *Nat. Rev. Drug Disc.* 2005, *4* (6), 477-488.

[10] (a)Thorson, J. S., Vogt, T. Glycosylated Natural Products. In *Carbohydrate-based Drug Discovery*. Wong, C.–H. ed., Wiley VCH, Weinheim, Germany, 2003. pp.803-829; (b) Wessjohann, L. A., Ruijter, E., Garcia-Rivera, D., Brandt, W. What can a chemist learn from nature's macrocycles? – A brief, conceptual view. *Mol. Div.* 2005, *9* (1-3), 171–186; (c) Quader, S.; Boyd, S. E.; Jenkins, I. D.; Houston, T. A. Multisite Modification of Neomycin B: A Combined Mitsunobu and Click Chemistry Approach. *J. Org. Chem.* 2007, *72* (6), 1962-1979.

[11] For examples of the recent failures of carbohydrate-based drugs during clinical trials see: Michael, A. Carbohydrates down, but not out. *Nature Biotechnol.* 1997, *15* (12),1233–1234.

[12] Holzer, C. T., von Itzstein, M., Jin, B., Pegg, M. S., Stewart, W. P., Wu, W. Y. Inhibition of sialidases from viral, bacterial and mammalian sources by analogues of 2-deoxy-2,3-didehydro-*N*-acetylneuraminic acid modified at the C-4 position. *Glycoconjugate J.* 1993, *10* (1), 40-44.

[13] Islam, T., Linhardt, R. J. Chemistry, Biochemistry, and Pharmaceutical potentials of Glycosaminoglycans and Related Saccharides. In *Carbohydrate-based Drug Discovery*. Wong, C.–H. ed., Wiley VCH, Weinheim, Germany, 2003. pp. 407-440.

[14] (a) Asano, N., Nash, R. J., Molyneux, R. J., Fleet, G. W. J. Sugar-mimic glycosidase inhibitors: natural occurrence, biological activity and prospects for therapeutic application. *Tetrahedron Asymm.* 2000, *11* (8), 1645-1680; (b) Tietze, L. F., Lieb, M., Herzig, T., Haunert, F., Schubert, I. A Strategy for Tumor-Selective Chemotherapy by Enzymatic Liberation of seco-duocarmycin SA-derivatives from Non-toxic Prodrugs. *Bioorg. Med. Chem.* 2001, *9* (7), 1929-1939.

[15] Masereel, B., Rolin, S., Abbate, F., Scozzafava, A., Supuran, C. T. Carbonic Anhydrase Inhibitors: Anticonvulsant Sulfonamides Incorporating Valproyl and Other Lipophilic Moieties. *J. Med. Chem.* 2002, 45, 312-320.

[16] For reviews on chemoselective ligation in glycopeptide synthesis, see: a) Hang, H.C.; Bertozzi, C.R. Chemoselective approaches to glycoprotein assembly. *Acc. Chem. Res.* 2001, *34* (9), 727-736; b) Peri, F., Nicotra, F., Chemoselective ligation in glycochemistry. *Chem. Commun.* 2004, (6), 623-627.

[17] (a) Huisgen, R. 1,3-Dipolar cycloadditions. *Angew. Chem.* 1963, *75* (13), 604-637; (b) Huisgen, R. 1,3-Dipolar cycloadditions. 76. Concerted nature of 1,3-dipolar cycloadditions and the question of diradical intermediates. *J. Org. Chem.* 1976, *41* (3), 403-419; (c) Gothelf, K. V., Jørgensen, K. A. Asymmetric 1,3-Dipolar Cycloaddition Reactions. *Chem. Rev.* 1998, *98* (2), 863-909.

[18] (a) Rostovtsev, V. V., Green, L. G., Fokin, V. V., Sharpless, K. B. A Stepwise Huisgen Cycloaddition Process: Copper(I)-Catalyzed Regioselective Ligation

of Azides and Terminal Alkynes. *Angew. Chem. Int. Ed.* 2002, *41* (14), 2596-2599; (b) Tornøe, C. W., Christensen, C., Meldal, M. Peptidotriazoles on Solid Phase: [1,2,3]-Triazoles by Regiospecific Copper(I)-Catalyzed 1,3-Dipolar Cycloadditions of Terminal Alkynes to Azides. *J. Org. Chem.* 2002, *67* (9), 3057-3064.

[19] Kolb, H. C., Finn, M. G., Sharpless, K. B. Click Chemistry: Diverse Chemical Function from a Few Good Reactions. *Angew. Chem. Int. Ed.* 2001, *40* (11), 2004-2021.

[20] (a) Sustmann, R. Simple model for substituent effects in cycloaddition reactions. I. 1,3 - Dipolar cycloadditions. *Tetrahedron Lett.* 1971, *29*, 2717-2720; (b) Bräse, S., Gil, C., Knepper, K., and Zimmermann, V. Organic Azides: An Exploding Diversity of a Unique Class of Compounds. *Angew. Chem. Int. Ed.* 2005, *44* (33), 5188–5240.

[21] (a) Himo, F., Lovell, T., Hilgraf, R., Rostovtsev, V. V., Noodleman, L., Sharpless, K. B., Fokin, V. V. Copper(I)-Catalyzed Synthesis of Azoles. DFT Study Predicts Unprecedented Reactivity and Intermediates. *J. Am. Chem. Soc.* 2005, *127* (41), 210-216; (b) Bock, V. D., Hiemstra, H., van Maarseveen, J. H. CuI-Catalyzed Alkyne–Azide "Click" Cycloadditions from a Mechanistic and Synthetic Perspective *Eur. J. Org. Chem.* 2006, 51–68.

[22] (a) Kolb, H. C., Sharpless, K. B. The growing impact of click chemistry on drug discovery. *Drug Disc. Today* 2003, *8* (24), 1128-1136; (b) Röper, S., Kolb, H. C. Click Chemistry for Drug Discovery. In *Fragment-based Approaches in Drug Discovery.* Jahnke, W., Erlanson, D. A. Eds. Wiley-VCH Verlag GmbH & Co. KGaA, Weinheim, Germany, 2006, pp. 113-139.

[23] Weber, L. In vitro combinatorial chemistry to create drug candidates. *Drug Disc. Today* 2004, *1* (3), 261-267.

[24] a) Oh, K., Guan, Z. A convergent synthesis of new β-turn mimics by click chemistry. *Chem. Commun.* 2006, (29), 3069–3071. b) Horne, W. S., Yadav, M. K., Stout, C. D., Ghadiri, M. R. Heterocyclic Peptide Backbone Modifications in an r-Helical Coiled Coil. *J. Am. Chem. Soc.* 2004, *126* (47), 15366-15367.

[25] For examples of the anti HIV-1 activity of 1,2,3-triazole containing compounds, see: (a) Whiting, M., Tripp, J. C., Lin, Y.–C., Lindstrom, W., Olson, A. J. Elder, J. H., Sharpless, K. B., Fokin, V. V. Rapid Discovery and Structure-Activity Profiling of Novel Inhibitors of Human Immunodeficiency Virus Type 1 Protease Enabled by the Copper(I)-Catalyzed Synthesis of 1,2,3-Triazoles and Their Further Functionalization. *J. Med. Chem.* 2006, *49* (26), 7697-7710; (b) Hosahudya N. Gopi, H. N., Tirupula, K. C., Baxter, S., Ajith, S., Chaiken, I. W. Click Chemistry on Azidoproline: High-Affinity Dual Antagonist for HIV-1 Envelope Glycoprotein gp120. *Chem MedChem.* 2006, *1* (1), 54–57; (c) Whiting, M., Muldoon, J., Lin, Y.–C., Silverman, S. M., Lindstrom, W., Olson, A. J., Kolb, H. C., Finn, M. G., Sharpless, K. B., Elder, J. H., Fokin, V. V. Inhibitors of HIV-1 protease by using in situ click chemistry. *Angew. Chem. Int. Ed.* 2006, *45* (9), 1435-1439.

[26] For examples of the antimicrobial properties of 1,2,3-triazole containing compounds, see: (a) Genin, M. J., Allwine, D. A., Anderson, D. J., Barbachyn, M. R., Emmert, D. E., Garmon, S. A., Graber, D. R., Grega, K. C., Hester, J. B., Hutchinson, D. K., Morris, J., Reischer, R. J., Ford, C. W., Zurenko, G. E., Hamel, J. C., Schaadt, R. D., Stapert, D., Yagi, B. H. Substituent Effects on the Antibacterial Activity of Nitrogen-Carbon-Linked (Azolylphenyl) oxazolidinones with Expanded Activity Against the Fastidious Gram-Negative Organisms *Haemophilus influenzae* and *Moraxella catarrhalis*. *J. Med. Chem.* 2000, *43* (5), 953–970; (b) Dabak, K., Sezer, Ö., Akar, A., Anaç, O. Synthesis and investigation of tuberculosis inhibition activities of some 1,2,3-triazole derivatives. *Eur. J. Med. Chem.* 2003, *38* (2), 215-218.

[27] Brockunier, L. L., Parmee, E. R., Ok, H. O., Candelore, M. R., Cascieri, M. A., Colwell, L. F., Deng, L., Feeney, W. P., Forrest, M. J., Hom, G. J., MacIntyre, D. E., Tota, L., Wyvratt, M. J., Fisher, M. H., Weber, A. E. Human β3-Adrenergic Receptor Agonists Containing 1,2,3-Triazole-Substituted Benzenesulfonamides. *Bioorg. Med. Chem. Lett.* 2000, *10* (20), 2111–2114.

[28] Cunha, A. C., Figueiredo, J. M., Tributino, J. L. M., Miranda, A. L. P., Castro, H. C., Zingali, R. B., Fraga, C. A. M., de Souza, M. C. B. V., Ferreirac, V. F., Barreiroa, E. J. Antiplatelet Properties of Novel *N*-Substituted-phenyl-1,2,3-triazole-4-acylhydrazone Derivatives. *Bioorg. Med. Chem.* 2003, *11* (9), 2051–2059.

[29] For examples of the anti-inflammatory activity of 1,2,3-triazole containing compounds, see: of (a) Tullis, J. S., Van-Rens, J. C., Natchus, M. G., Clark, M. P., De, B., Hsieh, L. C., Janusz, M. J. The Development of New Triazole Based Inhibitors of Tumor Necrosis Factor-α (TNF-α) Production. *Bioorg. Med. Chem. Lett.* 2003, *13* (10), 1665–1668; (b) Kim, D. K.; Kim, J.; Park, H. J. Synthesis and biological evaluation of novel 2-pyridinyl-[1,2,3]triazoles as inhibitors of transforming growth factor *b*1 type 1 receptor. *Bioorg. Med. Chem. Lett.* 2004, *14* (10), 2401–2405.

[30] (a) Matsuda, A., Sasaki, T. Antitumor activity of sugar-modified cytosine nucleosides. *Cancer Science* 2004, *95* (2), 105-111; (b) Al-Madhoun, A. S., Tjarks, W., Eriksson, S. The role of thymidine kinases in the activation of pyrimidine nucleoside analogues. *Mini-Rev. Med. Chem.* 2004, *4* (4), 341-350; (c) Ichikawa, E., Kato, K. Sugar-modified nucleosides in past 10 years, a review. *Curr. Med. Chem.* 2001, *8* (4), 385-423; (d) Gumina, G., Chong, Y., Choo, H., Song, G.–Y., Chu, C. K. L-nucleosides: antiviral activity and molecular mechanism. *Curr. Top. Med. Chem.* 2002, *2* (10), 1065-1086.

[31] For reviews concerning 1,3-dipolar cycloadditions on carbohydrates, see: (a) Gallos, J. K., Koumbis, A. E. 1,3-Dipolar cycloadditions in the Synthesis of Carbohydrate Mimics. Part 1: Nitrile Oxides and Nitronates. *Curr. Org. Chem.* 2003, *7* (5), 397-425; (b) Gallos, J. K., Koumbis, A. E., 1,3-Dipolar cycloadditions in the Synthesis of Carbohydrate Mimics. Part 2: Nitrones and Oximes. *Curr. Org. Chem.* 2003, *7* (6), 585-628; (c) Gallos, J. K., Koumbis, A. E., 1,3-Dipolar cycloadditions in the Synthesis of Carbohydrate Mimics. Part 3:

Azides, Diazo Compounds and Other Dipoles. *Curr. Org. Chem.* 2003, *7* (8), 771-793.

[32] (a) Lillelund, V. H., Jensen, H. H., Liang, X., Bols, M., Recent Developments of Transition-State Analogue Glycosidase Inhibitors of Non-Natural Product Origin. *Chem. Rev.* 2002, *102* (2), 515-553. (b) Houston, T. A.; Blanchfield, J. T. Back to (non)-Basics: Recent Developments in Neutral and Charge-Balanced Glycosidase Inhibitors. *Mini-Rev. Med. Chem.* 2003, *3*, 669-678.

[33] Nagstatin and heterocyclic mimics as a potent hexosamidase inhibitors, see: (a) Shanmugasundaram, B., Debowski, A. W., Dennis, R. J., Davies, G. J., Vocadlo, D. J., Vasella, A. Inhibition of *O*-GlcNAcase by a gluco-configured nagstatin and a PUGNAc-imidazole hybrid inhibitor. *Chem. Commun.* 2006, (42), 4372-4374; (b) Miroslav, T., Andrea, V. Synthesis of *N*-acetylglucosamine-derived nagstatin analogues and their evaluation as glycosidase inhibitors. *Helv. Chim. Acta.* 2005, *88* (1), 10-22. (c) Tatsuta, K., Shozo, M., Ohta, S., Gunji, H. Syntheses and glycosidase inhibiting activities of nagstatin analogs. *J. Antibiotics.* 1995, *48* (3), 286-288; (d) Tatsuta, K., Ikeda, Y., Miura, S. Synthesis and glycosidase inhibitory activities of nagstatin triazole analogs. *J. Antibiotics* 1996, *49* (8), 836-838.

[34] Li, R. K., Rinaldi, M. J. In Vitro Antifungal Activity of Nikkomycin Z in Combination with Fluconazole or Itraconazole. *Antimicrob. Agents Chemother.* 1999, *43* (6), 1401–1405.

[35] Kobayashi, Y. Chemistry and biology of trehazolins. *Carbohydr. Res.* 1999, *315* (1-2), 3-15.

[36] Rao, F. V., Douglas R. Houston, D. R., Boot, R. G., Aerts, J. M. F. G., Sakuda, S., van Aalten, D. M. F. Crystal Structures of Allosamidin Derivatives in Complex with Human Macrophage Chitinase. *J. Biol. Chem.* 2003, *278* (22), 20110-20116.

[37] (a) Dolhem, F., Al Tahli, F., Lièvre, C., Demailly, G. Efficient Synthesis of 1,2,3-Triazole-Fused Bicyclic Compounds from Aldoses. *Eur. J. Org. Chem.* 2005, *23*, 5019-5023; (b) Pearson, W. H., Lovering, F. E. Assembly of 3α-Arylperhydroindoles by the Intramolecular Cycloaddition of 2-Azaallyl Anions with Alkenes. Total Syntheses of (±)-Crinine, (±)-6-Epicrinine, (-)-Amabiline, and (-)-Augustamine. *J. Org. Chem.* 1998, *63* (26), 3607-3617.

[38] Rossi, L. L., Basu, A. Glycosidase inhibition by 1-glycosyl-4-phenyl triazoles. *Bioorg. Med. Chem. Lett.* 2005, *15* (15), 3596–3599.

[39] Périon, R., Ferrières, V., M. Isabel García-Moreno, M. I., Mellet, C. O. Duval, R., José M. Fernández, G., Plusquelleca, D. 1,2,3-Triazoles and related glycoconjugates as new glycosidase inhibitors. *Tetrahedron.* 2005, *61*, 9118-9128.

[40] Thomson, R., von Itzstein, M. *N*-Acetylneuraminic acid Derivatives and Mimetics as Anti-influenza Agents. In *Carbohydrate-based Drug Discovery.* Wong, C.–H., Ed.; Wiley VCH press, Weinheim, Germany, 2003, pp 831-861.

[41] Li, J., Zheng, M., Tang, W., He, P. –L., Zhu, W., Li, T. Syntheses of triazole-modified zanamivir analogues via click chemistry and anti-HIV activities. *Bioorg. Med. Chem. Lett.* 2006, *16* (21), 5009-5013.

[42] Smith, P. W., Whittington, A. R., Sollis, S. L., Howes, P. D., Taylor, N. R. Novel inhibitors of influenza sialidases related to zanamivir. Heterocyclic replacements of the glycerol sidechain. *Bioorg. Med. Chem. Lett.* 1997, *7* (17), 2239-2242.

[43] (a) Schmidt, R. R., Jung, K, -H., Glycosyl transferase inhibitors, in *Carbohydrate based drug discovery*. Vol. 2, Wong, C.–H., ed. Wiley-VCH press, Weinheim, Germany, 2003, pp. 609–659; (b) Palcic, M. M., Qian, X., Glycosyl transferase inhibitors, in *Carbohydrates in chemistry and biology*. Vol. 3, Ernst, B., Hart, G. W., Sinay, P. ed. Wiley-VCH press, Weinheim, Germany, 2000, pp. 293–312.

[44] Mitchell, M. L., Tian, F., Lee, L. V., Wong, C.–H. Synthesis and evaluation of transition-state analogue inhibitors of α-1,3- fucosyltransferase. *Angew. Chem. Int. Ed.* 2002, *41* (16), 3041-3044.

[45] Lee, L. V., Mitchell, M. L., Huang, S, -J., Fokin, V. V., Sharpless, K. B., Wong, C.–H. A Potent and Highly Selective Inhibitor of Human α-1,3-Fucosyltransferase via Click Chemistry. *J. Am. Chem. Soc.* 2003, *125* (32), 9588-9589.

[46] Sernee, M. F., Ralton, J. E., Dinev, Z., Khairallah, G. N., O'Hair, R. A., Williams, S. J., McConville, M. J. Leishmania β-1,2-mannan is assembled on a mannose-cyclic phosphate primer. *Proc. Nat. Acad. Sci. USA.* 2006, *103* (25), 9458-9463.

[47] (a) Naderer, T., McConville, M. J. Evidence that intracellular β1-2 mannan is a virulence factor in *Leishmania* parasites. *Mol. Biochem. Parasitol.* 2002, *125* (1-2), 147-161; (b) Garami, A., Ilg, T. Disruption of mannose activation in *Leishmania Mexica a*: GDP-mannose phosphorylase is required for virulence, but not viability. *EMBO J.* 2001, *20* (14), 3657-3666 (c) Garami, A. Mehlert, T. Ilg, *Mol. Cell. Biol.* 2001, *21* (23), 8168.

[48] van der Peet, P., Gannon, C. T., Ian Walker, I., Dinev, Z., Angelin, M., Tam, S., Ralton, J. E., McConville, M. J., Williams, S. J. Use of Click Chemistry to Define the Substrate Specificity of Leishmania β-1,2-Mannosyltransferases. *ChemBiochem.* 2006, *7* (9), 1384-1391.

[49] Supuran, C. T. Carbonic anhydrases: Catalytic and inhibition mechanism, distribution and physiological roles. In *Carbonic Anhydrase: Its Inhibitors and Activators*. Supuran, C. T., Scozzafava, A., Conway, J., Eds.; CRC Press: Boca Raton, FL, 2004; pp 1-24.

[50] For reviews on CA inhibitors see: (a) Pastorekova, S., Parkkila, S., Pastorek, J., Supuran, C. T. Carbonic Anhydrases: Current State of the Art, Therapeutic Applications and Future Prospects. *J. Enzyme Inhib. Med. Chem.* 2004, *19*(3), 199-229; (b) Scozzafava, A., Mastrolorenzo, A., Supuran, C. T. Carbonic anhydrase inhibitors and activators and their use in therapy. *Expert Opin. Ther. Patents* 2006, *16* (12), 1627-1664; (c) Clare, B. W., Supuran, C. T., A

perspective on quantitative structure-activity relationships and carbonic anhydrase inhibitors. *Expert Opin. Drug Metabol. Toxicol.* 2006, *2* (1), 113-137; (d) Thiry, A., Dogne, J. –A., Masereel, B., Supuran, C. T. Targeting tumor-associated carbonic anhydrase IX in cancer therapy. *Trends Pharm. Sci.* 2006, *27* (11), 564-573.

[51] Wilkinson, B. L., Bornaghi, L. F., Houston, T. A., Innocenti, A., Supuran, C. T., Poulsen, S.–A. A Novel Class of Carbonic Anhydrase Inhibitors: Glycoconjugate Benzene Sulfonamides Prepared by "Click-Tailing" *J. Med. Chem.* 2006, *49* (22), 6539-6548.

[52] (a) Wilkinson, B. L., Bornaghi, L. F., Houston, T. A., Innocenti, A., Vullo, D., Supuran, C. T., Poulsen, S.–A. Inhibition of membrane-associated carbonic anhydrase isozymes IX, XII and XIV with a library of glycoconjugate benzenesulfonamides. *Bioorg. Med. Chem. Lett.* 2007, *17* (4), 987-992; (b) Wilkinson, B. L., Bornaghi, L. F., Wright, A. D., Houston, T. A., Poulsen, S.–A. Antimycobacterial Activity of a Bis-Sulfonamide. *Bioorg. Med. Chem. Lett.* 2007, *17*(5), 1355-1357.

[53] Wilkinson, B. L., Bornaghi, L. F., Houston, T. A., Innocenti, A., Vullo, D., Supuran, C. T., Poulsen, S.–A. Carbonic anhydrase inhibitors: inhibition of isozymes I, II and IX with triazole linked *O*-glycosides of benzene sulfonamides. *J. Med. Chem.* 2007, *50*, 1651-1657.

[54] (a) Seitz, O. Glycopeptides and Glycoproteins: Synthetic Chemistry and Biology, in *Carbohydrate based drug discovery*. Vol. 2, Wong, C.–H., ed. 2003, Wiley-VCH press, Weinheim, Germany, pp. 169-209; (b) Buskas, T., Ingale, S., Boons, G.–J., Glycopeptides as versatile tools for glycobiology. *Glycobiology*. 2006, *16* (8), 113-136.

[55] For examples of enhanced biological activitiy of neoglycopeptides and glycol-engineering of peptide drugs, see: (a) Polt, R., Dhanasekaran, M., Keyari, C. M. Glycosylated neuropeptides: a new vista for neuropsychopharmacology? *Med. Res. Rev.* 2005, *25* (5), 557-585; (b) Susaki, H., Suzuki, K., Ikeda, M., Yamada, H., Watanabe, H. K. Synthesis and antidiuretic activities of novel glycoconjugates of arginine-vasopressin. Chem. Pharm. Bull. 1998, *46* (10), 1530; (c) Elliot, S., Lorenzini, T., Asher, S., Aoki, K., Brankow, D., Buck, L., Busse, L., Chang, D., Fuller, J., Grant, J., Hernday, N., Hokum, M., Hu, S., Knudten, A., Levin, N., Komorowski, R., Martin, F., Navarro, R., Osslund, T., Rogers, G., Rogers, N., Trail, G., Egrie, J. Enhancement of therapeutic protein *in vivo* activities through glycoengineering. *Nature Biotechnol.* 2003, *21* (4), 414-421.

[56] Dondoni, A., Marra, A. Methods for Anomeric Carbon-Linked and Fused Sugar Amino Acid Synthesis: The Gateway to Artificial Glycopeptides. *Chem. Rev.* 2000, *100* (12), 4395–4422.

[57] Wilkinson, B. L., Bornaghi, L. F., Poulsen, S.–A., Houston, T. A. Synthetic Utility of Glycosyltriazoles in Carbohydrate Chemistry. *Tetrahedron.* 2006, *62* (34), 8115–8125.

[58] Kuijpers, B. H. M., Groothuys, S., Keereweer, A., R., Quaedflieg, P. J. L. M., Blaauw, R. H., van Delft, F. L., Rutjes, F. P. J. T. Expedient Synthesis of Triazole linked Glycosyl Amino Acids and Peptides. *Org. Lett.* 2004, *6* (18), 3123-3126.

[59] Dondoni, A., Giovannini, P. P., Massi, A. Assembling Heterocycle-Tethered *C*-Glycosyl and α-Amino Acid Residues via 1,3-Dipolar Cycloaddition Reactions. *Org. Lett.* 2004, *6* (17), 2929-2932.

[60] Macmillan, D., Blanc, J. A novel neoglycopeptide linkage compatible with native chemical ligation. *Org. Biomol. Chem.* 2006, *4* (15), 2847-2850.

[61] Arosio, D., Bertoli, M., Manzonib, L., Scolastico, C. Click chemistry to functionalise peptidomimetics. *Tetrahedron Lett.* 2006, *47* (22), 3697–3700.

[62] Ermolat'ev, D. Dehaen, W., Van der Eycken, E. Indirect coupling of the 2(1H)-pyrazinone scaffold with various (oligo)-saccharides via "click chemistry": En route towards glycopeptidomimetics. *QSAR & Comb. Sci.* 2004, *23* (10), 915-918.

[63] Dörner, S., Westermann, B. A short route for the synthesis of ''sweet'' macrocycles via a click-dimerization–ring-closing metathesis approach. *Chem. Commun.* 2005, (22), 2852–2854.

[64] Fu, X., Albermann, C., Jiang, J., Liao, J., Zhang, C., Thorson, J. S. Antibiotic Optimisation via *in vitro* glycorandomisation. *Nature Biotechnol.* 2003, *21* (12), 1467-1469.

[65] Lin, H., Walsh, C. T. A Chemoenzymatic Approach to Glycopeptide Antibiotics. *J. Am. Chem. Soc.* 2004, *126* (43), 13998-14003.

[66] (a) Prescher, J. A., Bertozzi, C. R. Chemistry in living systems. *Nature Chem. Biol.* 2005, *1* (1), 13-21; (b) Prescher, J. A., Bertozzi, C. R. Chemical Technologies for Probing Glycans. *Cell.* 2006, *126* (5), 851-854.

[67] Prescher, J. A., Bertozzi, C. R. Chemical remodelling of cell surfaces in living animals. *Nature* 2004, *430* (7002), 873-877.

[68] Agard, N. J., Prescher, J. A., Bertozzi, C. R. A Strain-Promoted [3 + 2] Azide-Alkyne Cycloaddition for Covalent Modification of Biomolecules in Living Systems. *J. Am. Chem. Soc.* 2004, *126* (46), 15046-15047.

[69] (a) Kannagi, R. Molecular mechanism for cancer-associated induction of sialyl Lewis X and sialyl Lewis A expression-The Warburg effect revisited. *Glycoconjugate J.* 2004, *20* (5), 353-64; (b) Dube, D. H., Bertozzi, C. R. Glycans in cancer and inflammation - potential for therapeutics and diagnostics. *Nature Rev. Drug Disc.* 2005, *4* (6), 477-488; (c) Ohyama, C., Tsuboi, S., Fukuda, M. Dual roles of sialyl Lewis X oligosaccharides in tumor metastasis and rejection by natural killer cells. *EMBO J.* 1999, *18* (6), 1516-1525; (d) Fernandes B., Sagman U., Auger M., Demetrio M., Dennis J. W. Beta 1-6 branched oligosaccharides as a marker of tumor progression in human breast and colon neoplasia. *Cancer Res.* 1991, *51* (2), 718-723.

[70] Rabuka, D., Hubbard, S. C., Laughlin, S. T., Argade, S. P., Bertozzi, C. R. A Chemical Reporter Strategy to Probe Glycoprotein Fucosylation. *J. Am. Chem. Soc.* 2006, *128* (37), 12078-12079.

[71] Sawa, M., Hsu, T.–L., Itoh, T., Sugiyama, M., Hanson, S. R., Vogt, P. K., Wong, C. –H. Glycoproteomic probes for fluorescent imaging of fucosylated glycans in vivo. *Proc. Natl. Acad. Sci. USA.* 2006, *103* (33), 12371–12376.

[72] For reviews on the multivalent or glycoside cluster effect, see: (a) Kiessling, L., Pontrello, J. K., Schuster, M. C. Synthetic Multivalent Ligands as Effectors or Inhibitors of Biological Processes. In *Carbohydrate-based Drug Discovery.* Wong, C.–H., Ed.; Wiley-VCH, Weinheim, Germany, 2003, pp. 575-607.b) Kiessling, L. L., Gestwicki, J. E., Strong, L. E. Synthetic Multivalent Ligands as Probes of Signal Transduction. *Angew. Chem. Int. Ed.* 2006, *45* (15), 2348–2368. c) Kitov, P. I., Bundle, D. R. Thermodynamic Models of the Multivalency Effect. In *Carbohydrate-based Drug Discovery*; Wong, C.–H., Ed.; Wiley-VCH, Weinheim, Germany, 2003, pp. 541-573; d) Lundquist, J. J., Toone, E. J. The Cluster Glycoside Effect. *Chem. Rev.* 2002, *102* (2), 555-578.

[73] Collins, B. E., Paulson., J. C. Cell surface biology mediated by low affinity multivalent protein–glycan interactions. *Curr. Opin. Chem. Biol.* 2004, *8* (6), 617-625.

[74] For reviews detailing multivalent neoglycoconjugates as bacterial anti-adhesives: (a) Pieters, R. J. Intervention with Bacterial Adhesion by Multivalent Carbohydrates. *Med. Res. Rev.* 2006, 1-14; (b) Pieters, R. J. Interference with lectin binding and bacterial adhesion by multivalent carbohydrates and peptidic carbohydrate mimics. *Trends Glycosci. Glycotech.* 2004, *16* (90), 243-254; (c) Schengrund, C. "Multivalent" saccharides: development of new approaches for inhibiting the effects of glycosphingolipid-binding pathogens. *Biochem. Pharmacol.* 2003, *65* (5), 699-707.

[75] For reviews detailing multivalent carbohydrates as cancer vaccines: (a) Ragupathi, G., Coltart, D. M., Williams, L. J., Koide, F., Kagan, E., Allen, J.; Harris, C., Glunz, P. W., Livingston, P. O., Danishefsky, S. J. On the power of chemical synthesis: immunological evaluation of models for multiantigenic carbohydrate-based cancer vaccines. *Proc. Natl. Acad. Sci. USA.* 2002, *99* (21), 13699-13704; (b) Danishefsky, S. J. Synthetic Carbohydrate-Based Vaccines. In *Carbohydrate-based Drug Discovery.* Wong, C.–H., Ed.; Wiley-VCH, Weinheim, Germany, 2003, pp. 381-406.

[76] For reviews on glycopolymers as biomatierals see: (a) Klemm, D., Heublein, B., Fink, H, -P., Bohn, A. Cellulose: Fascinating Biopolymer and Sustainable Raw Material. *Angew. Chem. Int. Ed.* 2005, *44* (22), 3358-3393. (b) Heinze, T., Liebert, T., Unconventional methods in cellulose functionalisation. *Prog. Polym. Sci.* 2001, 1698-1762; (c) Müllera, F. A., Müllera, L., Hofmanna, I., Greila, P., Wenzelb, M. M., Staudenmaierb, R. Cellulose-based scaffold materials for cartilage tissue engineering. *Biomaterials.* 2006, *27*, 3955-3963; (d) Langer, R., David A. Tirrell, D. A. Designing materials for biology and medicine. *Nature* 2004, *428* (6982), 487-492.

[77] For reviews on glycopolymers as drug delivery and gene delivery, see: (a) Ladmiral, V., Melia, E., Haddleton, D. M. Synthetic glycopolymers: an overview. *Eur. Polymer J.* 2004, *40*, 431–449; (b) D'Souza, A. J. M., Topp, E.

M. Release from Polymeric Prodrugs: Linkages and Their Degradation. *J. Pharm. Sci.* 2004, *93* (8), 1962-1979; (c) Hoste, K., De Winne, K., Schacht, E. Polymeric prodrugs. *Int. J. Pharm.* 2004, *277*, 119–131; (d) Wang, W., Liu, X., Xie, Y., Zhang, H., Yu, W., Xiong, Y., Xie, W., Ma, X. Microencapsulation using natural polysaccharides for drug delivery and cell implantation. *J. Mater. Chem.* 2006, *16* (32), 3252-3267. (e) Green, D. W., Mann S., Oreffo, R. O. C. Mineralized polysaccharide capsules as biomimetic micro-environments for cell, gene and growth factor delivery in tissue engineering. *Soft Matter* 2006, *2* (9), 732-737.

[78] Duncan, R. The Dawning Era of Polymer Therapeutics. *Nat. Rev. Drug Disc.* 2003, *2* (5), 347-360.

[79] Wang, Q., Dordick, J. S., Linhardt, R. J. Synthesis and Application of Carbohydrate-Containing Polymers. *Chem. Mater.* 2002, *14* (8), 3232-3244.

[80] Ladmiral, V., Mantovani, G., Clarkson, G. J., Cauet, S., Irwin, J. L., Haddleton, D. M. Synthesis of Neoglycopolymers by a Combination of "Click Chemistry" and Living Radical Polymerization. *J. Am. Chem. Soc.* 2006, *128* (14), 4823-4830.

[81] Srinivasachari, S., Liu, Y., Zhang, G., Prevette, L., Reineke, T. M. Trehalose Click Polymers Inhibit Nanoparticle Aggregation and Promote pDNA Delivery in Serum. *J. Am. Chem. Soc.* 2006, *128* (25), 8176-8184.

[82] Hawker, C. J., Wooley, K. L. The Convergence of Synthetic Organic and Polymer Chemistries. *Science* 2005, *309* (5738), 1200-1205.

[83] (a) Oaksmith, J. M., Peters, U., Ganem, B. Three-Component Condensation Leading to β-Amino Acid Diamides: Convergent Assembly of β-Peptide Analogues *J. Am. Chem. Soc.* 2004, *126* (42), 13606-13607; (b) Wallace, G. A., Heathcock, C. H. Further Studies of the *Daphniphyllum* Alkaloid Polycyclization Cascade. *J. Org. Chem.* 2001, *66* (2), 450-454.

[84] Malkoch, M., Thibault, R. J., Drockenmuller, E., Messerschmidt, M., Voit, B., Russell, T. P., Hawker, C. J. Orthogonal Approaches to the Simultaneous and Cascade Functionalization of Macromolecules Using Click Chemistry. *J. Am. Chem. Soc.* 2005, *127* (42), 14942-14949.

[85] Liebert, T., Hänsch, C., Heinze, T. Click Chemistry with Polysaccharides. *Macromol. Rapid Commun.* 2006, *27* (3), 208–213.

[86] Keding, S. J. Danishefsky, S. J. Natural Product Synthesis Special Feature: Prospects for total synthesis: A vision for a totally synthetic vaccine targeting epithelial tumors. *Proc. Natl. Acad. Sci. USA.* 2004, *101* (33), 11937-11942.

[87] Wan, Q., Chen, J., Chen, G., Danishefsky, S. J. A Potentially Valuable Advance in the Synthesis of Carbohydrate-Based Anticancer Vaccines through Extended Cycloaddition Chemistry. *J. Org. Chem.* 2006, *71* (21), 8244-8249.

[88] Bao, X.; Duan, J.; Fang, X.; Fang, J. Chemical modifications of the (1→3)-β-D-glucan from spores of *Ganoderma lucidum* and investigation of their physicochemical properties and immunological activity. *Carbohydr. Res.* 2001, *336* (2), 127-140.

[89] (a) Sakurai, K., Shinkai, S., Molecular Recognition of Adenine, Cytosine, and Uracil in a Single-Stranded RNA by a Natural Polysaccharide: Schizophyllan. *J. Am. Chem. Soc.* 2000, *122* (36), 4520-4521; (b) Hasegawa, T., Fujisawa, T., Numata, M., Umeda, M., Matsumoto, T., Kimura, T., Okumura, S., Sakurai, K., Shinkai, S. Single-walled carbon nanotubes acquire a specific lectin-affinity through supramolecular wrapping with lactose-appended schizophyllan. *Chem. Commun.* 2004, (19), 2150-2151.

[90] Hasegawa, T., Umeda, M., Numata, M., Li, C., Bae, A. -H., Fujisawa, T., Haraguchi, S., Sakuraib, K., Shinkai, S. 'Click chemistry' on polysaccharides: a convenient, general, and monitorable approach to develop (1→3)-β-D-glucans with various functional appendages. *Carbohydr. Res.* 2006, *341* (1), 35–40.

[91] Cheshev, P., Marra, A., Dondoni, A. First synthesis of 1,2,3-triazolo-linked (1,6)-α-D-oligomannoses (triazolomannoses) by iterative Cu(I)-catalyzed alkyne–azide cycloaddition. *Org. Biomol. Chem.* 2006, *4* (17), 3225–3227.

[92] (a) Wu, X. Y., Bundle, D. R. Synthesis of Glycoconjugate Vaccines for *Candida albicans* Using Novel Linker Methodology. *J. Org. Chem.* 2005, *70* (18), 7381-7388; (b) Ni, J., Song, H., Wang, Y., Stamatos, N. M., Wang, L.–X. Toward a Carbohydrate-Based HIV-1 Vaccine: Synthesis and Immunological Studies of Oligomannose-Containing Glycoconjugates. *Bioconjugate. Chem.* 2006, *17* (2), 493-500; (c) Watt, J. A., Williams, S. J. Rapid, iterative assembly of octyl α-1,6-oligomannosides and their 6-deoxy equivalents. *Org. Biomol. Chem.* 2005, *3* (10), 1982-1992.

[93] Marmuse, L., Nepogodiev, S. A., Field, R. A. "Click chemistry" *en route* to pseudo-starch. *Org. Biomol. Chem.* 2005, *3* (12), 2225-2227.

[94] Mellet, C. O., Defaye, J., Fernándéz, J., M. G. Multivalent Cyclooligosaccharides: Versatile Carbohydrate Clusters with Dual Role as Molecular Receptors and Lectin Ligands. *Chem. Eur. J.* 2002, *8* (9), 1983-1990.

[95] (a) Loftsson, T., Masson, M. Cyclodextrins in topical drug formulations: theory and practice. *Int. J. Pharm.* 2001, *225*, 15-30; (b) Rao, V. M., Stella, V. J. When can cyclodextrins be considered for solubilization purposes? *J. Pharm. Sci.* 2003, *92* (5), 927-932; (c) Chourasia, M. K.; Jain, S. K. Polysaccharides for Colon Targeted Drug Delivery. *Drug Delivery* 2004, *11*(2), 129-148.

[96] Conteron, J. M., Vincent, C., Bosso, C., Penades, S. Glycophanes, cyclodextrin-cyclophane hybrid receptors for apolar binding in aqueous solutions. A stereoselective carbohydrate-carbohydrate interaction in water. *J. Am. Chem. Soc.* 1993, *115* (22), 10066-10076; See also reference 94 and references cited therein.

[97] Hoffmann, B., Bernet, B., Vasella, A. Oligosaccharide Analogues of Polysaccharides (Part 241): Synthesis of Cyclodextrin Analogues Containing a Substituted Buta-1,3-diyne or a 1,2,3-Triazole Unit and Analysis of Intramolecular Hydrogen Bonds. *Helv. Chim. Acta.* 2002, *85* (3), 265-286.

[98] Bodine, K. D., Gin, D. Y., Gin, M. S. Synthesis of Readily Modifiable Cyclodextrin Analogues via Cyclodimerization of an Alkynyl-Azido Trisaccharide. *J. Am. Chem. Soc.* 2004, *126* (6), 1638-1639.

[99] Bodine, K. D., Gin, D. Y., Gin, M. S. Highly Convergent Synthesis of C_3- or C_2-Symmetric Carbohydrate Macrocycles. *Org. Lett.* 2005, *7* (20), 4479-4482.

[100] Billing, J. F., Nilsson, U. J., C_2-Symmetric Macrocyclic Carbohydrate/ Amino Acid Hybrids through Copper(I)-Catalyzed Formation of 1,2,3-Triazoles. *J. Org. Chem.* 2005, *70* (12), 4847-4850.

[101] Hoogenboom, R., Moore, B. C., Schubert, U. S. Synthesis of star-shaped poly(ε-caprolactone) via 'click' chemistry and 'supramolecular click' chemistry. *Chem. Commun.* 2006, (38), 4010–4012.

[102] For reviews on the biological applications of glycodendrimers, see: (a) Cloninger, M. J. Biological applications of dendrimers. *Curr. Opin. Chem. Biol.* 2002, *6* (6), 742–748. (b) Gillies, E. R., Fréchet, J. M. J. Dendrimers and dendritic polymers in drug discovery. *Drug Disc. Today*. 2005, *10* (1), 35-43. (c) Bezouška, K. Design, functional evaluation and biomedical applications of carbohydrate dendrimers (glycodendrimers). *Rev. Mol. Biotechnol.* 2002, *90* (3-4), 269-290. (d) Benito, J. M., Gomez-Garcia, M., Ortiz Mellet, C., Baussanne, I., Defaye, J., Fernandez, J. M., Garcia. J. M. Optimizing saccharide-directed molecular delivery to biological receptors: design, synthesis, and biological evaluation of glycodendrimer-cyclodextrin conjugates. *J. Am. Chem. Soc.* 2004, *126* (33), 10355-10363. Bhadra, D., Yadav, A. K., Bhadra, S., Jain, N. K. Glycodendrimeric nanoparticulate carriers of primaquine phosphate for liver targeting. *Int. J. Pharm.* 2005, *295* (1-2), 221-233.

[103] (a) Vannucci, L., Fiserova, A., Sadalapure, K., Lindhorst, T. K., Kuldova, M., Rossmann, P., Horvath, O., Kren, V., Krist, P., Bezouska, K., Luptovcova, M., Mosca, F., Pospisil, M. Effects of N-acetyl-glucosamine-coated glycodendrimers as biological modulators in the B16F10 melanoma model in vivo. *Int. J. Oncol.* 2003, *23* (2), 285-296; (b) Roy R., Baek M.–G. Glycodendrimers: novel glycotope isosteres unmasking sugar coding. Case study with T-antigen markers from breast cancer MUC1 glycoprotein. *J. Biotechnol.* 2002, *90* (3-4), 291-309.

[104] (a) Pérez-Balderas, F., Ortega-Muñoz, M., Morales-Sanfrutos, J. Hernández-Mateo, F., Calvo-Flores, F. G., Calvo-Asín, J. A., Isac-García, J. Francisco Santoyo-González, F. Multivalent Neoglycoconjugates by Regiospecific Cycloaddition of Alkynes and Azides Using Organic-Soluble Copper Catalysts. *Org. Lett.* 2003, *5* (11), 1951-1954. (b) Calvo-Flores, F. G., Isac-García, J., Hernández-Mateo,F., Pérez-Balderas, F., Calvo-Asín, J. A., Sanchéz-Vaquero, E., Santoyo-González, F. 1,3-Dipolar Cycloadditions as a Tool for the Preparation of Multivalent Structures. *Org. Lett.* 2000, *2* (16), 2499-2502.

[105] Appukkuttan, P., Dehaen, W., Fokin, V. V., Van der Eycken, E. A Microwave-Assisted Click Chemistry Synthesis of 1,4-Disubstituted 1,2,3-Triazoles via a Copper(I)-Catalyzed Three-Component Reaction. *Org. Lett.* 2004, *6* (23), 4223-4225.

[106] Joosten, J. A. F., Niels T. H. Tholen, N. T. H., El Maate, F., Brouwer, A. J., van Esse, G. W., Rijkers, D. T. S., Liskamp, R. M. J., Pieters, R. J. High-Yielding

Microwave-Assisted Synthesis of Triazole linked Glycodendrimers by Copper-Catalyzed [3+2] Cycloaddition. *Eur. J. Chem.* 2005, *15*, 3182–3185.

[107] Bouillon, C., Meyer, A., Vidal, S., Jochum, A., Chevolot, Y., Cloarec, J.-P., Praly, J.–P., Vasseur, J.–J., Morvan, F. Microwave Assisted "Click" Chemistry for the Synthesis of Multiple Labeled-Carbohydrate Oligonucleotides on Solid Support. *J. Org. Chem.* 2006, *71* (12), 4700-4702.

[108] Deng, S. L., Gangadharmath, U., Chang, C.-W. Sonochemistry: A Powerful Way of Enhancing the Efficiency of Carbohydrate Synthesis. *J. Org. Chem.* 2006, *71* (14), 5179-5185.

[109] Chittaboina, S., Xie, F., Wang, Q. One-pot synthesis of triazole linked glycoconjugates. *Tetrahedron Lett.* 2005, *46* (13), 2331-2336.

[110] Wu, P., Malkoch, M., Hunt, J. N., Vestberg, R., Kaltgrad, E., Finn, M. G., Fokin, V. V., Sharpless, K. B., Hawker, C. J. Multivalent, bifunctional dendrimers prepared by click chemistry. *Chem. Commun.* 2005, (46), 5775–5777.

[111] Fernandez-Megia, E., Correa, J., Rodríguez-Meizoso, I., Riguera, R. A Click Approach to Unprotected Glycodendrimers. *Macromolecules.* 2006, *39* (6), 2113-2120.

[112] Fernandez-Megia, E., Correa, J., Riguera, R. Clickable PEG-Dendritic Block Copolymers. *Biomacromolecules.* 2006, *7* (11), 3104-3111.

[113] For reviews on galectins, see: (a) Leffler, H., Carlsson, S., Hedlund, M., Qian, Y., Poirier, F. Introduction to galectins *Glycoconjugate J.* 2004, *19* (7/8/9), 433–440; (b) Pieters, R. J. Inhibition and detection of galectins. *ChemBiochem.* 2006, *7* (5), 721-728.

[114] Giguère, D., Patnam, R., Bellefleur, M.–A., St-Pierre, C., Satob, S., Roy, R. Carbohydrate triazoles and isoxazoles as inhibitors of galectins-1 and -3. *Chem. Commun.* 2006, (22), 2379–2381.

[115] Tejler, J., Tullberg, E., Frejd, T., Lefflerb, H., Nilsson, U. J., Synthesis of multivalent lactose derivatives by 1,3-dipolar cycloadditions: selective galectin-1 inhibition. *Carbohydr. Res.* 2006, *341* (10), 1353–1362.

[116] Gao, Y. J., Eguchi, A., Kakehib, K., Lee, Y. C. Synthesis and molecular recognition of carbohydrate-centered multivalent glycoclusters by a plant lectin RCA_{120}. *Bioorg. Med. Chem.* 2005, *13* (22), 6151-6157.

[117] Chen, Q., Yang, F., Du, Y. Synthesis of a C_3-symmetric (1→6)-N-acetyl-β-D-glucosamine octadecasaccharide using click chemistry. *Carbohydr. Res.* 2005, *340* (16), 2476–2482.

[118] a) Ratner, D. M., Adams, E. W., Disney, M. D., Seeberger, P. H. Tools for Glycomics: Mapping Interactions of Carbohydrates in Biological Systems. *ChemBiochem.* 2004, *5* (10), 1375-1383; b) Dyukova V. I., Shilova N. V., Galanina O. E., Rubina A. Y., Bovin N. V. Design of carbohydrate multiarrays. *Biochim. Biophys. Acta* 2006, *1760* (4), 603-609; c) Smith, E. A. Thomas, W. D., Kiessling, L. L., Corn, R. M. Surface Plasmon Resonance Imaging Studies of Protein-Carbohydrate Interactions. *J. Am. Chem. Soc.* 2003, *125* (20), 6140-6148.

[119] Disney, M. D., Seeberger, P. H. The Use of Carbohydrate Microarrays to Study Carbohydrate-Cell Interactions and to Detect Pathogens. *Chem. Biol.* 2004, *11* (12), 1701-1707.

[120] Fazio, F., Bryan, M. C., Blixt, O., Paulson, J. C., Wong, C.-H. Synthesis of Sugar Arrays in Microtiter Plates. *J. Am. Chem. Soc.* 2002, *124* (48), 14397-14402.

[121] Bryan, M. C., Lee, L. V., Wong, C.–H. High throughput identification of fucosyltransferase inhibitors using carbohydrate microarrays. *Bioorg. Med. Chem. Lett.* 2004, *14* (12), 3185–3188.

[122] Huang, G-L., Liu, T.-C., Liu, M.-X., Mei, X.-Y. The application of quantum dots as fluorescent label to glycoarray. *Anal. Biochem,* 2005, *340* (1), 52-56.

[123] Bryan, M. C., Fazio, F., Lee, H.-K., Huang, C.–Y., Chang, A., Best, M. D., Calarese, D. A., Blixt, O., Paulson, J. C., Burton, D., Wilson, I. A., and Wong, C.–H., Covalent Display of Oligosaccharide Arrays in Microtiter Plates. *J. Am. Chem. Soc.* 2004, *126* (28), 8640-8641.

[124] Zhang, Y., Luo, S., Tang, Y., Yu, L., Hou, K.–Y., Cheng, J.–P., Zeng, X., and Wang, P. G. Carbohydrate-Protein Interactions by "Clicked" Carbohydrate Self-Assembled Monolayers. *Anal. Chem.* 2006, *78* (6), 2001-2008.

[125] Devaraj, N. K., Miller, G. P., Ebina, W., Kakaradov, B., Collman, J. P., Kool, E. T., Chidsey, C. E. D. Chemoselective Covalent Coupling of Oligonucleotide Probes to Self-Assembled Monolayers. *J. Am. Chem. Soc.* 2005, *127* (24), 8600-8601.

[126] Sun, X.–L., Stabler, C. L., Cazalis, C. S., and Chaikof, E. L. Carbohydrate and Protein Immobilization onto Solid Surfaces by Sequential Diels-Alder and Azide-Alkyne Cycloadditions. *Bioconjugate Chem.* 2006, *17* (1), 52-57.

In: Drug Design Research Perspectives
Editor: Stanley P. Kaplan, pp. 103-124
ISBN: 978-1-60021-823-1

Chapter V

Molecular Similarity Study in Compounds Comparison and Library Design

Aixia Yan[*]
State Key Laboratory of Chemical Resource Engineering,
Department of Pharmaceutical Engineering, P.O. Box 53,
Beijing University of Chemical Technology,
15 BeiSanHuan East Road, Beijing 100029, P. R. China

Abstract

In computer-aided drug design, molecular similarity study plays an important role in finding and designing new drugs from the known drugs or inhibitors for a given protein target. A list of compounds can be compared according to the similarity of their structures or properties. With regards to the comparison of several molecular structures, some similarity calculation methods based on 2D structural characteristics, 3D structural characteristics or 3D pharmacophore characteristics have been developed; as to the comparison of the similarity of the property of a set of compounds, some chemoinformatics tools for clustering compounds have been developed.

In addition, in the combinatorial library design, more attention is given to design target focusing library along with diversity and drug-like criteria. On the basis of the compounds similarity comparison study, the similarity of the two compounds library can be compared; furthermore, in high-throughput-screening, high quality compounds library, in particular, the focusing library of drug-likeness (or lead-likeness) for a given target could be built based on the results of molecular similarity study such as the 3D pharmacophore characteristics.

Molecular similarity study will continue to occupy an important role in drug discovery and library design.

[*] Corresponding author: E-mail: aixia_yan@yahoo.com or yanax@mail.buct.edu.cn; Tel.: +86 10 6442 1335; Fax. : +86 10 6441 6428

Keywords: Molecular similarity, similarity searching, 2D fingerprint based similarity, 3D pharmacophore (fingerprint) based similarity, Kohonen self-organizing map (SOM), Library design

1. Introduction

In computer-aided drug design, molecular similarity study plays an important role in finding and designing new drugs from the known drugs or inhibitors for a given protein target. From a set of known drugs or inhibitors, some common structural characteristics corresponding to a certain bioactivity could be found, which can be used for searching and identifying new molecules with the similar bioactivity.

A list of compounds can be compared according to the similarity of their structures or properties. With regards to the comparison of several molecular structures, some similarity calculation methods such as: 2D structure similarity, 3D structure similarity and 3D pharmacophore similarity have been developed; as to the comparison of the similarity of the property of a set of compounds, some chemoinformatics tools such as Kohonen self-organizing map (SOM) for clustering compounds have been developed.

In addition, in the combinatorial library design, more attention is given to design target focusing library along with diversity and drug-like criteria. On the basis of the compounds similarity comparison study, the similarity and dissimilarity (or diversity) of a compounds library can be evaluated and the similarity of the two compounds library can be compared; furthermore, in high-throughput-screening, high quality compounds library, in particular, the focusing library of drug-likeness (or lead-likeness) for a given target could be built based on the results of molecular similarity study such as the 3D pharmacophore characteristics.

This review will summary the recent molecular similarity comparison methods and their related applications from two parts: (1) molecular structure similarity comparison methods, which include 2D-graph based, 2D fingerprint based, 3D force field based, 3D pharmacophore and pharmacophore fingerprint based similarity methods, etc; and (2) molecular property similarity comparison and visualization using Kohonen self-organizing map (SOM). After that, based on molecular similarity comparison methods, the following questions such as evaluation of the diversity of a compounds library, comparison of two compounds libraries and design a diverse or focused library, will be discussed.

Molecular similarity study will continue to occupy an important role in drug discovery and library design.

2. Molecular Structures Comparison

Molecular similarity searching is a long-established yet active area of research. Similarity searching has the foundations in the Similar Property Principle, which states that structurally similar molecules are likely to have similar properties [1]. This is generally true, although there are some cases where minor changes of the structure lead to major changes of the property considered [2,3]. Thus, given a molecule of known biological activity, compounds that are structurally similar to it are likely to exhibit the same activity. This characteristic has been referred to as neighbourhood behaviour [4,5]. Basically, in a similarity searching, the molecule with the known desired activity can be used as a query (reference structure) to find some structurally similar molecules that display the same activity.

Molecular similarity has been extensively studied in recent years. Lots of methods and the related applications have been developed due to the complicate relationships between the molecular structures and their targets (or their bioactivities). A list of reviews for similarity searching methods and their applications in drug design have been reported in journals or in books [6-14].

A similarity measure comprises three components: the representation that is used to describe the molecules that are being compared; the weighting scheme that is used to assign differing degrees of importance to the various components of these representations; and the coefficient that is used to determine the degree of relatedness between two structural representations. (Thus far, there have been only limited studies of the effect of weighting on molecular similarity.) [12]. For each component, many choices can be made, their combination leading to a large number of possible similarity measures. So using a certain structural representation, the similarity of two molecules can be estimated by a similarity coefficient.

A molecule exhibits many different structural characteristics. As a whole, a molecule can be described from several aspects: its size, its chemical graph, its shape, its volume, etc.; a molecule can also be described by atoms or atom types, functional group (fragment), substructure, or scaffold that form the molecule. A molecular structural characteristic can also be reflected by its physicochemical properties such as atomic charges, atomic electronegativity, molecular electrostatic potential, hydrophobic property, hydrogen-bond donor and acceptor ability, and so on. Subsequently, based on the different structural representations, the similarity of two molecules can be compared by a list of different methods reflecting how similar of their structures and properties.

On the basis of the different molecular structural representations, there are several similarity coefficients (similarity index) that are used for quantifying the degree of similarity of two molecules. They include: Tanimoto Coefficient, Dice Coefficient, Cosine Coefficient and Euclidean Distance, etc, of which the most popular and best coefficient is the Tanimoto coefficient [6,12,15]. For continuous variables the Tanimoto coefficient $S_{A,B}$ between molecules A and B is represented as formula (1); for dichotomous variables (binary vectors), $S_{A,B}$ is defined as formula (2):

$$S_{A,B} = \frac{\sum_{j=1}^{j=n} x_{jA}x_{jB}}{\sum_{j=1}^{j=n}(x_{jA})^2 + \sum_{j=1}^{j=n}(x_{jB})^2 - \sum_{j=1}^{j=n} x_{jA}x_{jB}} \qquad (1)$$

where the molecules A and B are represented by vectors, X, of length n with the jth property having the value x_j.

$$S_{A,B} = \frac{c}{a+b-c} \quad (2)$$

where a is the number of bits "on" in A, b is the number of bits "on" in B, and c is the number of bits "on" in both A and B.

2.1. 2D Descriptors Based Similarity Methods

Molecules can be represented by all kinds of 2D descriptors, and then the intermolecular similarity can be calculated. 2D descriptors have been shown to be very effective in similarity comparison and in discriminating classes of molecules; in addition, they are fast to computer and they are, by nature, conformation-independent.

2.1.1. 2D Graphs Based Similarity

In two-dimension, molecules can be treated as topological graphs. The graph vertices represent the atoms, and the graph edges denote the bonds connecting each pair of covalently bonded atoms. Then we can map one molecule onto another and find the correspondence between atoms in the two molecules [16,17]. Thus the similarity between two molecules can be directly determined by comparing their corresponding graphs. In addition, the similarity of two molecules can also be calculated by measuring the similarity between two graphs by the 2D maximum common subgraphs (MCS) [18,19], which give an intuitively reasonable view of the similarity between two molecules.

Recently a type of 2D reduced graph based similarity method has also been developed. Reduced graphs are summary representations of chemical structures, where groups of connected atoms that share some characteristic are reduced into single nodes reflecting the connectivity of atoms in the original chemical graphs [20-26]. Various methods have been developed for quantifying the similarity of two structures based on their reduced graphs including mapping the reduced graph to a fingerprint [20,21], using graph matching techniques to compare the reduced graphs directly [22-24], and using the edit distance method [25,26]. A recent work of Barker et al. [24] has shown that graph matching, using a reduced graph representation of small molecules, is an effective and efficient similarity method, outperforming Daylight fingerprints. Use of the Bron-Kerbosch clique detection algorithm to find the MCIS between pairs of fully connected reduced graphs was the best graph-matching method of those investigated. The hits found using this method were more diverse than those found by other methods, clearly demonstrating that reduced graphs offer a powerful method for finding novel scaffolds [24]. The edit distance approach quantifies the degree of similarity of two reduced graphs based on the number and type of operations required to convert one graph to the other. In the work of Birchall et al. [26], a genetic algorithm (GA) was used for training the weights of the different edit distance operations. The method is applied to specific activity classes extracted from the MDDR database to derive activity-class specific weights. The GA-derived weights give substantially improved results in recall experiments as compared to using weights assigned on intuition [26].

2.1.2. 2D Fingerprint Based Similarity

Fingerprints, i.e., vectors where each element encodes some aspects of molecular structures, are commonly used as the input for similarity calculations since, once defined and generated, they can be quickly compared by using a similarity coefficient [13,14]. Molecular Fingerprint contains 2D Fingerprint, 3D Fingerprint, 3D pharmacophore Fingerprint. Molecular fingerprints are widely used for similarity-based virtual screening and drug discovery.

2D fingerprints, i.e. binary strings encoding the presence or absence of substructural fragments (or structural features) [13,14], for the measurement of inter-molecular structural similarity. 2D fingerprints are among the most popular similarity search tools because they transform compounds into a linear format, usually a bit string, and thus permit computationally efficient compound comparison. The commonly used 2D fingerprints include MDL keys [27], Daylight fingerprint [28], Unity 2D fingerprint [29], and many others [30]. The MDL products have exposed both 166 bit and 960 bit keysets based on 2D descriptors [27]. Daylight fingerprint [28] consist of 2048 bits, different molecular features are mapped to overlapping bit segments. Unity 2D fingerprint [29] is binary hashed fingerprint.

Kogej and coworkers [30] discussed the performance and the complementarity of nine 2D fingerprints (Daylight, Unity, AlFi, Hologram, CATS, TRUST, Molprint 2D, ChemGPS, and ALOGP) in retrieving active molecules by similarity searching against a set of query compounds. Four sets of compounds with binding affinity data against four protein families (GPCRs, kinases, ion channels, and proteases) were used. The Tanimoto index was used in the similarity searches. Their work suggests that a multifingerprint approach can be used as an efficient tool to balance the strengths and weaknesses of various fingerprints [30].

2D fingerprint based similarity searching method can be used for virtual screening, and the recent work at the University of Sheffield has been reviewed by Willett [12]. In his paper, a detailed comparison of a large number of similarity coefficients demonstrates that the well-known Tanimoto coefficient remains the method of choice for the computation of fingerprint-based similarity, despite possessing some inherent biases related to the sizes of the molecules that are being sought [12].

Xue and coworkers [31-33] proposed a kind of Mini-fingerprints (MFPs) for similarity searching. Mini-fingerprints (MFPs) are short binary bit string representations of molecular structure and properties, composed of few selected two-dimensional (2D) descriptors and a number of structural keys. MFPs were designed as short binary bit strings, consisting of only 50-60 bit positions or even less bits. Their work indicate that MFPs are effective in recognizing compounds with similar activity [31-33].

2.2. 3D Descriptors Based Similarity Methods

2.2.1. 3D Field-Based Similarity Methods

The 3D molecular field-based descriptors normally require alignment of the molecules to be compared. The field-based descriptors are derived from the effects of a test probe at a distance. This could be *e.g.* a positive charge, a water molecule or a methyl probe. Generally, additional properties such as atom centred partial charges have to be pre-computed before calculation of the field effect.

The 3D field-based properties such as Molecular Electrostatic Potential (MEP) [34-37] and shape [38-42] descriptors are calculated based on the application of a regular grid with many "grid points". The MEP similarity or the shape similarity can then be calculated based on some similarity coefficient such as Carbo index [43] as follows:

$$C_{AB} = \frac{\int P_A P_B dv}{(\int P_A^2 dv)^{1/2} (\int P_B^2 dv)^{1/2}} \tag{3}$$

Molecular similarity C_{AB} is determined from the 3D descriptor properties P_A and P_B of molecules A and B being compared over all space. Its structure is identical in construction to the cosine coefficient which is also utilized in 2D database similarity searching [6,14].

The Carbo index was originally utilized to measure the similarity of quantum-mechanically derived electron density between molecules [43]. Subsequently, variants of this index have also been introduced, in an attempt to increase the sensitivity of the formula to property magnitude [34, 44-45]. Later, similarity evaluations using the variants of the Carbo index have been extended to include the comparison of such MEP [34-37] and shape [38-42] descriptors. There are some other similarity coefficients such as Tanimoto coefficient (modified form), the Euclidean distance, etc. which are also widely used for 3D field-based molecular similarity comparison [7].

A variety of techniques have been developed in the calculations of MEP or shape descriptors and in the comparison of the molecular fields between molecules. They include numerical evaluation on rectilinear grids, analytical comparisons employing Gaussian function approximations, gnomonic projection and molecular volume measurement. These techniques have been reviewed recently [7].

2.2.2. 3D Maximum Common Subgraphs (MCS) Based Similarity

In a 3D chemical graph, the vertices of the graph denote the atoms and the edges indicate the geometric distance or range of distances between a pair of atoms (vertices). An edge in one graph is compatible with an edge in another graph if their two vertex endpoints and edge labels are compatible. The intermolecular similarity can be calculated on the basis of the 3D maximum common subgraphs (MCSs) methods [46,47], which have been used for similarity searching [48-51]. In a recent work of Raymond and Willett [51], they describe a new method for 3D MCS-based similarity searching. The method is based on a graph matching procedure that accommodates conformational flexibility by using distance ranges between pairs of atoms. These distance ranges are generated using triangle and tetrangle bound smoothing techniques from distance geometry. Simulated virtual screening experiments demonstrate that the method is broadly comparable in effectiveness to the 2D similarity searching (using Daylight 2D fingerprint and the 2D MCS measures) and retrieves many active molecules that are not identified by the 2D methods [51].

2.2.3. 3D Shape Based Similarity

Topomer shape similarity is a kind of 3D molecular shape based similarity method which is independent of molecular conformations [52-56]. A topomer is an invariant 3D representation of a molecular fragment, generated from its 2D topology by deterministic rules that produce absolute coordinates for each of its constituent atoms [52]. Topomers were originally devised to enable shape similarity searching of very large virtual libraries [53,54]. Cramer and colleagues applied topomer similarity as a predictor of similarity in biological activity [55,56]. In a prospective study, the degree of topomer similarity to the "nearest" of four A2 antagonist query structures correlated with the hit rate in the A2 assay, while correctly identifying a few unexpectedly active ligands [55]. In another recent work of them [56], they quantified the ability of topomer shape similarity to forecast a variety of biological similarities based on two extensive validation studies: (1) prospective "lead hopping", by biological testing of 308 compounds that were topomerically most similar, from among approximately 80000 candidates, to 14 query structures including 13 different kinds of reported biological activity, which yielded 11 successful "lead hops" in the 13 assays attempted; (2) retrospective comparison of topomer similarity with the standard "Tanimoto 2D Unity fingerprints", in their r tendencies to concentrate together potential and actual drugs reported to belong to the same "activity class", for twenty classes. Their work indicate that the topomer description of molecular structure is effective in predicting "lead hops" and topomer similarity could be used as a superior predictor of similar biological activities [56].

2.2.4. 3D Pharmacophore and Pharmacophore Fingerprints Based Similarity

The 3D pharmacophore hypothesis is based on the concept of the receptor and ligand as a 'lock and key' in which common interacting groups are found at similar distances apart [57]. A pharmacophore is a set of functional group types (features) in a certain spatial arrangement. The features of a pharmacophore can include: hydrophobic group, hydrogen-bond donor or acceptor, charge, similar functional groups, etc. A pharmacophore model can be built based on the alignment of the several inhibitors (on bioactive conformations or related) or based on the characteristics of active binding site of a protein target. The pharmacophore models could be used for identifying other molecules that share the same pharmacophore features. They are often referred to as multiple-point-pharmacophores: two-point pharmacophores (2PP) [58,59], which are known as atom pairs and represent all possible pairs of atoms in the molecule, three-point pharmacophores (3PP) [60-64], which allow for a more detailed representation of interatomic distances, and four-point pharmacophores (4PP) [65,66], which are able to distinguish between geometric isomers. On the basis of the molecular 3D pharmacophore models for representation of a set of molecules, their similarity can be estimated and other molecules that share the same pharmacophore features can be found.

Some work has incorporated whole-molecule information into pharmacophores on the basis of geometric distances between defined features and all ligand atoms [67]. In a recent work of Jenkins et al. [68], the 3D descriptions called FEPOPS (*fe*ature *po*int pharmacophores) was proposed. FEPOPS evaluates the regional correlation of additive physicochemical and pharmacophoric properties, benefiting from the advantages gained from both field-based similarity and pharmacophore-based queries. It was demonstrated that similarity searching with FEPOPS (for five drug targets) outperform the commonly used 2D similarity methods (DAYLIGHT, MACCS, Pipeline Pilot fingerprints) and a commercial 3D method (Pharmacophore Distance Triplets) in a scaffold hopping-oriented virtual screening protocol [68].

3D pharmacophore fingerprints have been introduced that store the information of pairs, triplets, or quartets of pharmacophore features for multiple conformations in the form of binned distance ranges, each of which is assigned to a bit in a bit-string or fingerprint [69-75]. The distances between features typically are binned to give a fixed number of possible pharmacophores. For flexible molecules, the "molecule fingerprint" is the union across all conformations. A "library fingerprint" typically is taken as the union across a set of molecules.

3D pharmacophore and pharmacophore fingerprints based similarity methods can be used for similarity searching, virtual screening and scaffold hopping. Zhang and Muegge [76] explored a broad range of 2D and 3D descriptors for similarity search and scaffold hopping for ligands of seven well-known drug targets; they focus on evaluating different ranking strategies to improve virtual screening performance for scaffold hopping. In a series of virtual screening studies involving 9969 MDDR compounds as negative controls it has been found that atom pair descriptors and 3D pharmacophore fingerprints combined with ranking, voting, and consensus scoring strategies perform well in finding novel bioactive scaffolds [76].

2.3. Evaluation of Molecular Similarity Methods

Evaluation the performance of molecular similarity methods should consider many factors such as query, size and quality of molecular database, property of binding target, molecular properties [or bioactivities), the descriptors used for molecular structural representation, and the similarity coefficient employed for similarity calculation, etc. It has been suggest that in order for a new method to be thoroughly tested, at least ten diverse sets should be used [8].

Given the same query and database, the effectiveness of the similarity searching methods can be measured using the cumulative recall measure or enrichment factors [77]. Recall is the fraction of the active structures that are retrieved in the search (a) over the total number of active structures in the database (A), i.e.

$$R = \frac{a}{A} \tag{4}$$

with regard to the 3D molecular similarity method dependent of molecular conformations, its performance is also influenced by the way of how to manage the flexibility of molecular conformations before molecular similarity calculations. Basically, it would be expected that descriptors contain conformational information would be more predictive of bioactivity than ones that do not, and that 3D descriptors would be better than 1D/2D descriptors. However, this has been difficult to demonstrate, and sometimes 2D descriptors actually outperform 3D ones. This may be caused by difficulties of the management the flexibility of the 3D molecular conformations.

3. Molecular Property Similarity Comparison and Visualization

With the methods for calculation the molecular structural similarity discussed above, for a given dataset, each pairwise molecular structural similarity can be calculated; the molecules could also be ranked in order of decreasing similarity to the query (reference molecule, eg. known drug). But how can one display and visualize the complex relationships of intermolecular similarities of all the molecules in a dataset simultaneously? One possible way is to use the cluster analysis. Cluster analysis, is the process of subdividing a group of chemical molecules into groups, or clusters, of molecules that exhibit a high degree of both intracluster similarity and intercluster dissimilarity [6]. For example, all the molecules in a whole dataset can be clustered by using some methods such as Ward's hierarchical clustering method [78], or the nonhierarchical nearest neighbor method of Jarvis and Patrick [79].

Here the Kohonen [80-83] self-organizing map (SOM) is taken as an example for molecular property similarity comparison and visualization. SOM is a kind of unsupervised neural network model. The basic processing elements, the artificial neurons, are arranged in an n-dimensional, usually two-dimensional, network. Objects, represented by m input variables are projected into this network. Usually, a 2D Kohonen network can be used to project a higher-dimensional (m-dimensional) space into a two-dimensional map. A 2D Kohonen layer commonly consists of a number of neurons arranged on a rectangular map. Each neuron has m weights assigned that match the m input variables (molecular descriptors) of the objects (chemical compounds). The architecture of a Kohonen network is illustrated in figure 1.

Learning in a Kohonen network is an unsupervised and competitive process. The value of weights are initialized by a random number. An object, a sample, s, represented by m input variables, $x_{si,}$ is projected into that neuron, c, that has weights, w_{ji}, most similar to the input variables (Eqn. (5)). This neuron is called the winning neuron.

$$out_c \leftarrow \min\left[\sum_{i=1}^{m}(x_{si} - w_{ji})^2\right] \tag{5}$$

In the next step, the weight of the winning neuron is corrected to make it even closer to the current sample s; the weights of the neighboring neurons are also updated to attract similar inputs. The weights are modified according to the following equation

$$w_{ji}^{new} = w_{ji}^{old} + \eta(t)\cdot a(d_c - d_j)(x_{si} - w_{ji}^{old}) \tag{6}$$

In which, w_{ji}^{new} and w_{ji}^{old} are the new and old weights, respectively,η is the learning rate, dependent on time or iteration cycle, t, the $a(d_c\text{-}d_j)$ is the neighborhood function dependent on the topological distance $d_c\text{-}d_j$ (topological distance between the winning neuron c and the current neuron j). After this adjustment of weights, a new object is taken and again a single neuron is selected as the winner with a corresponding adjustment of weights. In the end, each individual object has been assigned to a specific neuron.

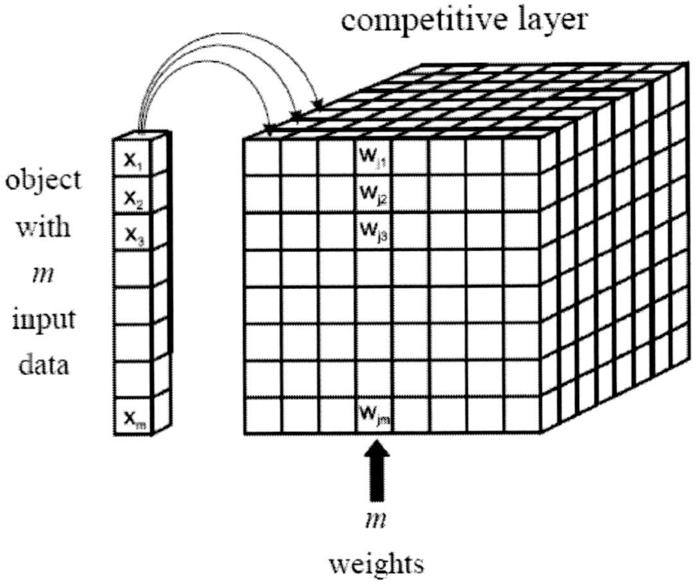

winning neuron $$out_c \leftarrow \min\left\{\sum_{i=1}^{m}(x_{si} - w_{ji})^2\right\}$$

Figure 1. Architecture of a Kohonen network. Here, a network with 8×8 neurons is depicted. The first subscripted index refers to the number of the neuron. During training, the m-dimensional input patterns iteratively enter the network and the responses of all neurons are calculated. The winning neuron having the minimum distance to the input pattern is determined.

The Kohonen self-organizing map (SOM) has the special property of effectively creating a spatially organized internal representation of various features of input signals and their abstractions. The perception of similarity of objects is an essential feature. That means, if two input data vectors are similar, they will be mapped into the same neuron or closely together in the output layer of the two-dimensional map. Thus the map can be used for visualizing similarity of the objects and for clustering the objects.

The map can be colored using a value (or a value range) of the corresponding property (or a biological activity) of each chemical compound. It should be noted that no knowledge on such property is used in determining the projection by training the SOM network. This is the essence of unsupervised learning.

The Kohonen self-organizing map (SOM) is an important tool that can simultaneously display a large dataset containing thousands of molecules and their properties. It can be used for the compounds pattern recognition and property comparison and visualization. A Kohonen network can be trained with some proper structural descriptors as the input vectors, and then the internal relationship of the objects (compounds) can be demonstrated on a two-dimensional map with different clusters. The clusters may be associated with different types of properties (different biological activities). For instance, in our former work [84-86], the solubility of organic compounds was clustered on a Kohonen map according to their structural characteristics represented by different structural descriptors. It is shown in figure 2, where 2084 organic compounds, when 18 2D descriptors (that are highly related to the solubility) were used as the input descriptors of Kohonen network, and the compounds were somehow orderly located into a 2D map according to the magnitude of their solubility. It is observed that compounds with a different range of solubility are projected into different areas. There is an island of molecules with high solubility, surrounded by a sea of intermediate solubility with some depressions of low solubility. The 18 input descriptors of SOM include logP, mean molecular polarizability, aromatic and aliphatic indicators of a molecule, highest hydrogen bond acceptor and donor potentials, number of hydrogen bond donor groups, number of atoms of element nitrogen, oxygen, fluorine, sulphur and chlorine, and five molecular 2D autocorrelation coefficients for the atomic properties of lone-pair electronegativity, π-electronegativity, σ-electronegativity, π-charge and partial atomic charges. They were used because that from our knowledge it can be assured that they must be highly associated with the solubility of organic compounds. Here using a 52×44 Kohonen map, the compounds were clustered in a two-dimensional map from the 18 dimensional input descriptors [86].

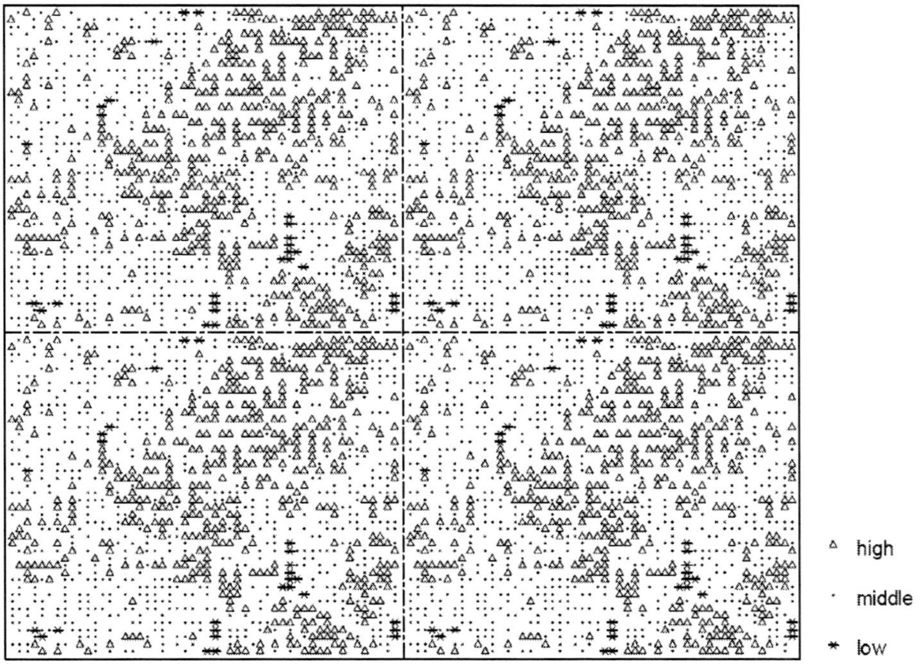

Figure 2. A four-fold toroidal KNN map for 2084 compounds by using the 18 topological input descriptors. "High" means compounds with high solubility where *logS* is in the range of [-2.045 ~ 2.346], "middle" means compounds with intermediate solubility where *logS* is in the range of [-6.436 ~ -2.046], and "low" means compounds with low solubility where *logS* is in the range of [-10.830 ~ -6.437]. Four identical KNN maps were put together like tiles in order to show the closed topology of the surface of a torus.

It was demonstrated that self-organizing maps can also be used as an important tool for distinguishing drug-like and nondrug-like molecules. In Lee and Schneider's recent work [87], using the SOM, the same dataset of drugs and nondrugs in Sadowski's work [88] was projected into a two-dimensional self-organizing map with the topological pharmacophore descriptors as the input vectors. The result of binary class distribution (as shown in figure 4b in ref. 87) clearly shown a drug and a nondrug region; while the binary prediction accuracy obtained (cc=0.33) is low compared to the much higher accuracy that can be yielded by the supervised neural networks and more problem-specific descriptors [88]. However, the self-organising neural networks here can be used to visualize the distribution of compound libraries of drugs and nondrugs from a 150-dimensional space into a two-dimensional preserving map (as shown in figure 4a in ref. 87). The results suggest that the topological pharmacophore descriptors used in this work can be used for distinguish drugs and nondrugs in some extent.

The other applications of Kohonen self-organizing map for clustering and property similarity comparison can be found in the references [89-92]. It must be emphasized that before performing a Kohonen self-organizing map projection, it is fundamentally important to select a suitable set of descriptors that exhibit a good discrimination ability in a proper hyperspace, i.e. allowing a good appraisal of molecular diversity (or highly associated with a certain biological activity), among all the compounds of the considered chemical database. Thus the complicated relationship between the molecular property and the structures can be properly display in a 2D Kohonen self-organizing map.

4. Library Design Based on Molecular Similarity Comparison

The goal of library design is to identify a small subset of the virtual space that best fits the design objectives [93]. Independently of the main objective to a diverse screening library or design a library focused around a lead series. A screening library based on many diverse chemical structures can be used as a general-purpose screening library to identify novel leads and not intended for any particular targets; while focused libraries are typically second-generation libraries, and represent an elaboration on a common structural theme. Based on kinds of the molecular similarity comparison methods discussed above, it is now possible to evaluate the diversity of a compounds library, compare two compounds libraries, and to generate large numbers of diverse or focused compound libraries.

4.1. Evaluation of the Diversity of a Compounds Library

How can one evaluate the diversity (or dissimilarity) of a compounds library? Basically, the diversity of a compounds library can be evaluated by the chemical space it covers. For example, some fundamentally important structural descriptors (or physicochemical properties) of the library, such as molecular weight, logP, number of hydrogen-bond donors and acceptors and some others can be easily calculated. The distribution of these structural descriptor values could reflect the diversity of a compounds library in some extent.

Each molecule in a library can also be described by a set of 2D or 3D fingerprints, where each feature within each structure is represented by a bit in a binary string. In this case, the simplest measure of diversity of a compounds library is simply to count the total number of features displayed by the entire library. The number of library union fingerprint, which is the union of all the individual bit-strings, gives total the number of features presented by the ensemble of molecules [94]. For instance, Daylight 2D substructure fingerprints were employed for diversity measurement and for comparing the number of molecular substructures presented by libraries of biopolymers, peptoids, and small molecule collections [95]. For three- or four-point pharmacophores fingerprints, the library union fingerprint includes all of the pharmacophores for all the conformations for all of the molecules in the library. This simple diversity measure is very quick to calculate for a large number of compounds. Subsequently, a somewhat diverse subset can be found rapidly [65].

4.2. Comparison of Two Compounds Libraries

When synthesis or testing a library of compounds, one needs to evaluate a library with another library, e.g. one wants to know if it covers more chemical space, how similar or dissimilar of the compounds in the two libraries. In this case, one needs to know the differences of the two libraries. How to compare two compounds library? An easy way for comparing two datasets libraries is just to compare the spread of activity values and the structural descriptor values of the two libraries. We can simply judge which dataset is more diverse. Basically, a library containing larger ranges of activity values and the structural descriptor values is more diverse than another one. More information can be obtained by comparing the detail distribution of the structural descriptor values in the two libraries.

Similarly, the diversity of two compounds library can also be compared on the basis of their corresponding 2D or 3D library union fingerprints. More information can be obtained by comparing the detail distribution of the exact fingerprints in the two libraries.

The Kohonen self-organizing map (SOM) has the characteristics of mapping the objects according to their similarity. SOM can also be used to investigate the similarity and dissimilarity of two different dataset libraries or determine the degree of overlap between the compounds in these two libraries [92]. The two library datasets can be compared according to the distribution of the objects in different datasets in a Kohonen map as shown in our former work [85]. In that work, the 2084 compounds in the Merck dataset and the 799 compounds in the Huuskonen dataset (excluding those already contained in the Merck dataset) were put together, and were fed into a 41×35 Kohonen map by using 18 2D descriptors (that are highly related to the solubility) as input descriptors. All the compounds were classified into three groups: only Merck dataset (1588 compounds) as shown by triangles, only Huuskonen dataset (799 compounds) represented by squares, and the overlap (the compounds existing in both datasets) as indicated by asterisks as shown in figure 3. The map was formed according to different datasets of the most frequently occupied neuron. It was observed that, apart from the removed compounds identical to the two datasets, 310 compounds of the reduced Huuskonen dataset occupy the same neurons as those of 376 compounds in Merck dataset. And one can see that the Merck dataset covers a large area and contains more information than the Huuskonen dataset. Thus, it can be concluded that the Merck dataset is more diverse than the Huuskonen dataset with respect to the chemical space of compound solubility.

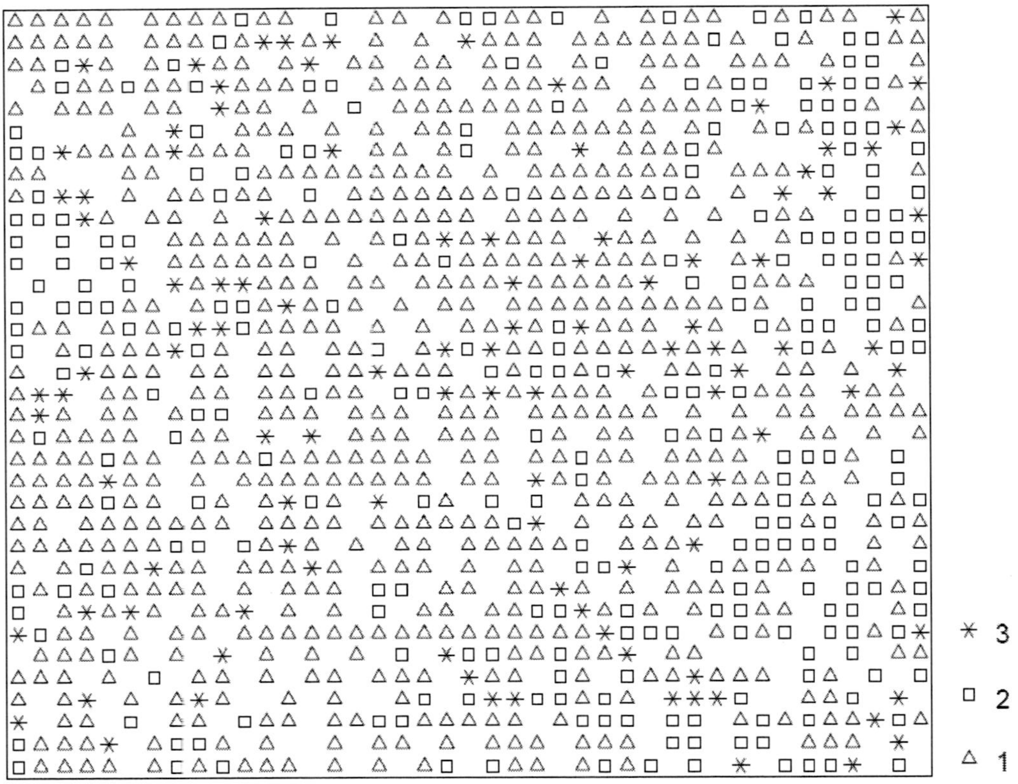

Figure 3. The distribution of the compounds of the two datasets in the same KNN (Kohonen's self-organizing neural network) map by using 18 topological descriptors as input descriptors. "triangle" represents the most frequently appearing 1588 compounds in the Merck dataset (excluding the same compounds with Huuskonen dataset); "square" represents the most frequently appearing 799 compounds in Huuskonen dataset (excluding the same compounds with Merck dataset), "star" represents the most frequently appearing overlap of Huuskonen dataset and Merck dataset.

4.3. Design a Diverse or a Focused Compounds Library

On the basis of kinds of molecular similarity comparison methods, one can evaluate the diversity of a compounds library, compare two compounds libraries, combined with some other tools sometime, it is possible for one to build an appropriate diverse or focused library.

For example, in order to obtain more possible leads with the similar bioactivity around a known inhibitor of a target, one may build a focused compound library with the same 2D or 3D scaffold as the known inhibitor. Furthermore, in high-throughput-screening, high quality compounds library, in particular, the focusing library of drug-likeness (or lead-likeness) for a given target could be built based on the results of molecular similarity study such as the 3D pharmacophore characteristics.

A compounds library can be designed based on 2D molecular similarity measure method such as using 2D descriptors for molecular structural representation. In a recent work Zheng et al. [96], a Focus-2D approach is used for the rational library design using Kier-Hall topological index based on measure of molecular similarity measure and simulated annealing as the optimization tool. Their work shows that this method correctly identifies building blocks found in active peptoids with adrenergic or opioid activities. Their work also suggested that Focus-2D can facilitate rational design of targeted chemical libraries in those cases when a lead molecule or several lead molecules are available [96].

A compound library can be designed based on 2D molecular similarity measure method such as using 2D fingerprint for molecular structural representation. Stahura et al. [97] recently provided an example of practical studies for lead identification aiming to inhibit a therapeutically relevant protein–protein interaction, where the focused compound libraries were generated by using a multiple fingerprint-based metric method. With the structure of the target known and one small molecular inhibitor, virtual screening (including 200,000 ACD molecules) at both the macromolecular level (docking) and small molecular level (similarity searching using fingerprints) was carried out. With the further aid of binary QSAR, 20 compounds were selected for testing and a total of five new synthetic inhibitors were identified. Based on one of the inhibitors, a small focused compound library of 139 compounds was generated by using a multiple fingerprint-based metric method. Different 2D mini-fingerprints are used to facilitate a similarity step for database mining, followed by a diversity step to assemble the final library. Testing of 139 compounds in this library yielded three other inhibitors and four active analogs of the template structure [97].

A compound library can also be designed based on 3D molecular similarity measure method such as on the basis of the 3D pharmacophore or pharmacophore fingerprints based similarity study. In a recent work of Eksterowicz and coworkers [98], the pharmacophores are generated from active site maps which are obtained from known protein structures with the ligands removed. All possible two-, three- and four-point pharmacophores are enumerated from the active site map and encoded as bit strings. The pharmacophores define a design space that can be used to select compounds using an informative library design tool [99]. Informative design chooses compounds which sample the design space pharmacophores with unique patterns. Iterative informative design was applied to select compounds for synthesis and screening. The method was evaluated against a collection of compounds assayed previously against a cyclin-dependent kinase target, CDK-2, starting with 23 X-ray co-crystal structures. During multiple iterations the method is able to focus in on the pharmacophores that are most relevant for activity, as demonstrated by the increase in the number of scaffolds selected and the higher enrichment in the later rounds of design. The method selects compounds from 12 out of the 15 active scaffolds from the CDK-2 library and outperforms a two-dimensional similarity search and docking calculations [98].

Beno and Mason have reviewed the design of combinatorial libraries using properties and 3D pharmacophore fingerprints [94]. The pharmacophore fingerprints can be used for diverse library design, screening set selection, and focused library design, giving a consistent approach [73,74]. Each of the pharmacophores present in the fingerprint represents a binding hypothesis, and focused libraries can be designed to produce compounds that cover as many of these hypotheses as possible, at any desired level of coverage [65]. Alternatively, coverage of 'biologically important' pharmacophores can be maximized. For example, the pharmacophores present in a large database of drug-like molecules, such as the MDDR, can be identified, and libraries could be designed to provide good coverage of that set of pharmacophores [74]. Focused libraries can also be designed to maximize the pharmacophoric similarity of the products with a single active compound (a known lead or leads) [100]. In the work of Pickett et al. [100], a Monte Carlo search procedure was performed for selection of a near-combinatorial subset in which all library members satisfy the design criteria while being as synthetically efficient as possible. Their work also illustrates that consideration of calculated log *P*, molecular weight, and polar surface area in the design of libraries can give compounds with improved absorption characteristics and provides a valuable tool to the medicinal chemist in lead optimization.

5. Conclusion

In summary, with all kinds of molecular similarity methods, we can perform molecular structural and property comparisons, similarity searching and design a diverse or focused compounds library from a known lead (or several leads) and the information of the target. Thus some structural and property similar molecules wound be identified or designed, an appropriate focused compounds library could be built, more hits are promised to be found. Molecular similarity study will continue to occupy an important role in drug discovery and library design.

Acknowledgment

This work was supported by the National Natural Science Foundation of China (20605003), National High Tech Project (2006AA02Z337), SRF for ROCS, and the "Special Funding for the Talent Enrollment" of Beijing University of Chemical Technology.

References

[1] Johnson, M.; Maggiora, G. M. *Concepts and Applications of Molecular Similarity*; John Wiley and Sons: New York, 1990.

[2] Kubinyi, H. Similarity and dissimilarity: A medicinal chemist's view. *Perspect. Drug Discov. Des.* 1998, *9-11*, 225-252.

[3] Kubinyi, H. Chemical similarity and biological activities. *J. Braz. Chem. Soc.* 2002, *13*, 717-726.

[4] Patterson, D.E.; Cramer, R.D.; Ferguson, A.M.; Clark, R.D.; Weinberger, L.E. Neighborhood Behavior: A Useful Concept for Validation of "Molecular Diversity" Descriptors. *J. Med. Chem.* 1996, *39,* 3049-3059.

[5] Horvath, D.; Jeandenans, C. Neighborhood Behavior of in Silico Structural Spaces with Respect to In Vitro Activity Spaces-A Benchmark for Neighborhood Behavior Assessment of Different in Silico Similarity Metrics, *J. Chem. Inf. Comput. Sci.* 2003, *43*, 691-698.

[6] Willett, P.; Barnard, J.M.; Downs, G.M. Chemical similarity searching. *J. Chem. Inf. Comput. Sci.* 1998, *38*, 983-996.

[7] Good, A.C.; Richards, W.G. Explicit calculation of 3D molecular similarity, *Perspect. Drug Discov. Des.*, 1998, *9-11*, 321-338.

[8] Sheridan, R.P.; Kearsley, S.K. Why do we need so many chemical similarity search methods? *Drug Disco. Today* 2002, *7*, 903-911.

[9] Nikolova, N.; Jaworska, J. Approaches to measure chemical similarity–a review. *QSAR Comb. Sci.* 2003, *22*, 1006-1026.

[10] Willett, P. The evaluation of molecular similarity and molecular diversity methods using biological activity data. *Methods Mol. Biol.* 2004, *275*, 51-63.

[11] Bender, A.; Glen, R. C. Molecular similarity: a key technique in molecular informatics. *Org. Biomol. Chem.* 2004, *2*, 3204-3218.

[12] Willett, P. Similarity-based virtual screening using 2D fingerprints. *Drug Discov. Today*, 2006, *11*, 1046-1053.

[13] Gasteiger, J. and Engel, T., eds (2003) *Chemoinformatics: A Textbook*, Wiley-VCH.

[14] R. Leach and V. J. Gillet, *An Introduction to Chemoinformatics*, Kluwer Academic Publishers, 2003.

[15] Whittle, M.; Gillet, V. J.; Willett, P.; Alex, A.; Loesel, J. Enhancing the Effectiveness of Virtual Screening by Fusing Nearest Neighbor Lists: A Comparison of Similarity Coefficients. *J. Chem. Inf. Comput. Sci.* 2004; *44*; 1840-1848.

[16] Hagadone, T.R. Molecular substructure similarity searching: efficient retrieval in two-dimensional structure databases. *J. Chem. Inf. Comput. Sci.* 1992, *32*, 515–521.

[17] Rarey, M. and Dixon, J.S. Feature trees: a new molecular similarity measure based on tree matching. *J. Comput.-Aided Mol. Design,* 1998, *12*, 471–490.

[18] Raymond, J.; Gardiner, E.; Willett, P. RASCAL: Calculation of Graph Similarity Using Maximum Common Edge Subgraphs. *Comput. J.* 2002, *45*, 631-644.

[19] Raymond, J.; Gardiner, E.; Willett, P. Heuristics for Rapid Similarity Searching of Chemical Graphs Using a Maximum Common Edge Subgraph Algorithm. *J. Chem. Inf. Comput. Sci.* 2002, *42*, 305-316.

[20] Gillet, V.J.; Willett, P.; Bradshaw, J. Similarity searching using reduced graphs. *J. Chem. Inf. Comput. Sci.* 2003, *43*, 338-345.

[21] Barker, E.J.; Gardiner, E.J.; Gillet, V.J.; Kitts, P.; Morris, J. Further development of reduced graphs for identifying bioactive compounds. *J. Chem. Inf. Comput. Sci.* 2003, *43*, 346-356.
[22] Rarey, M.; Dixon, J. S. Feature trees: A new molecular similarity measure based on tree matching. *J. Comput.-Aided Mol. Des.* 1998, *12*, 471-490.
[23] Takahashi, Y.; Sukekawa, M.; Sasaki, S. Automatic identification of molecular similarity using reduced-graph representation of chemicalstructure. *J. Chem. Inf. Comput. Sci.* 1992, *32*, 639-643.
[24] Barker, E. J.; Cosgrove, D. A.; Gardiner, E. J.; Gillet, V. J.; Kitts, P.; Willett, P. Scaffold-hopping using clique detection applied to reducedgraphs. *J. Chem. Inf. Model.* 2006, *46*, 503-511.
[25] Harper, G.; Bravi, G.S.; Pickett, S.D.; Hussain, J.; Green, D.V.S. The reduced graph descriptor in virtual screening and data- driven clustering of high-throughput screening data. *J. Chem. Inf. Comput. Sci.* 2004, *44*, 2145-2156.
[26] Birchall, K.; Gillet, V.J.; Harper, G.; Pickett, S.D. Training Similarity Measures for Specific Activities: Application to Reduced Graphs, *J. Chem. Inf. Model,* 2006, *46*, 577-586.
[27] Durant, J.L.; Leland, B.A.; Henry, D.R.; Nourse, J.G., Reoptimization of MDL Keys for Use in Drug Discovery, *J. Chem. Inf. Comput. Sci.*, 2002, *42*, 1273-1280.
[28] Daylight Chemical Information Systems Inc. http://www.daylight.com.
[29] UNITY; Tripos Inc. http://www.tripos.com.
[30] Kogej, T.; Engkvist, O.; Blomberg, N.; Muresan, S. Multifingerprint Based Similarity Searches for Targeted Class Compound Selection , *J. Chem. Inf. Model,* 2006, *46,* 1201-1213.
[31] Xue, L.; Godden, J. W.; Bajorath, J. Database Searching for Compounds with Similar Biological Activity Using Short Binary Bit String Representations of Molecules. *J. Chem. Inf. Comput. Sci.* 1999, *39*, 881-886.
[32] Xue, L.; Godden, J.W.; Bajorath, J. Evaluation of Descriptors and Mini-Fingerprints for the Identification of Molecules with Similar Activity, *J. Chem. Inf. Comput. Sci.,* 2000, *40*, 1227-1234.
[33] Xue, L.; Stahura, F.L.; Godden, J.W.; Bajorath, J. Mini-fingerprints Detect Similar Activity of Receptor Ligands Previously Recognized Only by Three-Dimensional Pharmacophore-Based Methods, *J. Chem. Inf. Comput. Sci.,* 2001, *4*, 394-401.
[34] Hodgkin, E.E. and Richards, W.G., Molecular similarity based on electrostatic potential and electric field. *Int. J. Quantum Chem. Quantum Biol. Symp.*, 1987, *14*, 105-110.
[35] Burt, C., Huxley, P. and Richards, W.G., The application of molecular similarity calculations. *J. Comp. Chem.*, 1990, *11*, 1139-1146.
[36] Richard, A.M., Quantitative comparison of MEPs for structure activity studies. *J. Comp. Chem.*, 1991, *12*, 959-969.

[37] Good, A.C., Hodgkin, E.E. and Richards, W.G., The utilisation of Gaussian functions for the rapid evaluation of molecular similarity. *J. Chem. Inf. Comput. Sci.*, 1992, *32*, 188-191.

[38] Meyer, A.M. and Richards, W.G., Similarity of molecular shape. *J. Comput.-Aided Mol. Design*, 1991, *5*, 426-439.

[39] van Geerestein, V.J., Perry, N.J., Grootenhuis, P.D.J. and Haasnoot, C.A.G., 3D database searching on the basis of shape using the SPERM prototype method. *Tetrahedron Comput. Methodol.*, 1992, *3*, 595-613.

[40] Good, A.C. and Richards, W.G., Rapid evaluation of shape similarity using Gaussian functions. *J. Chem. Inf. Comput. Sci.*, 1993, *33*, 112-116.

[41] Nilakantan, R., Bauman, N. and Venkataraghavan, R., New method for rapid characterization of molecular shapes: Applications in drug design. *J. Chem. Inf. Comput. Sci.*, 1993, *33*, 79-85.

[42] Hahn, M. Three-dimensional shape-based searching of conformationally flexible compounds. *J. Chem. Inf. Comput. Sci.*, 1997, *37*, 80-86.

[43] Carbo, R., Leyda, L. and Arnau, M., An electron density measure of the similarity between twocompounds. *Int. J. Quantum Chem.*, 1980, *17*, 1185-1189.

[44] Hodgkin, E.E. and Richards, W.G., Molecular similarity. *Chem. Br.*, 1988, 1141-1144.

[45] Petke, J.D., Cumulative and discrete similarity analysis of electrostatic potentials and fields. *J. Comp. Chem.*, 1993, *14*, 928–933.

[46] Willett, P. Matching of Chemical and Biological Structures Using Subgraph and Maximal Common Subgraph Isomorphism Algorithms. *IMA Math. Appl.* 1999, *108*, 11-38.

[47] Raymond, J.; Willett, P. Maximum Common Subgraph Isomorphism Algorithms for the Matching Of Chemical Structures. *J. Comput.-Aided Mol. Des.* 2002, *16*, 521-533.

[48] Brint, A.; Willett, P. Upperbound Procedures for the Identification of Similar Three-Dimensional Chemical Structures. *J. Comput.-Aided Mol. Des.* 1988, *2*, 311-320.

[49] Moon, J.B.; Howe, W.J. 3D Database Searching and De Novo Construction Methods in Molecular Design. *Tetrahedron Comput.Methodol.* 1990, *3*, 697-711.

[50] Ho, C.M.W.; Marshall, G.R. FOUNDATION: A Program to Retrieve All Possible Structures Containing a User-Defined Minimum Number of Matching Query Elements from Three-Dimensional Databases. *J. Comput.-Aided Mol. Des.* 1993, *7*, 3-22.

[51] Raymond, J.W.; Willett, P. Similarity Searching in Databases of Flexible 3D Structures Using Smoothed Bounded Distance Matrices, *J. Chem. Inf. Comput. Sci.,* 2003; *43*; 908-916.

[52] Cramer, R.D.; Clark, R.D.; Patterson, D.E.; Ferguson, A.M. Bioisosterism as a molecular diversity descriptor: steric fields of single topomeric conformers. *J. Med. Chem.* 1996, *39*, 3060-3069.

[53] Cramer, R.D.; Patterson, D.E.; Clark, R.D.; Soltanshahi, F.; Lawless, M.S. Virtual libraries: a new approach to decision making in molecular discovery research. *J. Chem. Inf. Comput.Sci.* 1998, *6*, 1010-1023.

[54] Andrews, K.M.; Cramer, R D. Toward General Methods of Targeted Library Design: Topomer Shape Similarity Searching with Diverse Structures as Queries. *J. Med. Chem.* 2000, *43*, 1723-1740.

[55] Cramer, R.D.; Poss, M.A.; Hermsmeier, M.A.; Caulfield, T.J.; Kowala, M C.; Valentine, M.T. Prospective Identification of Biologically Active Structures by Topomer Shape Similarity Searching. *J. Med. Chem.* 1999, *42*, 3919-3933.

[56] Cramer, R.D.; Jilek, R.J.; Guessregen, S.; Clark, S.J.; Wendt, B.; Clark, R.D., "Lead Hopping". Validation of Topomer Similarity as a Superior Predictor of Similar Biological Activities, *J. Med. Chem.;* 2004, *47*; 6777-6791.

[57] Marshall, C.R.; Barry, C.D.; Bosshard, H.E.; Dammkoehler R.A.; and Dunn, D.A. in The Conformational Parameter in Drug Design: The Active Analog Approach, *Computer-Assisted Drug Design, ACS Symposium Series 112*, ed. E. Olson and A. E. Christoffersen, ACS, Washington DC, USA, 1979.

[58] Sheridan, R.P.; Miller, M.D.; Underwood D.J.; Kearsley S.K. *J. Chem. Inf. Comput. Sci.*, 1996, *36*, 128.

[59] Schneider, G.; Neidhart, W.; Giller, T.; Schmid, G. "Scaffold-Hopping" by Topological Pharmacophore Search: A Contribution to Virtual Screening. *Angew. Chem., Int. Ed.* 1999, *38*, 2894-2896.

[60] Bemis, G.W.; Kuntz, I.D. A fast and efficient method for 2D and 3D molecular shape description, *J. Comput.-Aided Mol. Des.*, 1992, *6*, 607-628.

[61] Martin, Y.C. 3D database searching in drug design. *J. Med. Chem.*, 1992, *35*, 2145-2154.

[62] Nilakantan, R.; Bauman, N.; Venkataraghavan, R. New method for rapid characterization of molecular shapes: applications in drug design. *J. Chem. Inf. Comput. Sci.*, 1993, *33*, 79-85.

[63] Pickett, S.D.; Mason, J.S.; McLay, I.M. Diversity Profiling and Design Using 3D Pharmacophores: Pharmacophore-Derived Queries (PDQ). *J. Chem. Inf. Comput. Sci.*, 1996, *36*, 1214-1223.

[64] Mason, J. S.; Good A. C.; Martin, E. J. 3-D Pharmacophores in Drug Discovery. *Curr. Pharm. Des.*, 2001, *7*, 567-597.

[65] Mason, J.S.; Morize, I.; Menard, P.R.; Cheney, D.L.; Hulme, C.; Labaudiniere, R.F. New 4-point pharmacophore method for molecular similarity and diversity applications: Overview of the method and applications, including a novel approach to the design of combinatorial libraries containing privileged substructures. *J. Med. Chem.* 1999, *42*, 3251-3264.

[66] Duca, J.S.; Hopfinger, A.J. Estimation of Molecular Similarity Based on 4D-QSAR Analysis: Formalism and Validation. *J. Chem. Inf. Comput. Sci.*, 2001, *41*, 1367-1387.

[67] Makara, G.M. Measuring Molecular Similarity and Diversity: Total Pharmacophore Diversity, *J. Med. Chem.,* 2001, *44*, 3563-3571.

[68] Jenkins, J.L.; Glick, M.; Davies, J.W. A 3D Similarity Method for Scaffold Hopping from Known Drugs or Natural Ligands to New Chemotypes, *J. Med. Chem.*, 2004, *47*, 6144-6159.

[69] Good, A.C.; Cho, S.J.; Mason, J.S. Descriptors you can count on? Normalized and filtered pharmacophore descriptors for virtual screening. *J. Comput.-Aided Mol. Des.* 2004, *18*, 523-527.

[70] Mason, J.S.; Cheney, D.L. Library design and virtual screening using multiple 4-point pharmacophore fingerprints. *Pac. Symp. Biocomput.* 2000, 576-587.

[71] Mason, J.S.; Cheney, D.L. Ligand-receptor 3-D similarity studies using multiple 4-point pharmacophores. *Pac. Symp. Biocomput.* 1999, 456-467.

[72] Deng, Z.; Chuaqui, C.; Singh, J. Structural interaction fingerprint (SIFt): a novel method for analyzing three-dimensional protein-ligand binding interactions. *J. Med. Chem.* 2004, *47*, 337-344.

[73] McGregor, M. J.; Muskal, S.M. Pharmacophore Fingerprinting. 1. Application to QSAR and Focused Library Design. *J. Chem. Inf.Comput. Sci.* 1999, *39*, 569-574.

[74] McGregor, M.J.; Muskal, S.M. Pharmacophore Fingerprinting. 2. Application to Primary Library Design. *J. Chem. Inf. Comput. Sci.*2000, *40*, 117-125.

[75] Abrahamian, E.; Fox, P.C.; Naerum, L.; Christensen, I.T.; Thogersen, H.; Clark, R.D. Efficient Generation, Storage, and Manipulation of Fully Flexible Pharmacophore Multiplets and Their Use in 3-D Similarity Searching. *J. Chem. Inf. Comput. Sci.* 2003, *43*, 458-468.

[76] Zhang, Q.; Muegge, I. Scaffold Hopping through Virtual Screening Using 2D and 3D Similarity Descriptors: Ranking, Voting, and Consensus Scoring, *J. Med. Chem.* 2006, *49*, 1536-1548.

[77] Edgar, S.J.; Holliday, J.D.; Willett, P. Effectiveness of Retrieval in Similarity Searches of Chemical Databases: A Review of Performance Measures. *J. Mol. Graph. Model.* 2000, *18*, 343-357.

[78] Ward, J. H. Hierarchical Grouping to Optimize an Objective Function. *J. Am. Stat. Assoc.* 1963, *58*, 236-244.

[79] Jarvis, R. A.; Patrick, E. A. Clustering Using a Similarity Measure Based on Shared Nearest Neighbours. *IEEE Trans. Comput.* 1973, *C-22*, 1025-1034.

[80] Kohonen, T. *Self-Organization and Associative Memory*, Third Edition, Springer Verlag, Berlin, 1989.

[81] Kohonen, T. Self-Organizing Map (SOM], http://www.cis.hut.fi/research/som-research/; http://www.cis.hut.fi/research/som-research/nnrc-programs.shtml

[82] Zupan, J.; Gasteiger, J. *Neural Networks in Chemistry and Drug Design*, Second Edition, Wiley-VCH, Weinheim, 1999.

[83] Gasteiger, J.; Teckentrup, A.; Terfloth, L.; Spycher, S. Neural networks as data mining tools in drug design, *J. Phys. Org. Chem.*, 2003, *16*, 232-245.

[84] Yan, A.X.; Gasteiger, J. Prediction of Aqueous Solubility of Organic Compounds Based on a 3D Structure Representation, *J. Chem. Inf. Comput. Sci.*, 2003, *43*, 429-434.

[85] Yan, A.X.; Gasteiger, J. Prediction of Aqueous Solubility of Organic Compounds by Topological Descriptors , *QSAR Comb. Sci.*, 2003, *22*, 821-829.

[86] Yan, A.X.; Gasteiger, J.; Krug, M.; Anzali, S. Linear and nonlinear functions on modeling of aqueous solubility of organic compounds by two structure representation methods, *J. Computer-Aided Mol. Des.*, 2004, *18*,75-87.

[87] Lee, M.L.; Schneider, G. Scaffold architecture and pharmacophoric of natural products and trade drugs: application in design of natural product-based combinatorial libraries, *J. Comb. Chem.*, 2001, *3*, 284-289.

[88] Sadowski, J.; Kubinyi, H. A scoring scheme for discriminating between drugs and nondrugs. *J. Med. Chem.*, 1998, *41*, 3325-3329.

[89] Schneider, G. Neural networks are useful tools for drug design, *Neural Netw.*, 2000, *13*, 15-16.

[90] Lothar, T.; Gasteiger, J. Neural networks and genetic algorithms in drug design, *Drug Discov. Today*, 2001, *6*, Suppl., S102-S108.

[91] Polanski J. S elf-organizing neural networks for pharmacophore mapping. *Adv. Drug Deliv. Rev.*, 2003, *55*, 1149–1162.

[92] Yan, A.X., Application of Self-Organizing Maps in Compounds Pattern Recognition and Combinatorial Library Design, *Comb. Chem. High Throughput Screening*, 2006, *9*, 473-480.

[93] Agrafiotis, D.K. Multiobjective optimization of combinatorial libraries, *Mol. Divers.*, 2002, *5*, 209-230.

[94] Benc, B.R.; Mason, J.S. The design of combinatorial libraries using properties and 3D pharmacophore fingerprints. *Drug Discovery Today*, 2001, *6*, 251-258.

[95] Martin, E.J.; Blaney, J.M.; Siani, M.A.; Spellmeyer, D.C.; Wong, A.K.; Moos, W.H. Measuring diversity: Experimental design of combinatorial libraries for drug discovery. *J. Med. Chem.* 1995, *38*, 1431-1436

[96] Zheng, W.; Cho, S.J.; Tropsha, A. Rational Combinatorial Library Design. 1. Focus-2D: A New Approach to the Design of Targeted Combinatorial Chemical Libraries, *J. Chem. Inf. Comput. Sci.* 1998, *38*, 251-258.

[97] Stahura, F.L.; Xue, L.; Godden, J.W.; Bajorath, J. Methods for compound selection focused on hits and application in drug discovery. *J. Mol. Graph. Model.*, 2002, *20*, 439-446.

[98] Eksterowicz, J.E.; Evensen, E.; Lemmen, C.; Brady, G.P.; Lanctot, J.K.; Bradley, E.K.; Saiah, E.; Robinson, L.A.; Grootenhuis, P.D.J.; Blaney, J.M. Coupling structure-based design with combinatorial chemistry: application of active site derived pharmacophores with informative library design. *J. Mol. Graph. Model.*, 2002, *20*, 469-477.

[99] Teig, S.L. Informative libraries are more useful than diverse ones, *J. Biomol. Screening*, 1998, *3*, 85–88.

[100] Pickett, S.D.; McLay, I.M.; Clark, D.E. Enhancing the hit-to-lead properties of lead optimization libraries. *J. Chem. Inf. Comput. Sci.* 2000, *40*, 263–272.

In: Drug Design Research Perspectives
Editor: Stanley P. Kaplan, pp. 125-169
ISBN: 978-1-60021-823-1

Chapter VI

Perspectives of Microdialysis in the Evaluation of *In Vivo* Pharmacokinetics and Pharmacodynamics of New Chemical Entities

Christian Höcht*, Javier AW Opezzo, Marcos A. Mayer, Guillermo F. Bramuglia and Carlos A. Taira
Department of Pharmacology, Faculty of Pharmacy and Biochemistry,
University of Buenos Aires, Junín 956, (C1113AAD) Buenos Aires, Argentina

1. Abstract

An important trend in drug design is the evaluation of the *in vivo* pharmacokinetics and pharmacodynamics of new chemical entities (NCE). Although the pharmacokinetic properties of NCE can be studied by means of traditional blood sampling, regulatory authorities emphasize the evaluation of drug distribution at the target site. Another issue in the assessment of NCE is that the discovery of the existence of a large number of orphan receptors will increase the number of NCE to be tested. Therefore, a more rigorous selection process will be necessary at early stages of drug development. Integration of pharmacokinetic-pharmacodynamic (PK-PD) concepts through PK-PD modeling is a potential tool to enhance information gain and efficiency of drug selection during development. Therefore, the evaluation of drug concentrations at the biophase and their relationship with the pharmacological effect are two important trends in modern drug discovery.

New sampling techniques, including imaging techniques and microdialysis, are now available for the study of drug distribution at the target site. However, microdialysis seems to be the most adequate methodology for the evaluation of drug levels at the biophase in both preclinical and clinical research, considering that this technique samples

* Correspondence and address: Christian Höcht, Departamento de Farmacología, Facultad of Farmacia y Bioquímica, Universidad de Buenos Aires, Junín 956, (C1113AAD) Buenos Aires, ARGENTINA. Phone: +(54-11)-4964-8265 ; Fax: +(54-11)-4508-3645 ; e-mail: chocht@ffyb.uba.ar

extracellular free concentrations of drugs. Finally, microdialysis sampling might become a reference technique for PK-PD modeling by the fact that this technique allows the simultaneous determination of drug concentrations in one and more tissues and their effect on biochemical and clinical markers in the same subject and with good temporal resolution.

The aim of the chapter is to review the methodological and practical aspects of the microdialysis technique and its application for the study of drug distribution and PK-PD modeling in basic and clinical research of drugs and NCE.

2. Introduction

The demand for efficacious pharmacological agents is increasing due to higher lifestyle expectations and changes in demographic profiles (Debouck and Metcalf 2000). The discovery of the existence of a large number of orphan receptors will increase the number of new therapeutic agents to be tested. The availability of new techniques, such as the Human – Genome Project and the High throughput screening and combinatorial chemistry, will identify an exploding number of new molecular targets as well as potential new therapeutic agents (Curtis et. al 2000; Derendorf et al 2000; Toutain 2002). Therefore, a more rigorous selection process will be needed at early stages of drug development.

Integration of pharmacokinetic-pharmacodynamic (PK-PD) concepts through PK-PD modeling is a potential tool to enhance the information gain and the efficiency of drug selection during drug development (Elmquist and Sawchk 1997; Ben-Hun et al 1988).

In addition, another important trend of drug properties to evaluate at early stages of drug development is its capability to reach the target site at sufficient amount to exert its therapeutic effect. It has been estimated that up to 40% of NCE fail in the early phases of drug development because of inappropriate pharmacokinetic properties, such as inadequate drug distribution (Prentis et al 1988).

Although pharmacokinetic properties of NCE can be studied by means of traditional blood sampling, this technique is limited by the impossibility to evaluate variability in drug distribution at the target site. Therefore, in the last years, regulatory authorities have been emphasizing the evaluation of drug distribution at the target site during drug development (http://www.fda.gov/cder/guidance/2580dft.pdf).

New sampling techniques, including imaging techniques and microdialysis, are available for the study of drug distribution at the target site and PK-PD modeling. Use of these techniques in preclinical and early clinical drug development might reduce failing rates of NCE, also improving cost and the time required until a NCE reaches the market.

Considering that microdialysis technique samples the extracellular free concentration of drugs, it seems to be the most adequate methodology for the evaluation of drug levels at the biophase.

Taking into account these antecedents, the aim of this chapter is to: i) describe the concepts of drug distribution at the target site and PK-PD modeling; ii) compare the available techniques for the study of tissue distribution of drugs and PK-PD modeling and iii) to comment the methodological and practical aspects of microdialysis sampling and its application for the study of drug distribution and PK-PD modeling of drugs and NCE during basic and clinical research.

3. The Importance of Measuring Target Site Concentrations of Drugs in Drug Discovery

The selection of the adequate dose and dosing timing is a fundamental step to achieve therapeutic success with any pharmacological treatment. Selection of the best drug and dosing regimen does not only increase the chances of therapeutic success but also reduces the appearance of adverse drug reactions. However, the selection of the most adequate drug and dosing interval of the same is not an easy issue. For most therapeutic agents, drug and dose regimen are empirically selected based on trial and error and adjusting drug dosing taking into account patient weight. However, this strategy is subject to high clinical failure because of the large interindividual variability in pharmacokinetic and pharmacodynamic properties of therapeutic agents in patient population (Levy 1998; Ho and Kim 2005) (figure 1).

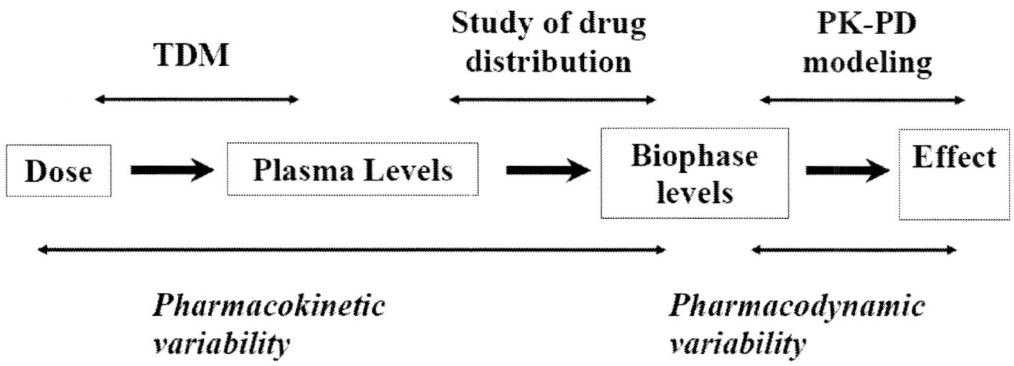

Figure 1. Sources of interindividual variability in drug response and methods available for its assessment.

Currently, therapeutic drug monitoring (TDM) (figure 1) is indicated for drugs with small therapeutic window, including antiepileptic drugs, antimicrobials, inmunesupressors, antiviral agents, teophylline and others (Gogtay 2001). TDM guides patient dosing according to plasma concentrations of therapeutic agents. In other words, measurement of plasma concentrations has the objective to consider the relationship between dose and attained plasma concentrations in the large interindividual variability of drug response in a treated population (Gogtay 2001). However, TDM has several limitations and its cost-benefit relationship is controversial (Schumacher and Barr 2002). It must be taken into account that most of the therapeutic agents exert their pharmacological effect in other tissues rather than the blood compartment, and tissue and plasma pharmacokinetics may differ substantially in patients and healthy volunteers (Schuck and Derendorf 2005; Höcht et al 2006a; Langer and Müller 2004). On the other hand, a large interindividual variability also exists between tissue concentrations and therapeutic response because of different receptor sensitivity along the patient population (Levy 1998). In conclusion, although TDM has been established as a valuable therapeutic tool for some drugs, measurement of plasma concentrations of drugs is a surrogate parameter with several limitations since it does not exclude interindividual variability in drug distribution at the target site and pharmacodynamic variability.

Based on the assumption that unbound plasma concentrations and free tissue levels are equal at equilibrium, it is frequently considered that total plasma concentrations and plasma protein binding can be used to predict free tissue levels of therapeutic agents. However, many studies have shown lower tissue unbound levels than plasma concentrations (Nordbring 1978; Fischman et al 1998; Lain 1994; Heikkinen 2000).

Tissue distribution is also affected by anatomic barriers, such as blood-brain barrier (BBB), presence of active transport systems, like P-glycoprotein, and tissue metabolism.

For instance, BBB prevents water soluble xenobiotics from entering the brain (Begley 2004). BBB is mainly formed by brain capillary endothelial cells, pericytes, astrocytes, and neuronal cells. Endothetial cells are different from peripheral cells because of the presence of tight junctions, which prevent paracellular transport of water-soluble compounds from blood to the brain (de Boer and Gaillard 2007).

It has also been discovered that many transport systems play an important role in determining tissue concentrations of drugs. These comprise carrier- and receptor-mediated transport systems, including cationic and anionic influx and efflux systems such as P-glycoprotein (Cordon-Cardo et al 1989), multidrug-resistance (MDR) proteins (de Boer et al 2003), nucleoside transporters, organic anion transporters, organic cation transporters, large amino acid transporter and the scavenger receptors SB-AI and SB-BI (Löscher and Potschka 2005a). These transporters modulate drug disposition process, with particular importance in terms of the clinical implications of transporters for drug-drug interactions, drug toxicity, interindividual variability in drug response, and disease (Ho and Kim 2005).

Efflux transporters, such as P-glycoprotein, are expressed in several tissues, especially in the central nervous system, hepatocytes, kidney and the small intestine. At these sites, efflux transporters reduce drug bioavailability, tissue distribution and enhance drug elimination in order to reduce drug exposure in a given patient (Fromm 2004). Moreover, in several pathologies, an overexpression of efflux transporters exist compromising therapeutic success because of limited distribution of the drug at the target site (Löscher and Potshcka 2005b).

For instance, reduction of central bioavailability of antiepileptic drugs by overexpression of efflux transporters has been established for several drugs in different models of experimental epilepsy as a mechanism of pharmacorresistance (Remy and Beck 2006; Lazarowski et al 2006; Löscher and Potschka 2005a). The need to study antiepileptic drug distribution at the target site is emphasized by the fact that overexpression of efflux transporters seems only to affect drug distribution in the biophase and not in other central nuclei. The finding that resistant epileptic patients have similar central adverse effect to drug treatment than responders (Löscher and Potschka 2005a) supports this conclusion.

Success of cancer treatment is also affected by the existence of an enhanced efflux transporter activity. In this case, efflux transporters reduce intracellular concentrations of cytotoxic drugs compromising the activity of chemotherapeutic agents (Lin 2006).

Tissue metabolism also affects the relationship between plasma unbound concentrations of therapeutic agents and their levels at the biophase. Although most part of drug biotransformation takes place in hepatocytes, other tissues present relative high metabolic activity, including small intestine, central nervous system and kidney (Gibbs and Hosea

2003). Therefore, tissue biotransformation could also influence target site concentrations and pharmacological effects.

Taking into account these antecedents, regulatory authorities encourage the study of tissue distribution of antimicrobial agents in unaffected and infected target sites and the relationship of unbound drug concentrations at the site of action to the *in vitro* susceptibility of the infecting microorganism (http://www.fda.gov/cder/guidance/2580dft.pdf).

In conclusion, the utility of therapeutic tools used to date for rational selection of the best drug and dosing regimen, such as TDM, is limited by the fact that this approach allows the determination of plasma drug concentrations rather than its levels at the target site. Several pharmacokinetic factors affect tissue distribution of drugs and their pharmacological effect. Therefore, an important fact in drug design is the evaluation of the *in vivo* pharmacokinetics of NCE at the biophase. In the following sections of the present chapter, available sampling techniques for the study of drug distribution at the target site, including imaging techniques and microdialysis, are discussed.

Variability in drug response is also due to pharmacodynamic factors (Levy 1998) (figure 1). Transporters activity, concentration of endogenous agonist and expression of receptors and their sensitivity to drug action vary among different patients and during the course of drug treatment on a single patient. Therefore, selection of the adequate dose of a therapeutic agent and its dose regimen also depends on pharmacodymanics of the drug at the target site. In the last years, PK-PD modeling have become an interestingly tool of pharmacological sciences for the study of factors influencing drug response (Derendorf et al 2000; Miller et al 2005; Meinbohm and Derendorf 2002) (table 1).

Table 1. Information obtained from pharmacokinetic-pharmacodynamic modeling during preclinical and clinical phases of drug development

Preclinical studies

- Precise definition of the dose–concentration–pharmacological effects and dose–concentration–toxicity relationship
- Determination of the appropriate dosing regimen for Phase I studies
- Identification of biomarkers and animal models for efficacy and toxicity
- Explore any dissociation between plasma concentration and duration and onset of pharmacological effect
- Provide information on drug effects that would be difficult to obtain in human subjects

Clinical studies

- Understanding the dose-concentration–pharmacological effects and dose–concentration–toxicity relationship in healthy volunteers.
- Characterization of pharmacokinetics and pharmacodynamics in special population
- Study of tolerance development
- Determination of the dosing regimens for Phase II studies.
- Confirms and explores the relationship between dose–concentration–effect in patients.
- Examination of a variety of therapeutic endpoints to understand the most adequate for further modeling.
- Determination of the dosing regimens for Phase III studies
- Prediction of the probability distribution of further clinical trial outcomes
- Assessment of pharmacokinetic and pharmacodynamic changes or relationship in the patients population

4. Utility of Pharmacokinetic-Pharmacodynamic Modeling in Drug Discovery

Two different approaches exist for characterizing the pharmacodynamics of a drug, namely the classical dose response trial and PK–PD modeling (Toutain 2002). PK–PD modeling has several advantages over classical dose–response studies. PK–PD modeling allows not only better pharmacodynamic characterization of drugs, but also permits screening and dosage–regimen selection (Toutain 2002). Information obtained from PK-PD modeling during preclinical and clinical drug development is resumed in table 1.

The principle of PK-PD modeling is described in figure 2. PK-PD relationships build a bridge between the time course of drug concentrations in the organism, as assessed by pharmacokinetics (PK), and the intensity of the observed pharmacological response, as quantified by pharmacodynamics (PD). The link between PK and PD of a drug is established by the use of mathematical models, allowing the estimation of parameters such as effective concentration to yield half-maximal response (EC_{50}) and maximal efficacy (E_{MAX}). PK-PD modeling also provides information about the onset, magnitude and duration of the therapeutic effect (Balant et al 1996).

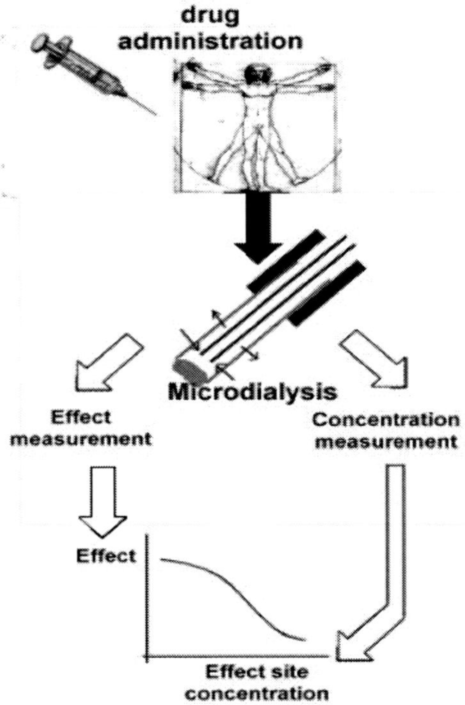

Figure 2. Representation of the applicability of microdialysis for PK-PD modeling. A microdialysis probe inserted into a specific tissue allows the simultaneous sampling of extracellular levels of the administered drug (concentration measurements) and endogenous compounds. Changes in the concentration of endogenous compounds or biochemical markers allow the measurements of drug response. Finally, the relationship between tissue concentrations of the drug and its effect on biochemical markers is determined by means of PK-PD modeling.

However, one disadvantage of the PK–PD modeling is the necessity of simultaneous measurement of drug tissue levels and its corresponding pharmacological effect at multiple time points (Toutain 2002). Blood sampling, which has traditionally been used for this purpose, has the disadvantage that the removal of samples by themselves may interfere with PK and PD drug behavior, especially in preclinical studies with small animals (Höcht et al 2004a). Furthermore, traditional sampling techniques allow the measurement of plasma concentrations of pharmacological agents rather than their levels at the target tissue. These limitations could be resolved by the application of new sampling techniques, such as *in vivo* microdialysis.

PK–PD modeling requires the simultaneous measurement of drug tissue levels and its corresponding pharmacological effects at multiple time points. Measurements of the active compound should be performed with fully validated analytical methods (Balant et al 1996). Although ideally concentrations of the therapeutic agent should be measured at the target site, in most situations this is not possible and frequent plasma sampling is the only alternative (Perez-Urizar et al 2000).

On the other hand, an accurate measurement of the intensity of the pharmacological effect of the active compounds is necessary for a PK-PD modeling design. A drug effect could be considered as any change in physiological parameters induced by the administration of a drug, compared to respective baseline values. Measurement of the effect should meet validation parameters, such as continuity, sensitivity, objectivity and repeatability (Perez-Urizar et al 2000).

To obtain the greatest precision in estimating PK-PD relationships, the number of measurements of drug tissue levels and its corresponding effect must be as large as possible [18]. However, multiple time point sampling is not always possible in the clinical setting. To overcome this limitation, population PK and PK-PD modeling is increasingly being introduced (Csajka and Verotta 2006).

PK-PD relationships have been described using diverse mathematical models depending on the nature of drug administration, the magnitude of the pharmacological effect and the time dependency of pharmacodynamics of the tested drug (Csajka and Verotta 2006; Derendorf et al 2000). Relative simple PK-PD models are needed for describing PK-PD relationship after multiple doses or long-term infusion, because the system is kinetically at steady state (Csajka and Verotta 2006). The most common mathematical equations employed at steady state condition are the linear, log linear and the E_{MAX} model. Although the linear and log linear model allow an easy parameter estimation, these models erroneously assume that the effect can increase with concentrations without limits (Perz-Urizar et al 2000). Therefore, the Emax model is the most broadly applied to characterise a myriad of pharmacological effects. This model derives from the classical theory of drug-receptor interaction, relating the effect to drug concentrations as in the following equation:

$$E = E_0 + \frac{E_{MAX} * C}{EC_{50} + C}$$

where E_{MAX} is the maximal effect and EC_{50} the effective concentration yielding half maximal response.

On the contrary, more complex PK-PD models are needed for describing PK-PD relationship after single dose administration or when time dependency in the pharmacodynamics of the drug is present (Csajka and Verotta 2006).

Plotting drug effects as a function of drug concentrations and connecting data in a chronological order allows the determination of possible delays between drug response and its tissue levels (Toutain 2002). A hysteresis loop appears in the plotting when the magnitude of an effect corresponds to more than one drug concentration. So, an anticlockwise hysteresis loop could be explained by the disequilibrium between biophase and plasma compartment (Sheiner et al 1979), appearance of active metabolites (Mandema et al 1992) or indirect mechanism of action (Dayneka et al 1993). On the other hand, tolerance in the pharmacological effect is suggested when a clockwise hysteresis loop is observed (Bauer et al 1997).

In theses cases, plasma concentrations can not be directly linked to drug effect, and more complex PK-PD models, such as an effect compartment model and a physiological indirect response model, are needed (Csajka and Verotta 2006).

The effect compartment model considers a hypothetical effect compartment as an additional compartment of a pharmacokinetic compartment model, representing the drug concentration at the effect site. The time-dependent aspects of the equilibrium between plasma concentration and effect are characterized by the first-order rate constant k_{e0} that represents the irreversible disappearance of the drug from the effect compartment and is in equilibrium with the rate constant of entrance of the drug to the effect compartment. The time course of drug concentration in the effect compartment is described by the following equation:

$$dCe/dt = ke_0 * (Cp-Ce)$$

where Ce and Cp represent the concentration in the effect compartment and plasma, respectively. This approach has been successfully applied to predict the PK-PD relationship of diverse drugs (Höcht et al. 2006b).

Physiologic indirect response models have been employed in a variety of studies regarding biological responses such as muscle relaxation, synthesis and secretion of endogenous compounds, mediator flux, cell trafficking, enzyme induction, or inactivation, among others (Balant et al 1996; Dayneka et al 1993). The physiological indirect response can be described as an inhibitory or stimulatory function applying the following equations:

$$I(t) = 1 - \frac{C(t)}{IC_{50} + C(t)}$$

$$S(t) = 1 + \frac{E_{MAX} * C(t)}{EC_{50} + C(t)}$$

where IC_{50} and EC_{50} are the concentrations eliciting 50% of inhibition or the maximal effect (E_{MAX}), respectively.

PK–PD modeling should cover the complete pharmacodynamic range of a drug after a single administration (Toutain 2002). Often, in clinical pharmacology, it is not possible to determine the maximal effect of a drug because of the appearance of adverse drug reactions, in which case an alternative PK–PD modeling must be applied (Schoemaker et al 1998). In this model, the authors replaced the parameter EC_{50} with S_0 in the E_{MAX} equation, which represents the initial sensitivity to the drug. The following equation describes the relationship between drug concentration and its pharmacological effect for this PK-PD model:

$$E = E_0 + \frac{S_0 * E_{MAX} * C}{E_{MAX} + S_0 * C}$$

Considering that the selection of an inadequate PK–PD model according to the PK–PD study characteristics might lead to an erroneous interpretation, it is extremely important to determine which PK–PD model is going to be applied for the analysis of the data.

When selecting the PK–PD model, the investigator must keep in mind experimental variables, including type of drug administration, type of pharmacological effect measured, the existence of time dependency in the pharmacological effect of a drug and the possibility to reach a maximal response in their experimental design.

5. Overview of Available Techniques for the Study of Drug Distribution and PK-PD Modeling

Target site concentrations of drugs and NCE can be evaluated by means of traditional techniques, including biopsies, skin blisters, or using more recently available methodologies, such as imaging techniques, ultrafiltration and microdialysis (Langer and Müller 2006).

Biopsy consists of a homogenized tissue sample that results in cellular lysis. Therefore, biopsy gives information about analyte concentrations in the homogenate without distinguishing between blood, intra- or extracellular drug levels. Moreover, it is an invasive method associated with the risk of infection due to cross-contamination and scarring. As tissue has to be removed, biopsy is not suitable for the study of the concentration–time course of a drug and, therefore, PK-PD modeling of drugs (Plock and Kloft 2005). Advantages of biopsy are that no special equipment is needed, there is no limitation with the size of the drug and no calibration is necessary (Plock and Kloft 2005).

Suction blisters have been an established method for more than 30 years for the study of pharmacokinetics in the dermis (Benfeldt et al 1999). The method principle relies on the separation of the epidermis from the dermis along the lamina lucida due to the application of prolonged suction to the skin surface (suction blister technique) or due to the adverse reaction effect of cantharidin (cantharides blister technique). The fluid drawn into the separated epidermis can be sampled and analyzed for the drug content (Plock and Kloft 2005). Skin blister technique has been compared to other sampling techniques, such as microdialysis, concluding that skin blister seems to be only valid for the dermal study of low protein bound drugs. Other limitations of skin blister sampling is the production of a great discomfort in the patients and is only applicable for studies at a certain time point not allowing continuous monitoring of drug skin penetration to the same test area.

In conclusion, traditional techniques have several drawbacks that limit their applicability for the evaluation of tissue distribution and PK-PD modeling during drug development. These methodologies have largely been replaced by modern analytical techniques, including ultrafiltration, imaging techniques and microdialysis sampling (table 2).

Table 2. Comparison of available modern sampling techniques

Technique	Advantages	Limitations
Membrane based techniques		
Ultrafiltration	• No *in vivo* calibration is needed • Simultaneous determination of drugs, metabolites and endogenous compounds • Cheap technique	• No applicable in tissues with limited flow rate and low replenishment of interstitial fluid. • Brain unsuitable for ultrafiltration sampling. • Scope of application of ultrafiltration technique limited in clinical pharmacology.
Microdialysis	• Simultaneous determination of the bioactive concentration of drugs, metabolites and endogenous compounds in the biophase. • Good temporal and spatial resolution. • Online coupling of analytical determination • Monitoring of drug concentrations in different tissues by multiprobe microdialysis. • Possibility of local drug administration. • Cheap technique.	• Semi-invasiveness technique • Diluting effect of the microdialysis procedure • Need of highly sensitive analytical methods • *In vivo* calibration of the microdialysis probe during the experiment • Sticking of lipophilic drugs to tubing and probe components • Low recovery of large molecules
Imaging techniques		
PET	• Non-invasive technique • High spatial and time resolution • No time consuming	• Short physical half-life of the most used radioisotope. • Did not discriminate between drugs and its metabolites. • Determinates total tissue concentrations of drugs. • Expensive technique.
MRS	• Allows discrimination between drugs and its metabolites. • Non-invasive technique • No time consuming	• Poor spatial resolution. • Low sensitivity • Expensive technique.

Abbreviations: MRS: magnetic resonance spectroscopy; PET: positron emission tomography.

Ultrafiltration is an alternative membrane sampling technique to microdialysis. This technique collects a sample by the application of negative pressure as driving force. The rate of fluid collection is determined by the amount of negative pressure applied, the membrane surface area and the hydraulic resistance (Vogt et al 2003). The mayor advantage of ultrafiltration with regards to membrane sampling techniques is that *in vivo* calibration of the technique is not necessary because the *in vivo* recovery is greater than 95% for small molecules (Vogt et al 2003). However, continuous tissue sampling with ultrafiltration depends on rapid replacement of interstitial fluid by blood vessels. In tissues with limited flow rate and low replenishment of interstitial fluid, such as subcutaneous tissue, only low sampling rates are possible (Vogt et al 2003). Brain is also unsuitable for ultrafiltration sampling because of the limited extracellular space. In addition, ultrafiltration does not allow the administration of a substance into the extracellular space through the probe.

On the other hand, although the applicability of ultrafiltration has been demonstrated for drug kinetics studies and glucose and lactate monitoring (Vogt et al 2003), the scope of application of ultrafiltration technique is limited in clinical pharmacology.

Several imaging techniques such as planar γ-scintigraphy, single photon emission computed tomography (SPET), positron emission tomography (PET), and magnetic resonance spectroscopy (MRS) have been developed for the study of drug distribution in basic and clinical settings (Langer and Müller 2004). PET is a new nuclear imaging technique that employs molecules labeled with positron-emiting radioisotopes (Nutt 2002). Advantages of this imaging technique reside in its non-invasive nature, high spatial resolution (1-5 mm) and time resolution (30 seg.). However, the physical half-life of the most used radioisotope (11C) (20.4 min) does not allow monitoring of tissue levels of radiolabeled drugs over several elimination half-lifes as desired in pharmacokinetic studies (Langer and Müller 2004). Another drawback of PET is that this methodology samples total tissue concentrations of drugs and their metabolites and, consequently, does not allow to differenciate them.

MRS utilizes nuclear magnetic resonance phenomenon, which consists in the emission of a radiosignal equal to the respective resonance frequency with amplitude proportionally to the number of nuclei present in the examinant objective (Langer and Müller 2004). MRS can be performed serially with a temporal resolution of minutes. Contrary to other imaging techniques, MRS is capable of identifying drug metabolites in tissues (Langer and Müller 2004). However, an important limitation of MRS is its low sensitivity, allowing only the evaluation of drug distribution of xenobiotics that are present in large concentrations. In addition, spatial resolution of this imaging methodology is low (Langer and Müller 2004).

Although imaging techniques are non-invasive, they are only applicable for a small group of compounds with special functional groups. Moreover, imaging techniques are very expensive and labor-intensive and, therefore, not suitable for clinical routine settings.

Finally, microdialysis is a membrane sampling technique that allows continuous monitoring of extracellular concentrations of NCE and drugs withouth fluid loss (Höcht et al 2004a). As discussed later, microdialysis has several advantages compared to other sampling techniques, including the capability to monitor unbound tissue concentrations of drugs and their metabolites, the simultaneous study of drug concentrations in several tissues and the possibility of concomitant sampling of endogenous compounds or biochemical markers (Höcht et al 2006b). However, it must be pointed out that microdialysis is a semi-invasive methodology that often requires calibration during the experiment (Höcht et al 2004a).

In conclusion, to date, modern techniques exist for the evaluation of tissue distribution of drugs at the target site and simultaneous PK-PD modeling, providing information that has been inaccessible to previously available methods. Considering advantages of microdialysis sampling over imaging techniques regarding study of drug distribution, in the next section of this chapter its principles and applications will be reviewed.

6. Principles of Microdialysis Sampling

In the past twenty years, the microdialysis technique has become a method of choice for the study of tissue concentrations of both endogenous and exogenous substances. Microdialysis is a powerful and versatile sampling technique based on the dialysis principle (figure 3) (Lönnroth et al 1987). In this technique, a probe that is inserted into a tissue mimics the function of a capillary blood vessel (figure 3). The probe has a hollow fiber that is permeable to water and small molecules, and when the perfusate solution passes through the dialysis membrane, molecules diffuse into (recovery) or out of (delivery) the perfusion fluid (figure 3) depending on the concentration gradient. Thus, microdialysis can be used for both collecting a substance in the dialysate as well as delivering it into the periprobe fluid. The latter is referred to as reverse microdialysis (see later) (Höcht et al 2007a). Then, dialysate samples are analyzed using quantitative techniques like liquid chromatography (LC) or capillary electrophoresis (CE).

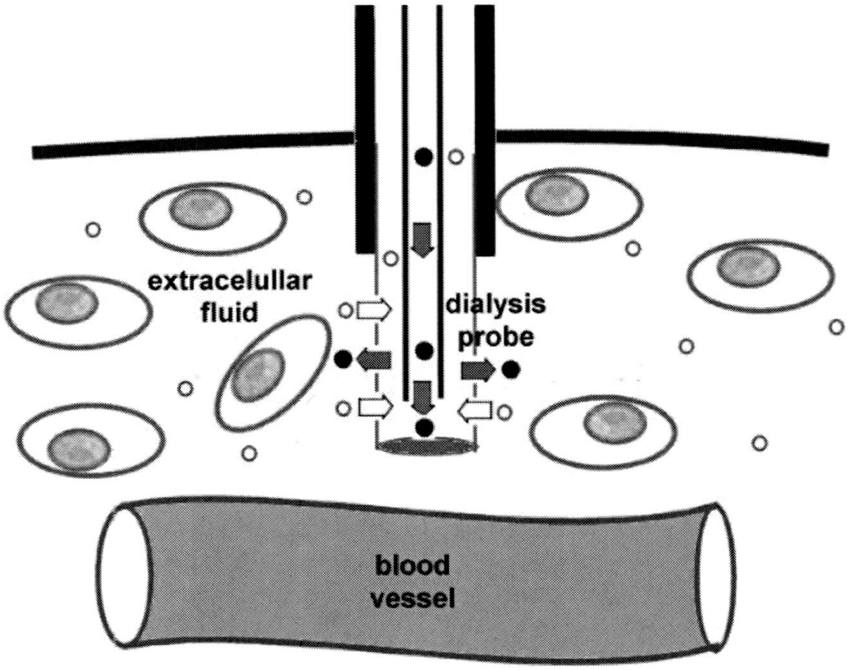

Figure 3. Representation of the microdialysis principle. A concentric microdialysis probe inserted in the extracellular space of a tissue is perfused with a solution containing an exogenous compound (black circles) to administer. The solution is pumped through the inner cannula allowing the diffusion of the exogenous compounds to the surrounding tissue. Simultaneously, endogenous compounds or drugs (open circles) diffuse from the extracellular space into the microdialysis probe and are collected in the dialysate for further analysis.

The basic setup for a microdialysis experiment consists of a microdialysis probe, a perfusion pump, and an analytical method with the required sensitivity to quantify small concentrations of substances (Höcht et al 2004a).

Microdialysis sampling is not performed under equilibrium conditions because the perfusate is constantly being pumped through the probe, and therefore concentrations of the drug in the sample are some fraction of that in the surrounding tissue. Therefore, limit of quantification of analytical methods should be extremely low (Davies et al 2000).

On the other hand, to obtain true tissue concentrations the factor by which these concentrations are interrelated needs to be determined. This factor, which is obtained during an *in vivo* or an *in vitro* calibration procedure, is called relative recovery (Höcht et al 2004a).

Microdialysis sampling offers several advantages over conventional methods and imaging methods of studying the pharmacokinetics and pharmacodynamics of drugs. Microdialysis technique allows continuous tissue sampling without removing liquid, achieving higher temporal resolution than traditional techniques, without interfering with the PK and PD behavior of the drug. An economical and ethical advantage is that 5-10 times fewer animal experiments have to be performed to determine the time profile of a drug (figure 4) (Höcht et al 2004a).

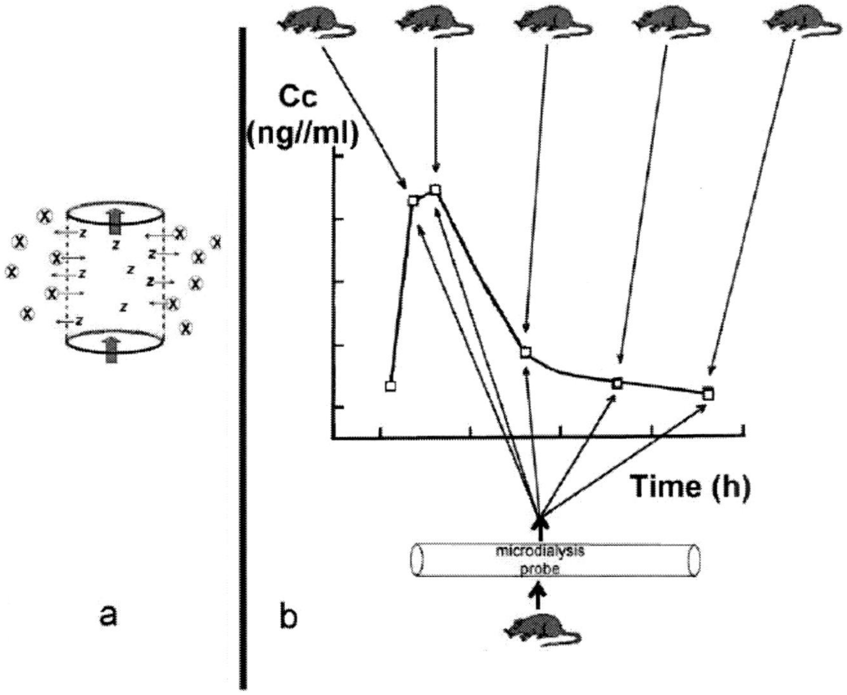

Figure 4. Representation of advantages of microdialysis sampling. Traditional blood sampling in small laboratory animals must be undertaken using several experimental subjects for each sampling time point. Conversely, microdialysis sampling allows the measurement of all sampling time points in a single experimental subject considering that no fluid loss occurred.

Placement of multiple microdialysis probes in different tissues allows monitoring of the drug time course in different organs in the same animal, supplying information about the distribution process of xenobiotics.

The fact that the microdialysis technique provides protein free samples permits the analysis of the sample without pretreatment and the liquid chromatography (LC) or capillary electrophoresis (CE) on-line coupling. Conversely, traditional blood sampling requires clean up procedures prior to analysis with the possibility of analyte loss during protein precipitation and the need of an internal standard for an accurate determination of the drug. On the other hand, tissue enzymes are also excluded from the dialysate sample without further enzymatic degradation of the drug.

The microdialysis technique not only allows sampling of extracellular levels of drugs but also endogenous compounds such as neurotransmitters, metabolites, glucose, lactate and low molecular wheighing peptides. Therefore, this technique is highly useful for the simultaneous study of the pharmacokinetic behavior of drugs and their effects on extracellular levels of endogenous compounds (Höcht et al 2006b).

However, the microdialysis sampling technique has some disadvantages compared to traditional sampling methods. Drug tissue concentrations often are in the picomolar to micromolar range and the limit of detection of the analytical methods should be for some drugs extremely low. On the other hand, *in vivo* recovery of the microdialysis probe must be determined during the experiment to permit conversion of microdialysis concentration into extracellular levels if an absolute concentration is needed (Höcht et al 2004a). The calibration aspects of the microdialysis probe will be discussed in following sections.

In addition, a minimal lesion of the tissue surrounding the probe is produced by implantation of the microdialysis probe, causing tissue responses such as a compromise BBB (Benveniste and Hüttermeier 2000; de Lange et al 1997) and acute inflammation in different tissues (Ault et al 1994; Palsmeier and Lunte 1994; Davies and Lunte 1995). However, several studies have demonstrated that baseline conditions are reached following periods of 60 min after probe implantation.

7. Methodological Aspects of Microdialysis Sampling

In this section, technical aspects of the microdialysis setup, including microdialysis probes, perfusion fluid, microdialysis probe implantation and microdialysis probe sterilization, are discussed (table 3). Calibration of microdialysis probes is discussed in a separate section due to its importance for the study of drug distribution.

Table 3. Technical components of a microdialysis protocol

Component	Examples	Comments
Microdialysis probes	Differences in design (concentric, linear and shunt) and rigidity (flexible versus rigid)	Selection of microdialysis probe depends on sampling tissue.
Microdialysis membrane	Low MW cutt-off membranes (Cellulose, polyacetronitrile /sodium methallyl sulfonate, polycarbonate / polyether) and high MW cutoff membranes (acrylic copolymer, polyethylenesulphone)	Selection of dialysis membrane depends on MW of the compounds of interest and the possibility of physical and chemical interactions between membrane and the interest compound.
Perfusion fluid	Saline solution, Ringer Solution, Krebs Ringer solution, Mock-cerebrospinal fluids, Bile salt Ringer's, Modified Krebs-Henseleit solution	Composition of perfusion media is determinate by the extracellular fluid of the sampled tissue and the dialysis membrane (for high MW cutoff membranes)
Analytical techniques	Non-separation based technique (biosensors, immunoassays and mass spectrometry) versus separation based techniques (LC-UV, LC-EC, LC-MS, LC-FL, CE-EC, CE-LIF).	Selection of the analytical methodology depends on the number of compounds to be determined in each sample and the sensitivity required for the compound detection.

Abreviattions: CE: capillary electrophoresis; EC: electrochemical detection; FL; fluorometric detection; LC: liquid chromatography; LIF: laser induced fluorescence detection; MS: mass spectrometry detection; MW: molecular weight; UV: ultraviolet detection.

7.1. Microdialysis Probes

The microdialysis probe is perhaps the nucleus of the microdialysis experiment. Many types are described and they are used in different experiments. The different geometry of microdialysis probes enables their use in virtually any tissue and any fluid of the body (de Lange et al 2000).

Many laboratories have designed their own probe considering that its construction takes no longer than several minutes. To date, there are also many commercial approved probes for studies in soft tissues and brain of humans and animals.

In general, probes will have a longitudinal, a semicircular or an I-shape design. Various designs have been described, including concentric cannula probe, linear probe, and shunt probe (figure 5). In addition, several modified probe designs have been reported, such as spinal loop dialysis catheter (Marsala et al 1995), flexible intravenous probe (Evrard et al 1996), and shunt intraarterial microdialysis probe (Höcht et al 2003).

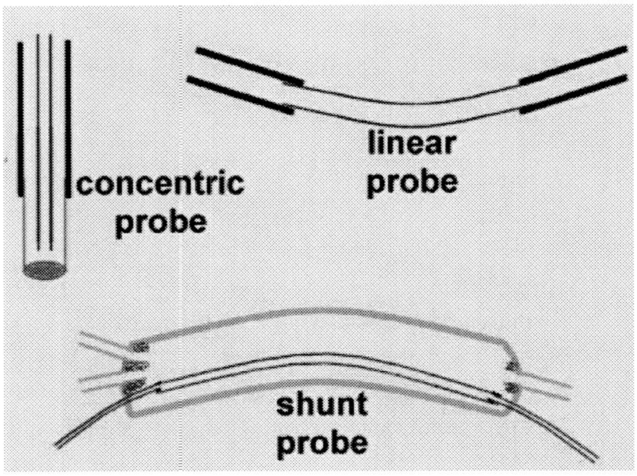

Figure 5. Available designs of microdialysis probes.

For soft peripheral tissues like muscle, skin, liver, tumor, and fluids like blood and bile, flexible probes can be used. Linear probes are useful for monitoring transdermal drug delivery. However, an increase in skin blood flow, erythema and skin thickness was demonstrated after insertion of a microdialysis probe (Groth and Serup 1998). Local anaesthesia prior to insertion reduces the vascular effects of needle insertion trauma.icrodialysis in adipose tissue and resting skeletal muscle can be performed with an especially developed commercial catheter to achieve excellent diffusion characteristics. Insertion into the tissue is easily achieved with help of a unique slit cannula introducer that leaves the catheter in place when withdrawn (Islinger et al 2004).

Some probes can be rigid, as in example, the concentric cannula used intracerebrally. In this case, guide cannulas can also be implanted, opening up the possibility to insert the probe after surgical recovery, thereby decreasing negative effects of anesthesia.

The catheter tip of the CMA 70 Brain Microdialysis Catheter® (Solna, Sweden) has a gold tip, which makes it visible on a CT-scan in order to locate the exact position of the catheter in the brain (Hlatky et al 2002). Another catheter for the study of the human brain is the CMA 71 High Cut-Off Brain Microdialysis Catheter® (Solna, Sweden). The most important characteristic of it is the dialyzing membrane that allows diffusion of large molecules such as cytokines. The membrane has a cut-off of around 100.000 Da (Hutchinson et al 2005).

7.2. Microdialysis Membrane

The choice of the membrane type is an essential element to optimize the probe for a particular microdialysis protocol (table 3). Conventional microdialysis probes are constructed with 20 kDa molecular weight cut-off membranes enabling measurement of small molecules. Common substances used as membrane materials are cuproamonic rayon, celluloses, polycarbonate, polyethersulfone or cuprophan and the effective membrane dialysis length ranges from 1 to 30 mm (Höcht et al 2004a).

Recently, a 100 kDa molecular weight cut-off microdialysis catheter have been used to sample larger molecules such as cytokines (Hutchinson et al 2005). These probes can be constructed with polyethylenesulphone (Sjögren et al 2002). One of the technical problems encountered with the 100 kDa microdialysis membrane is the issue of poor sample volume retrieval, considering that it enables passage of high molar mass molecules. One way of addressing this problem is to equalize the pressure differences between in-flowing and out-flowing perfusate at the membrane by increasing the colloidal pressure of the perfusate using dextran (Rosdahl et al 1997).

For selection of the most adequate dialysis membrane, molecular weight of the substance of interest must be taken into consideration. As discussed previously, only substances with a molecular weight lower than the membrane cut-off are able of passing the membrane. However, even if the molar mass falls below the molecular weight cutoff, an acceptable relative recovery will only be attained with substances having a molar mass lower than approximately one-fourth of the membrane cutoff (Plock and Kloft 2005).

The size of the microdialysis membrane can also influence the relative recovery. According to Fick's law, perfusion across a membrane is proportional to its area. Therefore, increasing the length of the microdialysis membrane will lead to an increase in relative recovery (Plock and Kloft 2005). In addition, increasing the outer diameter of the inner cannula may also enhance relative recovery of the probe (Wisniewski and Torto 2002; Torto et al 1999).

An important key is that the membrane does not interact with the surrounding tissue or with the perfusate. Low recoveries of acid aminoacids have been described due to the presence of surface charge (Sandberg and Lindström 1983). Recovery of neuropeptides can vary as much as 20% with different dialysis membranes (Kendrick 1989).

7.3. Perfusion Solutions

Different perfusion media have been used in microdialysis experiments and they vary widely in their composition and pH (table 3). Composition, ion strength, osmotic value and pH of the perfusion solution should be as close as possible to those of the extracellular fluid of the dialyzed tissue.

In most experiments, the perfusate is an aqueous solution of sodium and potassium salts and other ions in a minor proportion, without proteins or a very small concentration of them. Addition of dextran to the perfusion solution prevents fluid loss. Rosdahl et al. (1997) demonstrated that the estimated concentrations of dextran in the perfusion solution at which no net loss of perfusion fluid occurred produces a colloid osmotic pressure similar to the reported values for plasma. This implies that the plasma colloid osmotic pressure contributes to a mass transfer of fluid from microdialysis catheter to capillaries. In some cases, proteins should be added to the perfusion medium to prevent sticking of drugs to the microdialysis probe and tubing connection (Maidment et al 1989).

7.4. Microdialysis Probe Esterilization

When microdialysis probes are implanted into a tissue, it is necessary to keep asepsis in the area immediately around the site of the insertion. Huff et al. (2003) reported a preliminary evaluation of several disinfection/sterilization techniques used with experimental microdialysis probes. These authors found that two disinfection methods′, 70% ethanol and a commercial contact lens solution, and two sterilization methods, hydrogen peroxide plasma, and e-beam radiation, did not affect the functionallity of the probes. However, hydrogen peroxide plasma and contact lens solution groups reduced extraction efficiencies of microdialysis probes.

Gamma irradiation has also been used for sterilization of microdialysis probes (Ekstrøm et al 1997). Probes were sterilized by gamma irradiation (32 kGy) prior to use. Subsequent *in vitro* experiments confirmed that the irradiated probes behaved identically to native probes from the supplier. In this work, no sign of infection at the implantation sites was found, probably due to the strict attention to sterility during the probe implantation as well as to prophylactic treatment with a broad-spectrum antibiotic (Ekstrøm et al 1997). However, one of the most popular methods to sterilize microdialysis probes is the gas sterilization method with ethylene oxide (Mantovani et al 2002; Castejon et al 1999).

7.5. Analytical Methods

An important aspect of microdialysis in pharmacokinetic studies is the selection of an adequate analytical method for drug concentration determination. Microdialysis generates small volume samples (1-10 μl), because of the need of slow perfusion rates (0.1-2 μl/min) to obtain high recoveries of the drug maintaining an adequate temporal resolution. In addition, microdialysis samples often contain the analyte at low concentration (pM-μM range). Therefore, another important key of microdialysis sampling in pharmacokinetic studies is the selection of the most sensitive analytical method.

A wide range of analytical methods can be used for the analysis of microdialysis samples. Non-separation-based methods allow the detection of one analyte at a time, in contrast to separation-based methods that can be used for the detection of multiple analytes (original drug and metabolites) in each sample.

Available non-separation-based methods for analysis of microdialysis samples include biosensors, immunoassays and mass spectrometry (Davies et al 2000). These methods, compared to separation-based methods, have the advantage of no sample dilution.

Biosensors have been combined with microdialysis for direct on-line monitoring both *in vivo* and *in vitro*. The apparatus consists of a syringe pump, a microdialysis probe and an enzyme reactor, followed by an electrochemical (EC) detector. The enzyme is present to provide selectivity and to generate a redox-active product that can be detected by the EC detector. This system has been used mainly for the monitoring of endogenous compounds, such as lactate, glucose and glutamate (Davies et al 2000) and has no special utility for pharmacokinetic studies of drugs.

Immunoassay coupled to microdialysis sampling has been employed for measuring peptides and some drugs (Rocha et al 1997; Mackie et al 1997). It seems to be the most adequate method to analyze drugs with peptidic structure because of its high sensitivity and requirement of small sample volumes. Peptides cross the dialysis membrane much slower than smaller molecular weight compounds, leading to low recovery values and diluted samples. However, the assay itself is time-consuming and antibodies often exhibit crossreactivity leading to erroneous results.

More important in the drug discovery research is the use of mass spectrometric (MS) detection coupled to microdialysis. MS detection is particularly useful for the analysis of microdialysis samples because of the high selectivity, the ability to identify the analytes based on their molecular weight and the handling of small sample volumes. The main disadvantage of this method is that the high ionic strength of microdialysis samples generates high background noise. However, there are several methods of reducing or removing the dialysis salts, including the use of off-line C18 solid phase extraction (Prokai et al 1998), on line sample dilution with LC or a flow injection analysis carrier stream (Davies et al 2000), off-line sample dilution with organic solvents (Davies et al 2000), and parallel by-pass columns (Michelsen and Pettersson 1994).

Separation-based methods available for analysis of microdialysis samples include liquid chromatography (LC), microbore LC, capillary LC and capillary electrophoresis (CE). A wide range of different detectors can be coupled to the separation method. Immunoassay, ultraviolet absorbance (UV), EC, fluorescence (FL) and MS are the most common detectors for LC analysis. For CE, EC and laser induced fluorescence (LIF) detection have been most commonly employed.

The great diversity of detection methods provides to the researcher the chance to select the more adequate analytical system for the determination of the compound of interest. The most important aspect of the analytical system is its sensitivity in order to improve the temporal resolution of a microdialysis protocol.

The drawback of the separation-based methods is the dilution of the microdialysis sample. In comparison to the non-separation-based methods, these methods allow the detection of many compounds in each sample allowing simultaneous quantification of drug metabolites.

To improve the sensitivity of the LC method, it is also important to select the column type. Important aspects of chromatography columns that determine the sensitivity are its length, particle size and internal diameter. Microbore columns have become popular for the analysis of microdialysis samples, considering that its use results in a lower sample dilution because of the smaller length and inner diameter (Davies et al 2000). When using microbore columns, it is important to reduce at minimum dead volumes in the injector, microbore column and detector connections to preserve peak resolution (Cheng and Kuo 1995). The requirement of low flow rates (usually 0.1 $ml.min^{-1}$) for microbore chromatography is other important aspect. To resolve this issue there are commercially available pumps with low flow rates as well as flow splitting systems (Davies et al 2000).

Capillary LC has been also employed for the analysis of microdialysis samples. One advantage of capillary LC columns is that relatively large volumes of analytes can be injected on the column and then eluted in a very small volume using gradient elution (Boyd et al 2000).

Despite the advantage of microbore and capillary LC, conventional columns are still more commonly employed, considering that they do not need special instrumentation such as flow rate pumps, low dead volume injectors or reduced volume detection cells.

Another approach of the LC method is the on-line analysis. This system avoids sample handling problems associated with off-line collection such as liquid evaporation of the sample, improving also temporal resolution. One drawback of the on-line analysis of samples is that the temporal resolution will be limited by the analysis time. This disadvantage could be overcome using systems with multiple sampling valves (Steele and Lunte 1995).

CE is becoming increasingly popular as an analysis method for microdialysis samples. In spite of the fact that CE only needs injection volumes of 1-10 nl for off-line analysis, 1-5 μl samples are generally required due to difficulties with physical manipulation. One disadvantage of CE is that high ionic strength of the microdialysis sample causes lower detection sensitivity (Davies et al 2000).

7.6. Tissue Damage by Microdialysis Probe Implantation

A minimal lesion of the surrounding tissue is produced by implantation of the microdialysis probe, causing tissue responses such as a compromise BBB (Benveniste and Hüttermeier 2000; de Lange et al 1997) and acute inflammation of different tissues (Ault et al 1994; Palsmeier and lunte 1994; Davies and Lunte 1995). However, several studies have demonstrated that baseline conditions are reached following period of 60 min after probe implantation.

In the central nervous system, tissue disturbance originated by microdialysis probe implantation initially induces a release of neurotransmitter in the surrounding area of probe implantation. However, neurotransmitters rapidly return to basal levels. There is also an initial period of disturbed tissue function that lasts from 30 min to 24 h, characterized by increased glucose metabolism, decreased blood flow and disturbed neurotransmitters release (Ungerstedt, 1991). After chronically implantation of a microdialysis probe for 2 days, histological studies have shown the presence of oedema, minor haemorrhages and leukocytes accumulation in the surrounding tissue of the probe (Benveniste, and Hansen, 1991).

Therefore, researchers must consider these acute and chronic tissue alterations in the design of a microdialysis study.

7.7. Anaesthetized Versus Awake Animal Model

Microdialysis experiments can be realized in anaesthetized or freely moving animals. When using anaesthetized animals, the effect of the anaesthetic preparation on the PK and PD behavior of the drug must be evaluated (Claassen, 1994). In general, anaesthetized animals are more suitable for experiments with more than one microdialysis probe, local injection cannulas and in experiments requiring dissections (Ungerstedt, 1991).

In microdialysis experiments conducted in awake animals, the researcher should consider that stress could introduce artefacts in the measurements. In accordance, environmental stress needs to be evaluated (Bourne, 2003).

8. Importance of Microdialysis Probe Calibration

Use of microdialysis sampling for determination of target site concentrations of drugs needs accurate calibration of the microdialysis probe, considering that the desired information is the absolute drug concentration in the tissue. The relationship between concentration of the drug in the dialysate and its concentration in the sample matrix may be thought of as the recovery of the probe. Several factors influence recovery of the microdialysis probe, including temperature, perfusate flow rate, chemical and physical properties of the dialysis membrane, probe geometry, membrane surface area, chemical properties of the drug and the diffusion rate of the drug within the matrix (Johnson and Justice 1983). All of these parameters remain constant during the experiment under normal conditions. Few parameters, such as membrane surface area and perfusate flow rate, can be modified by the researcher to improve recovery of the microdialysis probe. If high microdialysis probe recovery is needed to obtain concentrated samples, the researcher can opt among increasing the size of the surface area or decreasing the perfusate rate. However, increase of membrane surface area implies loss of spatial resolution and decrease of the perfusate rate implies loss of temporal resolution, because of prolongation of the sampling interval.

Recovery of the microdialysis probe can be determined by both *in vitro* and *in vivo* assays. The *in vitro* estimation of the probe recovery, serves to prove if the microdialysis probe works, since *in vitro* recovery values are often an overestimation of the *in vivo* recovery. This is explained by the fact that the nature of the tissue to be sampled and its interactions with the drug diminishes the recovery (Ståhle et al 1990; Benveniste et al 1989a). In most cases drug diffusion through the tissue is the rate limiting step in its recovery, especially in hard tissues. Therefore, determination of *in vivo* recovery of microdialysis probe is essential for an accurate estimation of drug concentration in different tissues.

In vitro recovery of the probe also serves to evaluate the possibility of drug sticking in probe connections, to determine optimal perfusion rate for *in vivo* experiments and to validate *in vivo* recovery determination by the retrodialysis method.

In vivo recovery of the microdialysis probe can be determined through different methods, including flow-rate or stop- flow method (Jacobson et al 1985), zero-net-flux method (Lonnroth et al 1987), retrodialysis method (Wang et al 1991) and delivery method (Benveniste 1989b). Because the zero-net-flux method and the flow rate method require that examination of the study subject under steady state conditions prior to the experiment, the total study time is extended. This limits the application of these methods for pharmacokinetics purposes.

Determination of the *in vivo* recovery with the retrodialysis or delivery method is more appropriately. Both methods are based on the principle that recovery is independent of the direction of analyte diffusion. Therefore, loss of the analyte from the perfusion media to the extracellular matrix should be equal to gain of the analyte from the tissue to the perfusate. This equality must be determined by the determination of *in vitro* recovery of the analyte both by loss and gain methods.

The retrodialysis method consists in the diffusive loss of molecules of a calibrator from the perfusate into the environment surrounding the probe. The dialysance or permeability-area product (PeA) of the calibrator must be similar to that of the compound of interest (Höcht et al 2004a). The *in vivo* recovery of the calibrator is calculated with the following equation:

$$R = (Cin - Cdial) / Cin$$

where R is the calibrator *in vivo* recovery, Cin is the concentration of the calibrator in the perfusate and Cout is the concentration of the calibrator in the dialysate. Unbound concentrations of the compound of interest in the surrounding environment (C) are calculated using the following equation:

$$C = Cout / R$$

where Cout is the compound of interest concentration in the dialysate and R is the *in vivo* recovery of the microdialysis probe.

The delivery method consists in the determination of the *in vivo* recovery of the compound of interest by retrodialysis before the actual experiment. *In vivo* recovery of the compound of interest is determined before drug administration by perfusing microdialysis probe with a solution of the compound of interest, taking the proportion of loss across the dialysis membrane as an estimate of the recovery.

Comparing the last two methods, a limitation of the retrodialysis is the necessity of an analyte with similar physico-chemical characteristic as the compound of interest. On the other hand, if the analyte is added to the perfusate at high concentrations, it could introduce unwanted perturbations in the surrounding tissue area.

The delivery method overcomes the uncertainty introduced by using a standard to mimic the compound of interest. The shortcoming of this approach is that recovery changes resulting from the experiment are not detected. However, recovery of the microdialysis probe generally remains constant in pharmacokinetic studies (Höcht et al 2004a).

Another issue of microdialysis sampling in pharmacokinetic studies is the fact that microdialysis generates data that are the integral of the concentration surrounding the probe during the sampling interval, conversely to tissue samples, which provides a point measurement during the same interval. To estimate pharmacokinetic parameters, microdialysis data must to be transformed from a series of integrals to a series of points corresponding to the same time of tissue samples collection.

It is necessary to calculate first the time point during the sample interval at which the mean microdialysis samples concentration is attained (T) using the following equation developed by Ståhle (1992):

$$T = [\ln (k\Delta t) - \ln (1 - e^{-k\Delta t})]/k$$

where Δt is the sample interval and k the elimination rate constant.

Next, to calculate the concentration of drug in the microdialysis sample at the end time of the sample interval, the following equation must to be used:

$$C(t) = C(0)\ e\ \text{-}k(\Delta t - T)$$

where C(t) is the concentration of the drug at the end of the sample interval, C(0) is the mean concentration of the compound of interest in the microdialysis sample, Δt is the sample interval and T is the time point during the sample interval at which the mean microdialysis samples concentration is attained.

9. Current Applications of Microdialysis Sampling for Evaluation of Tissue Distribution and PK-PD Modeling of Drugs

Microdialysis sampling has been used for the study of target site distribution and PK-PD modeling of established drugs and NCE in both preclinical and clinical protocols. Although pharmacokinetic properties and PK-PD relationship of different drug classes have been evaluated by means of microdialysis sampling, most of microdialysis studies addressed drug distribution and PK-PD modeling of antiinfective, antineoplastic and central acting drugs (Höcht et al 2006a; Höcht et al 2006b). In addition, microdialysis sampling is useful for PK-PD modeling of cardiovascular drugs and for the evaluation of dermal pharmacokinetics of drugs after its topical application (Höcht et al 2006a; Höcht et al 2006b). In this section, most important microdialysis studies in this area are discussed. Principal applications of microdialysis sampling in drug development are resummed in table 4.

Table 4. Principal microdialysis applications in drug development

Therapeutic class	Application
Antimicrobial agents	• Antibiotic penetration in non-infected soft tissues • Distribution of antimicrobial agents in infected tissues • *In vivo* PK- *in vitro* PD modeling of antibiotics • *In vivo* evaluation of prodrugs
Antineoplastic drugs	• Anticancer drug distribution in solid tumors • Compartmental distribution within a tumor • Pharmacodynamics of antineoplastic drugs on biochemical markers • *In vivo* PK-PD modeling of anticancer agents
Anticonvulsivant agents	• Brain penetration of anticonvulsant agents • Distribution of anticonvulsivants at the epileptic focus • Evaluation of central pharmacokinetics of developmental anticonvulsivant drugs • Evaluation of efflux transporters inhibitors as a therapeutic strategy to overcome antiepileptic drug pharmacoresistance • *In vivo* PK-PD modeling of antiepileptic drugs

Table 4. (Continued).

Therapeutic class	Application
Psychotropic drugs	• *In vivo* evaluation of neurochemical effects of antidepressants and antipsychotic agents.
Antiinflamatory agents	• *In vivo* evaluation of effects on inflammatory mediators. • *In vivo* determination of COX-1/COX-2 selectivity of NSAIDs
Cardiovascular drugs	• *In vivo* PK-PD modeling of antihypertensive agents
Lipolytic agents	• Pharmadynamics by monitoring glicerol in mmicrodialysis samples

9.1. Antiinfective Drugs

Traditionally, pharmacokinetic assessment of antimicrobial agents was largely based on the measurement of total plasma concentrations. However, the use of plasma antibiotic levels is not ideal, because most infections occur in tissue sites, and therefore the ability of antibiotics to reach the target site is a key determinant of clinical outcome. It is considered that total plasma concentrations and plasma protein binding can be used to predict free tissue levels of antibiotics, based on the assumption that unbound plasma concentrations and free tissue levels are equal at equilibrium, considering that tissue distribution is generally mediated only by passive diffusion. However, many studies have shown lower tissue unbound levels than plasma concentrations (Nordbring 1978; Fischman et al 1998; Jain 1994; Heikkinen 2000). As discussed above, tissue distribution is also affected by anatomic barriers, such as BBB, presence of active transport systems and tissue metabolism. On the other hand, time to reach equilibrium between plasma and tissue concentrations of antibiotics may range from minutes to days (Liu and Derendorf 2003). Therefore, measurement of unbound drug concentrations in the interstitial fluid of the target tissue should be considered a gold standard for improvement of antimicrobial therapy and dose adjustment.

Microdialysis has been used to measure various antimicrobials agents in human and laboratory animals tissues, including aminoglucosides, penicillins, cephalosporines, fosfomycin, fluoroquinolones and antiviral agents (for review see Joukhadar et al 2001; Schuck and Derendorf 2005; Brunner et al 2005). These studies have served to evaluate drug distribution in several organs, including infective tissues, and to develop *in vivo* PK- *in vitro* PD models at the target site using the same parameters calculated in plasma: time (T) above the minimum inhibitory concentration (MIC) (T>MIC), the ratio of the maximum concentration of drug in serum (Cmax) to the MIC (Cmax/MIC), the area under the inhibitory curve or the area under the curve (AUC)/MIC ratio (Brunner et al 2005).

Using microdialysis sampling, the relevance of protein binding in therapeutic failure was demonstrated. A relationship was found between the failure in cefoperazone treatment in serious illness and the degree of drug bounded to proteins (Wise 1986). In addition, cefpodoxime, with lower protein binding than cefixime (25 vs 65%) showed higher peak concentration and tissue penetration than cefixime (Liu et al 2000), suggesting a favorable efficacy of cefpodoxime, and this is supported by clinical trial data provided by a study in paediatric acute otitis media (Cohen et al 1994).

Microdialysis sampling was also used for the determination of antimicrobial agents in the interstitial fluid of non-infected soft tissues of healthy volunteers and rats. Several studies have demonstrated that tissue concentrations of different antibiotics, including ertapenem (Burkhardt et al 2006a), gemifloxacin (Islinger et al 2004), telithromycin, (Graninger et al 2004), and imipenem (Marchand et al 2005) among others, are similar in soft tissues with regards to unbound plasma concentrations, indicating a good distribution of antimicrobial in non-infected tissues.

In addition, drug distribution of antimicrobial drugs in infected tissues was also evaluated by means of microdialysis sampling both in basic research and clinical setting. Experiments conducted in rats showed that lung and peritoneal penetration of imipenem is not reduced in rats with pneumonia and peritonitis, respectively (Dahyot et al 2006; Lefeuvre et al 2006). In another study, lung penetration of meropenem was determined in patients with pneumonia and metapneumonic pleural emphysema by means of microdialysis (Tomaselli et al 2004). Although meropenem rapidly penetrated infected lung tissue, interstitial lung fluid levels were lower than serum concentrations with target site drug concentrations above the MIC90 threshold for many clinically relevant pathogens for up to 6 hours (Tomaselli et al 2004). In another microdialysis clinical protocol, moxifloxacin exhibited similar penetration into the interstitial space fluid in normal subcutaneous tissue and infected decubitus ulcers in patients with spinal cord injury (Burkhardt et al 2006b).

Conversely, penetration of the fluorquinolone levofloxacin into tissues appears to be unaffected by local inflammation. Bellman et al (2004) observed that administration of a standard dose of levofloxacin reached adequate levels at target site, although the extent of tissue penetration showed a high interindividual variability. Zeitlinger et al. (2007) have also demonstrated a reduced distribution of levofloxacin in infected lung with regards to interstitial fluid of non-infected soft tissues obtained from healthy volunteers.

Microdialysis sampling was also extensively employed for PK-PD modeling of antiinfective agents (for review see Brunner et al 2005; Schuck and Derendorf 2005). A three step approach has been used for the *in vivo* PK- *in vitro* PD modeling by means of microdialysis. First, interstitial fluid concentrations of the antibacterial drug at the target site, e.g. site of infection, are measured by means of microdialysis. Then, time versus drug concentration profile measured *in vivo* is simulated in an *in vitro* setting on bacterial cultures. In a third step, unbound antibiotic concentrations are linked to bacterial kill rates by means of a PK-PD model (Delacher et al 2003).

Delacher et al. (2003) have demonstrated a significant correlation between the maximal bactericidal effect and several pharmacokinetic surrogate parameters, such as AUC/MIC, Cmax/MIC and T>MIC. The authors concluded that the therapeutic success or failure in antibacterial therapy depends on the target site concentrations of the antimicrobial agent. Moreover, *in vivo* PK- *in vitro* PD modeling provides valuable guidance for drug and dose selection of antibacterial drugs (Delacher et al 2003).

In vivo PK – *in vitro* PD modeling of antimicrobial drugs was also studied in critical ill patients by means of microdialysis. Zeitlinger et al (2003) have applied an in-vivo pharmacokinetic/in-vitro pharmacodynamic method to simulate bacterial killing in plasma and the interstitium of skeletal muscle tissue after intravenous administration of cefpirome and fosfomycin alone and in combination to patients with sepsis. The *in vitro* simulation of *in vivo* plasma and tissue pharmacokinetics of cefpirome and fosfomycin has shown that both antimicrobial agents kill Staphylococcus aureus and Pseudomonas aeruginosa strains after single dose administration, observing a synergic antimicrobial effect by the combined use. Therefore, this data confirms antimicrobial strategies of simultaneous administration of cefpirome and fosfomycin in patients with severe soft tissue infection (Zeitlinger et al 2003).

However, it must be pointed out that all PK–PD studies of antimicrobial drugs by means of microdialysis have used a combined *in vivo* PK–*in vitro* PD simulation without applying mathematical PK–PD models in their analysis. Recently, Liu et al. (2005) demonstrated that a PK–PD model based on unbound antibiotic concentrations at the site of infection, and a sigmoid Emax relationship, effectively described the antimicrobial efficacy of both cefpodoxime and cefixime. This approach offers a more detailed information than the MIC does about the time course of antibacterial efficacy of antibiotics (Liu et al 2005).

In summary, several microdialysis studies have demonstrated that anti-infective levels at the target site are sub-inhibitory, although effective concentrations are attained in plasma. Therefore, study of target site concentrations and PK-PD modeling of anti-infective agents in clinical drug development and in critically ill patients by means of microdialysis should enhance knowledge on adequate antibiotic drug dosing and improve patient's outcome.

9.2. Antineoplastic Drugs

Tumor drug exposure, a marker linked to clinical outcome, may be dramatically reduced due to diffusion barriers in solid tumors (Lin 2006). Differences in tumor drug distribution do not allow to predict the antineoplastic response from plasma profiles (Presant et al 1994), thus measurements of drug exposure into tumor interstitium by microdialysis may help to develop clinical PK-PD models with the aim of individualize drug therapy (Brunner and Müller 2002).

Microdialysis has been employed for the characterization of different antineoplastic drugs. Methotrexate, cisplatin, capecitabine, 5-fluorouracil, dacarbacine and melphalan have been measured into the tumor using clinical microdialysis in several types of malignancies such as breast cancer, melanoma, osteosarcoma and malignant fibrous histiocitoma (for review see Brunner and Müller 2002; Zhouh and Gallo 2005).

Measurements of plasma drug concentration have been employed during high-dose methotrexate treatment (Evans et al 1998). Although this practice has served to identify individuals with impaired renal function in order to avoid serious side effects, it assumed a relationship between pharmacological response and circulating levels of drug. Müller et al (1998) have studied interstitial tumor pharmacokinetics and plasma-to-tumor transfer rates of methotrexate in breast cancer patients by insertion of microdialysis probes into the primary tumor and the periumbilical subcutaneous adipose layer of previously chemotherapy-naive breast cancer patients. Absence of correlation between plasma AUC and the AUC in the interstitial space of tumor tissue was found, together with a high interindividual variability in transendothelial methotrexate transfer. Therefore, plasma levels of methotrexate were not predictive of intratumor levels.

Similar results were obtained for 5-fluorouracil (5-FU), showing that plasma or subcutaneous levels of 5-FU failed to predict tumor response. Conversely, high interstitial tumor concentrations of 5-FU were associated with increased tumor response. This information may explain drug resistance in some patients and help to optimize dosing and administration schedules (Müller et al 1997).

Microdialysis has also been used to assess the pharmacodynamics of chemotherapeutic agents (Zhou and Gallo 2005). Castejon et al. (1999) have determined plasma concentrations of serotonin and 5-hydroxyindoleacetic acid during cisplatin treatment by means of microdialysis. Serotonin is involved in the production of emesis associated to antineoplastic treatment. Microdialysis has also been used for the monitoring of extracellular levels of growth factors, such as the vascular endothelial growth factor (VEGF), during treatment with tamoxifen in a mouse model of human breast cancer (Garvin and Dabrosin 2003).

Although to date scarce PK-PD modeling studies applying microdialysis were made, integration of the pharmacological response with tumor PK profiles of the corresponding drug would help to define PK-PD relationship, which is essential for the rational design of drug administration regimens in cancer patients. Thompson et al. (2001) studied the clinical and biochemical responses to the time course of melphalan in the subcutaneous interstitial space and in tumor tissue from patients with various limb malignancies. The authors showed a significant correlation between the melphalan mean concentration in subcutaneous microdialysate and tumour response (Thomson et al 2001).

In another PK-PD protocol, Müller et al. (2000) have determined the unbound, interstitial drug pharmacokinetics of 5-fluorouracil and methotrexate in solid tumour lesions of patients by means of *in vivo* microdialysis. The authors then made a pharmacodynamic simulation of the time versus drug concentration profile in an *in vitro* setting by exposing breast cancer cells to interstitial tumour concentration of the antineoplastic drugs. The authors concluded that *in vivo* PK–*in vitro* PD models might provide a rational approach for describing and predicting pharmacodynamics of cytotoxic drugs at the target site (Müller et al 2000).

In summary, high interindividual variability in intratumoral drug distribution exists, indicating lack of correlation between plasma concentration of antineoplastic drugs and their interstitial tumoral levels. Plasma measurements do not serve as surrogates for intratumoral concentration and microdialysis may help to design optimal treatment schedules and to select appropriate drug, doses and dosing intervals for antineoplastic agents. Surprisingly, in the last three years, only few clinical microdialysis protocols were conducted for the study of intratumoral distribution of anticancer drugs. A possible explanation is that microdialysis experiments in cancer patients must be conducted in strict compliance with regulatory demands and need to be based on appropriate ethical conditions. In addition, there are other factors that limit the use of microdialysis in cancer research. Puncture of solid tumors by microdialysis catheter implantation may induce metastasis. However, the estimated incidence of metastasis by puncture ranged from 0.003% to 0.005 % and there is no evidence that puncture of tumor lesions affected the course or prognosis of the underlying disease (Lundstedt et al 1991).

Another limitation of the applicability of microdialysis sampling in oncology is the fact that the majority of the antineoplastic drugs act within cells. The relationship between extracellular drug concentrations and intracellular drug levels remains unknown. This drawback could be overcome by applying an attractive approach consisting in the simultaneous study of drug distribution with microdialysis and PET. PET measures total (intracellular, extracellular, and intravascular) concentrations of radiolabeled drugs in tissue, and microdialysis determines unbound drug concentrations in the extracellular space fluid of tissue. Therefore, combination of both techniques allows the description of the intracellular drug pharmacokinetics. Langer et al. (2005), using (18)F-labeled-ciprofloxacin as a model drug, have found that *in vivo* intracellular ciprofloxacin pharmacokinetics was in accordance with previous *in vitro* data describing cellular ciprofloxacin uptake and retention. Therefore, PET/microdialysis combination might be useful during research and development of new drugs, for which knowledge of intracellular concentrations is of interest.

Finally, some antineoplastic drugs, such as 5-fluorouracil, requires intracellular enzymatic conversion in order to exert their cytotoxic activity. In addition, other aspects like tumor location and accessibility for microdialysis probe implantation, and the possibility of variation in interstitial concentrations of cytotoxic drugs in different metastases in a patient restrict the utility of microdialysis for studies of antineoplastic drug distribution (Höcht et al 2006a).

9.3. Central Acting Drugs

Knowledge of brain uptake is an important issue for drugs acting in the central nervous system, such as anticonvulsivants, antidepressants, anaesthetics and anticancer drugs. Concentration-time profile of the drug in the central nervous system determines the intensity and duration of the effects of a drug with a central action. As in the clinical setting, the measurement of brain concentrations of the drug is highly restricted; alternative determination of lumbar or ventricular cerebrospinal fluid is sometimes used. However, study of drug concentrations in the cerebrospinal fluid provides only limited information with respect to drug distribution into the brain, because brain parenchyma distribution of a central acting agent is determined by multiple factors, including active biotransformation, active transport at the BBB and intracellular-extracellular exchange (de Lange and Danhof 2002). Intracerebral microdialysis would be a significant improvement, considering that it allows the study of the pharmacokinetic profile of the unbound drug fraction at a specific region within the brain. However, as discussed previously, microdialysis is a semi-invasive technique and may bear risks. Microdialysis can be used in the early phase of clinical trials to establish brain penetration of drugs. A phase II trial of topiramate in severe traumatic brain injury demonstrated different free-drug concentrations in the extracellular space from that measured in cerebrospinal fluid (CSF) (Alves et al 2003).

An interesting field of intracerebral microdialysis in the clinical setting is the study of brain distribution of anticonvulsivant agents. Despite the existence of a large variety of antiepileptic drugs (AED), almost 30% of epileptic patients are resistant to treatment (Regesta and Tanganelli 1999). There is increasing evidence that over-expression of multidrug transporters such as P-glycoprotein is involved in the generation of pharmacoresistance to antiepileptic drugs because of a greater efflux from the brain of the central acting agent (Löscher 2002). Therefore, monitoring of target site concentration of

AED by means of intracerebral microdialysis could elucidate the involvement of the efflux transport in the generation of the pharmacoresistance to antiepileptic drugs.

In the clinical setting, microdialysis protocols have demonstrated that intracerebral concentrations of valproic acid (Lindberger et al 2001) and phenytoin (Scheyer et al 1994) were slightly lower than plasma levels, suggesting that both anticonvulsivant drugs may be substrate for an energy-dependent carrier transport out of the central nervous system. Conversely, carbamazepine and carbamazepine-10,11-epoxide concentrations in the extracellular brain fluid closely mirror their unbound serum concentrations (Scheyer et al 1994b).

However, clinical study of the involvement of drug carriers in pharmacoresistance to antepileptic drugs is restricted by the limitation in obtaining suitable control tissue to compare with epileptogenic brain tissue of patients (Treiman and Woodbury 1995). This limitation was reflected in a recent study, in which authors found a high interindividual variability in extracellular levels of AED of cortical regions of patients with intractable epilepsy (Rambeck et al 2006). Eventhough authors found lower extracellular concentration of several antiepileptic drugs in the cerebrospinal fluid of patients with intractable epilepsy, in the absence of data from non-epileptic tissues it was not possible to determine if their findings were related to overexpression of multidrug transporters in the brain (Rambeck et al 2006). Therefore, animal models of epilepsy may be useful to test the transporter hypothesis.

Microdialysis sampling has been extensively used for the study of antiepileptic drug concentrations at the target site in non-epileptic animals and experimental models of refractory epilepsy. Involvement of drug efflux transporters in brain parenchyma distribution of antiepileptic drugs was demonstrated in non-epileptic animals by enhancement of drug central levels by the application of drug efflux transporters inhibitors (Löscher and Potschka 2005a; Remy 2006).

Only few studies evaluated central pharmacokinetics of antiepileptic drugs in refractory epilepsy models with conflicting results. Whilst Rizzi et al. (2002) using the Kainate model of epilepsy showed that overexpression of P-glycoprotein in the hippocampus is associated with reduced brain concentrations of phenytoin compared with control animals, Potschka and Löscher (2002) did not find altered levels of phenytoin in hippocampus and amygdala of kindled when compared to non-kindled rats. Recently, we have demonstrated a critical role of P-gp overexpression in the development of pharmacoresistance to phenytoin in a model of epilepsy induced by 3-mercaptopropionic acid chronic administration, suggesting that administration of efflux transporters inhibitors could be an effective strategy to decrease pharmacoresistance to phenytoin antiepileptic treatment (Höcht et al 2007b).

Intracerebral microdialysis was also used for measurement of brain concentration of other central acting drugs. Ederoth el al (2003) studied the BBB transport of morphine in patients with severe brain trauma by simultaneous microdialysis of morphine levels in "better" and "worse" brain tissue. The authors found that unbound brain morphine levels were lower than plasma concentrations suggesting the existence of an efflux transport system for morphine across the human BBB. In addition, the results suggested an increase of BBB permeability to morphine in the "worse" brain tissue.

Brain penetration of anticancer drugs was also evaluated using microdialysis sampling in basic research. Zhuang et al (2006) have shown that P-glycoprotein and breast cancer resistance protein 1 (BCRP-1) limit topotecan distribution in brain parenchyma. Microdialysis sampling also allows the study of drug distribution to brain tumors. A significant increase in the penetration of methotrexate into the tumor tissue was found (Devineni et al 1996). A hybrid physiologically-based pharmacokinetic model was used by the authors to characterize the mechanisms that may be responsible for the high methotrexate tumor concentration.

Brain microdialysis has also been used for PK-PD modeling of therapeutic agents at the central nervous system. PK–PD modeling allows the study of the mechanism responsible for the time delay of central actions of drugs. In an elegant study, Bouw et al. (2001) have simultaneously determined blood and brain concentrations of morphine-6-glucuronide and its antinociceptive effect by means of microdialysis sampling. By applying a PK–PD model with an effect compartment, the authors found a greater delay in the onset of the effect when antinociception was related to plasma morphine-6-glucuronide concentrations with regard to brain levels. Therefore, it was concluded that half of the effect delay could be explained by transport across the blood–brain barrier, suggesting that the remaining delay is a result of drug distribution in the brain parenchyma.

Microdialysis was used to describe the relationship between norfloxacin concentrations in the CNS and its adverse reactions, such as convulsive effect (Chenel et al 2004). Brain extracellular concentrations of norfloxacin by means of microdialysis and a quantitative electroencephalogram (EEG) were simultaneously determined. Blood samples were also collected to determine norfloxacin plasma levels. Although norfloxacin brain concentrations peaked early after its intravenous administration, the effect on the EEG measurement was delayed. By applying a PK–PD model with an effect compartment, the authors demonstrated that the delayed EEG effect of norfloxacin is not due to BBB transport.

In conclusion, microdialysis sampling is a promising technique for drug development of central acting drugs, especially in the preclinical phase. The most attractive application of brain microdialysis seems to be the evaluation of brain parenchyma distribution of developing antiepileptic drugs and its ability to overcome pharmacoresistance in chronic epileptic animal models. In addition, impact of therapeutic strategies to enhance brain penetration of antiepileptic, such as efflux transporters inhibitors, could also be evaluated during preclinical phase of drug development by means of microdialysis sampling.

9.4. Antihypertensive Drugs

A poor concentration-response relationship with regard to the blood pressure effect of antihypertensive drugs has been found. Taking into account the pharmacodynamic properties of antihypertensive drugs, the suggestion of absence of relationship between plasma levels of antihypertensive drugs and its blood pressure lowering effect could reflect an inadequacy or failure in the approaches designed to detect such correlation. A number of factors have hampered the possible identification of a correlation, including failure to study individual patients, inability to collect sufficient pharmacodynamic data, failure to identify and account for temporal delay in the onset of the pharmacological effect, the use of restricted concentration ranges and the use of dose rather than concentration (Meredith 1997; Brynne et al. 1998).

Using a 'shunt' intra-arterial microdialysis probe, a good relationship was found between metoprolol concentration in the effect compartment and its hypotensive and chronotropic effect (Höcht et al 2004b; Höcht et al. 2005; Höcht et al 2006c). Moreover, the maximal response was significantly greater in hypertensive animals, such as spontaneously hypertensive rats and animals with aortic coarctation, with regards to their respective control animals. Therefore, this data suggests that the proposed lack of relationship between plasma levels of β-blockers and its antihypertensive effect is probably a consequence of an inadequate experimental design and data analysis.

In conclusion, microdialysis sampling is a powerful tool for PK–PD modeling of cardiovascular drugs in basic research, taking into account that it allows continuous and simultaneous determination of antihypertensive drugs plasma levels and their corresponding effect on blood pressure and heart rate in the same animal. Moreover, PK–PD modeling not only allows a better pharmacodynamic characterization of blood pressure lowering agents, but also permits the study of the physiopathological mechanisms of the hypertensive stage in different experimental models.

9.5. Microdialysis for the Evaluation of Drug Distribution and PK-PD Modeling of New Chemical Entities

To date, most of the microdialysis protocols have been conducted to evaluate the PK and PD of established drugs. Only few works have described tissue distribution and PK-PD modeling of NCE by means of microdialysis sampling. Among the reasons for the little use of microdialysis in drug development, it must be mentioned that microdialysis is a relative new technique; regulatory authorities have only recently emphasized the importance of studying drug distribution in drug development and knowledge of PK-PD modeling is limited.

Pharmacodynamics and PK-PD modeling of central acting NCE has been evaluated using intracerebral microdialysis. Neurochemical effects of desvenlafaxine, a novel antidepressant drug, on cathecholamine hypothalamic levels were studied in freely moving rats. Administration of desvenlafaxine induced an increase in noradrenaline levels in hypothalamic dialysate withouth any effect on dopamine and serotonin concentrations, suggesting its utility in a variety of central nervous system-related disorders (Deecher et al 2006).

In another report, using *in vivo* microdialysis, the protective effect of NGP1-01, a dual blocker of neuronal voltage- and ligand-operated calcium channels, was evaluated by monitoring choline release during NMDA infusion as a measure of excitotoxic membrane breakdown (Kiewert et al 2006). Intraperitoneal administration of NGP1-01 reduced NMDA-induced membrane breakdown demonstrating that NGP1-01 simultaneously blocks both major neuronal calcium channels and is sufficiently brain-permeable. Therefore, NGP1-01 is a promising lead structure for a new class of dual-mechanism neuroprotective agents.

Microdialysis was also used to describe, by means of PK–PD modeling, the effect of drug candidates on dopaminergic activity at different nuclei of the CNS (Weikop et al 2004; Raje et al 2005). The effect of benzatropine analogues on dopamine concentration in the nucleus accumbens after its intravenous administration was evaluated (Raje et al 2005). The authors fitted plasma concentration of the analogues and their effects on extracellular dopamine levels to two different PK–PD models, such as an effect compartment model and a model with indirect physiological response. The authors demonstrated that the indirect model is more suitable for PK–PD modeling of benzatropine analogues than the linked PK–PD model. These results are in accordance with the mechanism of action of the analogues because these drugs bind to the dopamine transporter inhibiting the dopamine re-uptake and consequently elevate dopamine extracellular levels.

Finally, ocular pharmacokinetics of prodrug candidates of antiviral agents has been evaluated by means of microdialysis sampling. It was found that dipeptide ester derivates of acyclovir and ganciclovir enhance ocular absorption of the respective antiviral agent and seem to be promising candidates for the treatment of ocular infections (Anand et al 2006; Majumdar et al 2006).

Considering the methodological aspects of microdialysis, it is expected that the use of this technique for pharmacokinetic and pharmacodynamic assessment of NCE would increase in the next years.

10. Application of Reverse Microdialysis for the Asssessment of Pharmacodynamics of NCE

As previously mentioned in this chapter, another application of the microdialysis technique is reverse microdialysis (Höcht et al 2007a). Based on the dialysis principle, this application allows the local administration of a substance by means of the microdialysis probe, at the same time that it samples the extracellular levels of endogenous compounds (figure 3). The reverse microdialysis technique is based on the principle that the inclusion of a high amount of drug in the perfusate will allow the drug to diffuse through the microdialysis membrane into the tissue, and thus, will permit its local administration in a controlled manner.

One advantage of reverse microdialysis with respect to other drug delivery methods is the possibility to manipulate and estimate the amount of drug that diffuses into the extracellular space. It can be calculated by the determination of the compounds concentration in the perfusate and in the dialysate. The difference of these concentrations represents the amount of drug delivered into the tissue during the sampling time. Therefore, the sum of the difference of concentrations between perfusate and dialysate of the drug for each sampling interval represents the total amount of drug delivered during the reverse microdialysis experiments (Chan and Chan, 1999). In the reverse microdialysis technique, drug administration into the perfusion site is of zero order, similar to an intravenous constant rate infusion. Consequently, constant levels at the site of perfusion are attained when the diffusion rate of the drug from the tissue surrounding the microdialysis probe, due to distribution in other tissues, is similar to the diffusion rate of the drug from the probe to the extracellular space (Höcht et al. 2007a).

Reverse microdialysis is an interesting alternative to other drug delivery methods, such as microinfusion through an implanted intracranial cannula. Drug administration by microinfusion has several disadvantages compared to reverse microdialysis. For instance, in microinfusion studies the compound is administered by the introduction of a determined volume of fluid into the extracellular space of the tissue of interest. As volume expansion occurs, this technique is rather useful for chronic drug administration, especially for chronic infusion into brain parenchyma (Bazzett et al 1991), considering that microinfusion may produce non-specific tissue damage due to the volume of fluid introduced. On the contrary, reverse microdialysis delivers drugs without net gain of fluid, making this technique suitable for chronic drug administration. However, it must be pointed out that during chronic procedures the recovery of the microdialysis probe might change because of alterations in the surrounding tissue, such as gliosis or inflammation.

Another important difference between reverse microdialysis and microinfusion is that in the last one, after local administration of a drug, it diffuses rapidly out of the site of injection into the peripheral bloodstream, making interpretation of results extremely difficult (Evans et al 1975). With the reverse microdialysis technique, the drug is chronically perfused into the tissue, maintaining more constant drug concentrations at the target tissue. In addition, drug targets a larger volume of the tissue, given the fact that it diffuses to the interstitial space from the whole microdialysis membrane. Therefore, an important aspect in reverse microdialysis is the selection of the adequate membrane size in order to ensure drug administration to a great volume of the target tissue.

Considering that solutions injected by microinfusion through an intracranial implanted cannula are assumed to enter the brain as a spherical drop, any variation in the drop size, resulting from pressure injection delivery, may produce variability in the diffusion area (Rice et al 1985).

Another advantage of reverse microdialysis drug delivery is the potential to monitor the proper operation of the microdialysis probe throughout a period of chronic drug administration. In intracranial injection studies, incomplete expulsion due to cannula obstruction has been reported (Hargraves and Freed 1987). On the other hand, reverse microdialysis is more suitable for dose–response studies in the same animal, as there is no need to reimplant the microdialysis probe for further dose administrations.

However, an important drawback of reverse microdialysis with regards to other delivery methods is the greater cost of the equipment, and this might negatively influence at the moment of selecting the most appropiate technique in order to conduct an experiment.

Reverse microdialysis has been extensively employed for the study of local effects of pharmacological and toxicological agents in basic research, especially at the central nervous system, where it has been used for the study of the effects on neurotransmission at different central nuclei of diverse pharmacological and toxicological agents, such as antidepressants, antipsychotics, antiparkinsonians, hallucinogens, drugs of abuse and experimental drugs (for review see Höcht et al 2007a).

In the clinical setting, reverse microdialysis has been used for the study of local effects of drugs in the adipose tissue, skeletal muscle and dermis (Höcht et al 2007a). Reverse microdialysis was extensively employed for the study of lipolytic and anti-lipolytic drugs. Microdialysis has some advantages with regard to *in vitro* lipolysis studies, considering that it reflects the metabolism of adipose tissue in its natural environment, and offers the possibility to perform local manipulation and simultaneously study of metabolism (Kolehmainen et al., 2000).

An important issue is that the recovery is dependent on local tissue blood flow. To eliminate this variability, blood flow must be simultaneously measured using ethanol dilution technique (Kolehmainen et al., 2000). In this method, ethanol is added to the perfusion media and ethanol concentrations are analyzed from the perfusate and dialysate. Changes in ethanol concentration, expressed as dialysate/perfusate ratio, describe the changes in the local blood flow (Kolehmainen et al., 2000). Lipolytic effects are determined by the measurement of glycerol dialysate concentrations previously and during administration of the lipolytic or anti-lipolytic drug through the microdialysis catheter.

Another advantage of the use of reverse microdialysis for the study of drug effects on lipolysis is the fact that this technique allows the evaluation of the concentration–response relationship in a single subject by perfusing the microdialysis catheter with different concentrations of the lipolytic agent.

Reverse microdialysis technique has been also used to study drug effects on muscle metabolism and adverse drug reactions at the dermis, but these issues exceed the purpose of the present chapter (Höcht et al 2007a).

In conclusion, reverse microdialysis is a promising tool for the assessment of local pharmacodynamics of NCE, especially to evaluate neurochemical effects of central acting drugs and effects on lipolysis of lipolytic and anti-lipolytic agents.

11. Conclusion

Evaluation of drug concentrations at the biophase is an essential issue in development of highly efficient therapeutic agents, especially for antimicrobial drugs, anticancer agents and central acting drugs. On the other hand, PK–PD modeling has several advantages compared with dose–response trials, allowing a better pharmacodynamic characterization and screening of dosage–regimen selection of therapeutic agents. However, only few PK–PD studies are described in both preclinical and clinical settings. The need for simultaneous measurement of drug tissue levels and corresponding pharmacological effects at multiple time points probably limits the feasibility of PK–PD studies.

Traditional sampling techniques, such as biopsy and skin blister, have several limitations for the study of drug distribution at the target site, including single point determination, measurement of total drug concentration at the biophase and its invasive nature. However, more sophisticated technologies are now available, improving evaluation of tissue pharmacokinetics of drugs and NCE. PET and microdialysis are two of the most applied methodologies for the study of this trend. These methods are capable of characterizing drug disposition in humans in a non-invasive or minimally invasive manner and thus provide information that has been inaccessible to previously available methods. Comparing these two techniques, PET is a non-invasive imaging technique that allows determination of total tissue concentrations of drugs without discerning between the original drug and its metabolites. Another drawback of PET is its high cost. Conversely, microdialysis sampling is able to monitor free drug concentrations at the biophase allowing the simultaneous determination of the drug, drug metabolites and endogenous compounds in a single protocol. However, the most important drawback of microdialysis is its semi-invasive nature. Combining PET and microdialysis sampling in the same experimental subject appear to be an interesting approach for the measurement of intracellular drug levels of drugs acting at cytoplasmatic and nuclear receptors.

In addition, microdialysis technique allows the simultaneous determination of unbound concentrations of drugs at several tissues and their pharmacological effect on biochemical and clinical markers. Therefore, microdialysis overcomes major experimental limitations of PK–PD modeling and might become a reference technique for PK–PD studies.

The applicability of microdialysis to study drug distribution and PK–PD modeling during drug development would be restricted to the preclinical and early clinical phases because of its low throughput, its semi-invasive nature and the need for technical expertise and additional laboratory equipment.

Although the majority of basic and clinical applications reported so far have studied already marketed drugs, microdialysis approach now offers the possibility of gaining early pharmacokinetic information on novel drug candidates. Whether used alone or in combination, microdialysis sampling will exert an increasingly important impact on future drug development, and will make an important contribution to providing us with safer and more efficacious doses of novel drugs.

References

Alves, O.L., Doyle, A.J., Clausen, T., Gilman, C., and Bullock, R. (2003). Evaluation of topiramate neuroprotective effect in severe TBI using microdialysis. *Ann. N. Y. Acad. Sci., 993,* 25-34.

Anand, B.S., Katragadda, S., Gunda, S., and Mitra, A.K. (2006) *In vivo* ocular pharmacokinetics of acyclovir dipeptide ester prodrugs by microdialysis in rabbits. *Mol. Pharm., 3,* 431-440.

Ault, J.M., Riley, C.M., Meltzer, N.M., and Lunte, C.E. (1994) Dermal microdialysis sampling *in vivo*. *Pharm. Res*., *11*, 1631-1939.

Rambeck, B., Jürgens, U.H., May, T.W., Pannek, H.W., Behne, F., Ebner, A., Gorji, A., Straub, H., Speckmann, E.J., Pohlmann-Eden, B., and Löscher, W. (2006) Comparison of Brain Extracellular Fluid, Brain Tissue, Cerebrospinal Fluid, and Serum Concentrations of Antiepileptic Drugs Measured Intraoperatively in Patients with Intractable Epilepsy, *Epilepsia*, *47,* 681-694.

Balant, L.P., Gex-Fabry, M., and Balant-Gorgia, E. (1996) Drug metabolism as a confounding factor in PK/PD population approaches. *Therapie, 51,* 390-398.

Bauer, J.A., Balthasar, J.P., and Fung, H.L. (1997) Application of pharmacodynamic modeling for designing time-variant dosing regimens to overcome nitroglycerin tolerance in experimental heart failure. *Pharm. Res*., *14,* 1140-1145.

Bazzett, T.J., Becker, J.B., and Albin, R.L. (1991). A novel device for chronic intracranial drug delivery via microdialysis. *J. Neurosci. Methods*, *40,* 1-8.

Begley, D.J. (2004) Delivery of therapeutic agents to the central nervous system: the problems and the possibilities. *Pharmacol. Ther.*, *104*, 29–45.

Bellmann, R., Kuchling, G., Dehghanyar, P., Zeitlinger, M., Minar, E., Mayer, B.X., Muller, M., and Joukhadar, C. (2004) Tissue pharmacokinetics of levofloxacin in human soft tissue infections. *Br. J. Clin. Pharmacol.*, *57,* 563-568.

Benfeldt, E., Serup, J., and Menne, T. (1999) Microdialysis vs. suction blister technique for *in vivo* sampling of pharmacokinetics in the human dermis. *Acta Derm Venérelo*, *79*, 338–342.

Ben-Hun, J., Cooper, R.L., Cringle, S.J., and Constable, I.J. (1988) Ocular dialysis. A new technique for *in vivo* intraocular pharmacokinetic measurement. *Arch. Ophthalmol.*, *106,* 254-259.

Benveniste, H., Hansen, A.J., and Ottosen, N.S. (1989a) Determination of brain interstitial concentrations by microdialysis. *J. Neurochem.*, *52*, 1741-1750.

Benveniste, H. (1989b) Brain microdialysis. *J. Neurochem.*, *52*, 1667-1679.

Benveniste, H., and Hüttemeier, P.C. (1990) Microdialysis – theory and application. *Prog. Neurobiol.*, *35*, 195-215.

Benveniste, H., and Hansen, A.J. (1991). Practical aspects of using microdialysis for determination of brain interstitial concentrations. In: Robinson, T.E., Justice Jr., J.B. (Eds.), Microdialysis in the Neurosciences. (pp. 23-43). Amsterdam: Elsevier.

Bourne, J.A. (2003). Intracerebral microdialysis: 30 years as a tool for the neuroscientist. *Clin. Exp. Pharmacol. Physiol.*, *30,* 16-24.

Bouw, M.R., Xie, R., Tunblad, K., and Hammarlund-Udenaes, M. (2001) Blood–brain barrier transport and brain distribution of morphine-6-glucuronide in relation to the antinociceptive effect in rats – pharmacokinetic/pharmacodynamic modeling. *Br. J. Pharmacol.*, *134*, 1796-1804.

Boyd, B.W., Witowski, S.R., and Kennedy, R.T. (2000) Trace-level amino acid analysis by capillary liquid chromatography and application to *in vivo* microdialysis sampling with 10-s temporal resolution.*Anal. Chem.*, *72*, 865-871.

Browne, T.R., and Holmes, G.L. (2001) Epilepsy. *N. Engl. J. Med.*, *344,* 1145-1151.

Brunner, M., and Müller, M. (2002) Microdialysis: an *in vivo* approach for measuring drug delivery in oncology. *Eur. J. Clin. Pharmacol.*, *58,* 227-234.

Brunner, M., Derendorf, H., and Müller, M. (2005) Microdialysis for *in vivo* pharmacokinetic/pharmacodynamic characterization of anti-infective drugs. *Curr. Opin. Pharmacol.*, *5,* 495-499.

Brynne, L., Karlsson, M.O., and Paalzow, L.K. (1998) Concentration-effect relationship of l-propranolol and metoprolol in spontaneous hypertensive rats after exercise-induced tachycardia. *J. Pharmacol. Exp. Ther.*, *286,* 1152–1158.

Burkhardt, O., Brunner, M., Schmidt, S., Grant, M., Tang, Y., and Derendorf, H. (2006a) Penetration of ertapenem into skeletal muscle and subcutaneous adipose tissue in healthy volunteers measured by *in vivo* microdialysis. *J. Antimicrob. Chemother.*, *58,* 632-636.

Burkhardt, O., Derendorf, H., Jager, D., Kumar, V., Madabushi, R., Rohl, K., and Barth, J. (2006b) Moxifloxacin distribution in the interstitial space of infected decubitus ulcer tissue of patients with spinal cord injury measured by *in vivo* microdialysis. *Scand. J. Infect. Dis.*, *38,* 904-908.

Castejon, A.M., Paez, X., Hernandez, L., Luigi, X., and Cubeddu, L.X. (1999) Use of Intravenous Microdialysis to Monitor Changes in Serotonin Release and Metabolism Induced by Cisplatin in Cancer Patients: Comparative Effects of Granisetron and Ondansetron. *J. Pharmacol. Exp. Ther.*, *291*, 960–966.

Chan, S.H.H., and Chan, J.Y.H. (1999). Application of reverse microdialysis in the evaluation of neural regulation of cardiovascular functions. *Anal. Chim. Acta, 379,* 275-279.

Chenel, M., Marchand, S., Dupuis, A., Lamarche, I., Paquereau, J., Pariat, C., and Couet, W. (2004) Simultaneous central nervous system distribution and pharmacokinetic–pharmacodynamic modeling of the electroencephalogram effect of norfloxacin administered at a convulsivant dose. *Br. J. Pharmacol., 142,* 323-330.

Cheng, F.C., and Kuo, J.S. (1995) High-performance liquid chromatographic analysis with electrochemical detection of biogenic amines using microbore columns. *J. Chromatogr. Biomed. Appl.*, *665,* 1-13.

Claassen, V. (1994). Anaesthesia. In: Claaseen, V. (Ed.), Neglected factors in pharmacology and neuroscience research, pp 382-421, Amsterdam: Elsevier.

Cohen, R., de La Rocque, F., Boucherat, M. et al. (1994) Cefpodoxime proxetil vs cefixime for painful febrile acute otitis media in children. *Medecine et Maladies Infectieuses*, 24, 844-51.

Cordon-Cardo, C.J., O'Brien, J.P., Casals, D., Rittman-Grauer, L., Biedler, J.L. Melamed, M.R., and Bertino, J.R. (1989) Multidrug-resistance gene P-glycoprotein is expressed by endothelial cells at blood-brain barrier sites. *Proc. Natl. Acad. Sci. USA.*, *86,* 695–98.

Csajka, C., and Verotta, D. (2006) Pharmacokinetic-pharmacodynamic modeling: history and perspectives. *J. Pharmacokinet. Pharmacodyn.*, *33*, 227-279.

Curtis, G., Colburn, W., Heath, G., Lenehan, T., and Kotschwar, T. (2000) Faster drug development. *Appl. Clin. Trials., 9*, 52-55.

Dahyot, C., Marchand, S., Pessini, G.L., Pariat, C., Debaene, B., Couet, W., and Mimoz, O. (2006) Microdialysis study of imipenem distribution in skeletal muscle and lung extracellular fluids of Acinetobacter baumannii-infected rats. *Antimicrob Agents Chemother., 50,* 2265-2267.

Davies, M.I., and Lunte, C.E. (1995) Microdialysis sampling for hepatic metabolism studies. Impact of microdialysis probe design and implantation technique on liver tissue. *Drug Metab. Dispos., 23,* 1072-1079.

Davies, M.I., Cooper, J.D., Desmond, S.S., Lunte, C.E., and Lunte, S.M. (2000) Analytical considerations for microdialysis sampling. *Adv. Drug Deliv. Rev.*, *45*, 169-188.

Dayneka, N.L., Garg, V., and Jusko, W. (1993) Comparison of four basic models of indirect pharmacodynamic responses. *J. Pharmacokinet. Biopharm., 21,* 457-478.

de Boer, A.G., van der Sandt, I.C., and Gaillard, P.J. (2003) The role of drug transporters at the blood-brain barrier. *Annu. Rev. Pharmacol. Toxicol., 43*, 629–56.

de Boer, A.G., and Gaillard, P.J. (2007) Drug Targeting to the Brain. *Annu. Rev. Pharmocol. Toxicol., 47,* 323-355.

de Lange, E.C., Danhof, M., de Boer, A.G., and Breimer, D.D. (1997) Methodological considerations of intracerebral microdialysis in pharmacokinetic studies on drug transport across the blood-brain barrier. *Brain Res. Rev., 25,* 27-49.

de Lange, E,. de Boer, A.G., and Breimer, D. (2000) Methodological issues in microdialysis sampling for pharmacokinetic studies. *Adv. Drug. Deliv. Rev., 45*, 125-48.

de Lange, E.C.M., and Danhof, M. (2002) Considerations in the Use of Cerebrospinal Fluid Pharmacokinetics to Predict Brain Target Concentrations in the Clinical Setting. Implications of the Barriers Between Blood and Brain. *Clin. Pharmacokinet., 41,* 691-703.

Debouck, C., and Metcalf, B. (2000) The impact of genomics on drug discovery. *Ann. Rev. Pharmacol. Toxicol., 40,* 193-208.

Deecher, D.C., Beyer, C.E., Johnston, G., Bray, J., Shah, S., Abou-Gharbia, M., and Andree, T.H. (2006) Desvenlafaxine succinate: A new serotonin and norepinephrine reuptake inhibitor. *J. Pharmacol. Exp. Ther., 318,* 657-665.

Delacher, S., Derendorf, H., Hollenstein, U., Brunner, M., Joukhadar, C., Hofmann, S., Georgopoulos, A., Eichler, H.G., and Muller, M. (2000) A combined *in vivo* pharmacokinetic–*in vitro* pharmacodynamic approach to simulate target site pharmacodynamics of antibiotics in humans. *J. Antimicrob. Chemother., 46,* 733-739.

Derendorf, H., Lesko, L.J., Chaikin, P., Colburn, W.A., Lee, P., Miller, R., Powell, R., Rhodes, G., Stanski, D., and Venitz, J. (2000) Pharmacokinetic/Pharmacodynamic modeling in drug research and development. *J. Clin. Pharmacol., 40,*1399-1418.

Devineni, D., Klein-Szanto, A., Gallo, J.M. (1996) *In vivo* microdialysis to characterize drug transport in brain tumors: analysis of methotrexate uptake in rat glioma-2 (RG-2)-bearing rats. *Cancer Chemother. Pharmacol., 38*, 499-507.

Ederoth, P., Tunblad, K., Bouw, R., Lundberg, C.J., Ungerstedt, U., Nordstrom, C.H., and Hammarlund-Udenaes, M. (2004) Blood–brain barrier transport of morphine in patients with severe brain trauma. *Br. J. Clin. Pharmacol., 57,* 427-435.

Ekstrøm, P.O., Andersen, A., Saeter, G., Giercksky, K.E., and Slørdal, L. (1997) Continuous intratumoral microdialysis during high-dose methotrexate therapy in a patient with malignant fibrous histiocytoma of the femur: a case report. *Cancer Chemother. Pharmacol., 39,* 267–72.

Elmquist, W.F., and Sawchuk, R.J. (1997) Application of microdialysis in pharmacokinetic studies. *Pharm. Res., 14,* 267–287.

Evans, W.E., Relling, M.V., Rodman, J.H., Crom, W.R., Boyett, J.M., and Pui, C.H. (1998) Conventional compared with individualized chemotherapy for childhood acute lymphoblastic leukemia. *N. Engl. J. Med., 338,* 499-501.

Evans, B.K., Armstrong, S., Singer, G., Cook, R.D., and Burnstock, G. (1975). Intracranial injection of drugs: comparison of diffusion of 6-OHDA and guanethidine. *Pharmacol. Biochem. Behav., 3,* 205-217.

Evrard, P.A., Cumps, J., and Verbeeck, R.K. (1996) Concentration-dependent plasma protein binding of flurbiprofen in the rat: an *in vivo* microdialysis study. *Pharm. Res., 13,* 18-22.

Fischman, A.J., Babich, J.W., Bonab, A.A., Alpert, N.M., Vincent, J., Callahan, R.J., Correia, J.A., and Rubin R.H. (1998) Pharmacokinetics of [18F]trovafloxacin in healthy human subjects studied with positron emission tomography. *Antimicrob. Agents Chemother., 42,* 2048-2054.

Fromm, M.F. (2004) Importance of P-glycoprotein at blood–tissue barriers. *Trends Pharmacol. Sci., 25,* 423-429

Garvin, S., and Dabrosin, C. (2003) Tamoxifen inhibits secretion of vascular endothelial growth factor in breast cancer *in vivo*. *Cancer Res., 63*, 8742-8748.

Gattringer, R., Urbauer, E., Traunmuller, F., Zeitlinger, M., Dehghanyar, P., Zeleny, P., Graninger, W., Muller, M., and Joukhadar, C. (2004) Pharmacokinetics of telithromycin in plasma and soft tissues after single-dose administration to healthy volunteers. *Antimicrob. Agents Chemother., 48,* 4650-4653.

Gibbs, M.A., and Hosea, N.A. (2003) Factors Affecting the Clinical Development of Cytochrome P450 3A Substrates. *Clin. Pharmacokinet., 42,* 969-984.

Gogtay, N.J., Kshirsagar, N.A., and Dalvi, S.S. (2001) Therapeutic drug monitoring in a developing country: an overview. *Br. J. Clin. Pharmacol.,52,*103S-108S.

Groth, L., and Serup, J. (1998) Cutaneous Microdialysis in Man: Effects of Needle Insertion Trauma and Anaesthesia on Skin Perfusion, Erythema and Skin Thickness. *Acta Derm. Venereol. (Stockh), 78,* 5–9.

Potschka, H., and Löscher, W. (2002) A comparison of extracellular levels of phenytoin in amygdala and hippocampus of kindled and non-kindled rats. *Neuroreport, 13,* 167-170.

Hargraves, R., and Freed, W.J. (1987) Chronic intrastriatal dopamine infusions in rats with unilateral lesions of the sustancia nigra. *Life Sciences, 40,* 959-966.

Heikkinen, T., Laine, K., Neuvonen, P.J., and Ekblad, U. (2000) The transplacental transfer of the macrolide antibiotics erythromycin, roxithromycin and azithromycin. *Br. J. Obstet. Gynaecol., 107,* 770-775.

Hlatky, R., Furuya, Y., Valadka, A.B., Goodman, J.C., and Robertson, C.S. (2002) Comparison of microdialysate arginine and glutamate levels in severely head-injured patient. *Acta Neurochir. Suppl., 81,* 347-349.

Ho, R.H., and Kim, R.B. (2005) Transporters and drug therapy: implications for drug disposition and disease. *Clin. Pharmacol. Ther., 78,* 260–277.

Höcht, C., Opezzo, J.A.W., and Taira, C.A. (2003) Validation of a new intraarterial microdialysis shunt probe for the estimation of pharmacokinetic parameters. *J. Pharm. Biomed. Anal., 31,* 1105-13.

Höcht, C., Opezzo, J.A.W., and Taira, C.A. (2004a) Microdialysis in drug discovey. *Curr. Drug Discov. Technol., 1,* 269-285.

Höcht, C., Di Verniero, C., Opezzo, J.A.W., and Taira, CA. (2004b) Pharmacokinetic-pharmacodynamic properties of metoprolol in chronic aortic coarctated rats. *Naunyn-Schmiedeberg's Arch. Pharmacol., 370,* 1–8.

Höcht, C., Di Verniero, C., Opezzo, J.A.W., and Taira, CA. (2005) Applicability of microdialysis as a technique for pharmacokinetic–pharmacodynamic (PK–PD) modeling of antihypertensive beta-blockers. *J. Pharmacol. Toxicol. Methods, 52,* 244–250.

Höcht, C., Opezzo, J.A.W., Bramuglia, G.F., and Taira, C.A. (2006a) Application of microdialysis in clinical pharmacology. *Curr. Clin. Pharmacol., 1,* 163-184.

Höcht, C., Opezzo, J.A.W., Bramuglia, G.F., and Taira, CA. (2006b) Application of microdialysis for PK-PD modelling. *Expert Opin. Drug Discov., 1* ,289-301.

Höcht, C., Di Vereniero, C., Opezzo, J.A.W., Bramuglia, G.F., and Taira, C.A. (2006c) Pharmacokinetic-pharmacodynamic (PK-PD) modeling of cardiovascular effects of metoprolol in spontaneously hypertensive rats: a microdialysis study *Naunyn-Schmiedeberg's Arch. Pharmacol., 373,* 310-318.

Hocht, C., Opezzo, J.A., and Taira, C.A. (2007a) Applicability of reverse microdialysis in pharmacological and toxicological studies. *J. Pharmacol. Toxicol. Methods, 55,* 3-15.

Höcht, C., Lazarowski, A., Gonzalez, N.N., Auzmendi, J., Opezzo, J.A., Bramuglia, G.F., Taira, C.A., and Girardi, E. (2007b) Nimodipine restores the altered hippocampal phenytoin pharmacokinetics in a refractory epileptic model. *Neurosci. Let. 2007; 413,* 168-172.

Huff, J.K., Bresnahan, J.F., and Davies, M.I. (2003) Preliminary evaluation of several disinfection/sterilization techniques for use with microdialysis probes. *Life. Sci., 73,* 275-287.

Hutchinson, P.J., O'Connell, M.T., Nortje, J., Smith, P., Al-Rawi, P.G., Gupta, A.K., Menon, D.K., and Pickard, J.D. (2005) Cerebral microdialysis methodology evaluation of 20 kDa and 100 kDa catheters. *Physiol. Meas., 26,* 423–428.

Islinger, F., Bouw, R., Stahl, M., Lackner, E., Zeleny, P., Brunner, M., Muller, M., Eichler, H.G., and Joukhadar, C. (2004) Concentrations of Gemifloxacin at the Target Site in Healthy Volunteers after a Single Oral Dose. *Antimicrob. Agents. Chemoter., 48,* 4246-4249.

Jacobson, I., Sandberg, M., and Hamberg, A. (1985) Mass transfer in brain dialysis devices--a new method for the estimation of extracellular amino acids concentration. *J. Neurosci. Meth., 15,* 263-268.

Jain, E.K. (1994) The next frontier of molecular medicine: delivery of therapeutics. *Nat. Med., 4,* 655-657.

Johnson, R.D., and Justice, J.B. (1983) Model studies for brain dialysis. *Brain Res. Bull., 10,* 567-71.

Joukhadar, C., Derendorf, H., and Müller, M. (2001) Microdialysis: A novel tool for clinical studies of anti-infectives agents. *Eur. J. Clin. Pharmacol., 57,* 211-19.

Kendrick, K.M. (1989) Use of microdialysis in neuroendocrinology. *Methods. Enzymol., 168,* 182-205.

Kiewert, C., Hartmann, J., Stoll, J., Thekkumkara, T.J., Van der Schyf, C.J., and Klein, J. (2006) NGP1-01 is a brain-permeable dual blocker of neuronal voltage- and ligand-operated calcium channels. *Neurochem. Res., 31,* 395-399.

Kolehmainen, M., Ohisalo, J.J., Kaartinen, J.M., Tuononen, V., Paakkonen, M., Poikolainen, E., Alhava, E., and Usitupa, M.I. (2000). Concordance of *in vivo* microdialysis and *in vitro* techniques in the studies of adipose tissue metabolism. *Int. J. Obes. Relat. Metab. Disord., 24,* 1426-1432.

Langer, O., and Müller, M. (2004) Methods to Assess Tissue-Specific Distribution and Metabolism of Drugs. *Curr. Drug. Metabolism, 5,* 463-481.

Langer, O., Karch, R., Muller, U., Dobrozemsky, G., Abrahim, A., Zeitlinger, M., Lackner, E., Joukhadar, C., Dudczak, R., Kletter, K., Muller, M., and Brunner, M. (2005) Combined PET and microdialysis for *in vivo* assessment of intracellular drug pharmacokinetics in humans. *J. Nucl. Med., 46,* 1835-1841.

Lazarowski, A., Czornyj, L., Lubieniecki, L., Vazquez, S., D'Giano, C., Sevlever, G., Taratuto, A.L., Brusco, A., and Girardi, E. (2006) Multidrug-Resistance (MDR) Proteins Develops Refractory Epilepsy Phenotype: Clinical and Experimental Evidences, *Curr. Drug Ther, 1,* 291-309.

Lefeuvre, S., Marchand, S., Lamarche, I., Mimoz, O., and Couet, W. (2006) Microdialysis study of imipenem distribution in the intraperitoneal fluid of rats with or without experimental peritonitis. *Antimicrob. Agents Chemother., 50,* 34-37.

Levy, G. (1998) Predicting Effective Drug Concentrations for Individual Patients. Determinants of Pharmacodynamic Variability. *Clin. Pharmacokinet., 34,* 323-333.

Lin, J.H. (2006) Tissue Distribution and Pharmacodynamics: A Complicated Relationship. *Curr. Drug Metab., 7,*39-65.

Lindberger, M., Tomson, T., Wallstedt, L., and Stahle L. (2001) Distribution of valproate to subdural cerebrospinal fluid, subcutaneous extracellular fluid, and plasma in humans: a microdialysis study. *Epilepsia, 42,* 256-61.

Liu, P., Müller, M., Grant, M., Webb, A., Obermann, B., and Derendorf, H. (2000) Interstitial tissue concentrations of cefpodoxime. *J. Antimicrob. Chemoter., 50,* 19-22.

Liu, P., and Derendorf, H. (2003) Antimicrobial tissue concentrations. *Infect. Dis. Clin. N. Am.,17,* 599-613.

Liu, P., Rand, K.H., Obermann, B., and Derendorf, H. (2005) Pharmacokinetic–pharmacodynamic modelling of antibacterial activity of cefpodoxime and cefixime in *in vitro* kinetic models. *Int. J. Antimicrob. Agents, 25,* 120-129.

Lönnroth, P., Jansson, P.A., and Smith, U. (1987) A microdialysis method allowing characterization of intercellular water space in humans. *Am. J. Physiol., 253,* E228-E231.

Löscher, W. (2002) Current status and future directions in the pharmacotherapy of epilepsy. *Trends Pharmacol. Sci, 23,* 113-118.

Löscher, W., and Potschka, H. (2005a) Drug resistance in brain disease and the role of drug efflux transporters. *Nature Rev. Neurosci., 6,* 591-602.

Löscher, W., and Potschka, H. (2005b) Role of drug efflux transporters in the brain for drug disposition and treatment of brain diseases. *Prog. Neurobiol., 76,* 22–76.

Lundstedt, C., Stridbeck, H., Andersson, R., Tranberg, K.G., and Andren-Sandberg, A. (1991) Tumor seeding occurring after fine-needle biopsy of abdominal malignancies. *Acta Radiol., 32,* 518-20.

Mackie, C.E., English, H.E., Lelievre, E., Gordon, B.H., Genissel, P., and Robinson, B.V.J. (1997) Radioimmunoassay for the measurement of S9788 in serum and microdialysis samples. *J. Pharm. Biomed. Anal., 15,* 917-928.

Maidment, N.T., Brumbaugh, D.R., Rudolph, V.D., Erdelyi, E., and Evans, CJ. (1989) Microdialysis of extracellular endogenous opioid peptides from rat brain *in vivo*. *Neuroscience, 33,* 549-557.

Majumdar, S., Kansara, V., and Mitra, A.K. (2006) Vitreal pharmacokinetics of dipeptide monoester prodrugs of ganciclovir. *J. Ocul. Pharmacol. Ther., 22,* 231-241.

Mandema, J.W., Tuck, B., van Stevenink, A.L., Breimer, D.D., Cohen, A.F., and Danhof, M. (1992) Pharmacokinetic–pharmacodynamic modeling of the central nervous system effects of midazolam and its main metabolite α-hydroxy-midazolam in healthy volunteers. *Clin. Pharmacol. Ther., 51,* 715-728.

Mantovani, V., Kennergren, C., Berglin, E. Moratti, R., Lonnroth, P., Hamberger, A., and Vigano, M. (2002) Intramyocardial Troponin-T Monitoring with Microdialysis in Coronary Artery Bypass Surgery. *Scand. Cardiovasc. J., 36,* 308–12.

Marchand, S., Dahyot, C., Lamarche, I., Mimoz, O., and Couet, W. (2005) Microdialysis study of imipenem distribution in skeletal muscle and lung extracellular fluids of noninfected rats. *Antimicrob. Agents Chemother., 49,* 2356-2361.

Marsala, M., Malmberg, A., and Yaksh, T. (1995) The spinal loop dialysis catheter: characterization of use in the unanesthetized rat. *J. Neurosci. Methods, 62,* 43-53.

Meibohm, B., and Derendorf, H. (2002) Pharmacokinetic/pharmacodynamic studies in drug development. *J. Pharm. Sci., 91,* 18-31.

Meredith, PA. (1997) Clinical relevance of optimal pharmacokinetics in the treatment of hypertension. *J. Hypertens., 15,* S27–S31.

Michelsen, P., and Pettersson, G. (1994) An automated liquid chromatography/mass spectrometry system coupled on-line with microdialysis for the *in vivo* analysis of contrast agents. *Rapid Commun. Mass Spectrom., 8,* 517-520.

Miller, R., Ewy, W., Corrigan, B.W. Ouellet, D., Hermann, D., Kowalski, K.G., Lockwood, P., Koup, J.R., Donevan, S., El-Kattan, A., Li, C.S., Werth, J.L., Feltner, D.E., and Lalonde, R.L. (2005) How modeling and simulation have enhanced decision making in new drug development. *J. Pharmacokinet. Pharmacodyn., 32,* 185-197.

Müller, M., Brunner, M., Schmid R., Mader, R.M., Bockenheimer, J., Steger, G.G., Steiner, B., Eichler, H.G., and Blochl-Daum, B. (1998) Interstitial methotrexate kinetics in primary breast cancer lesions. *Cancer Res., 58,* 2982-2985.

Müller, M., Mader, R.M., Steiner, B., Steger, G.G., Jansen, B., Gnant, M., Helbich, T., Jakesz, R., Eichler, H.G., and Blochl-Daum, B. (1997) 5-fluorouracil kinetics in the interstitial tumor space: clinical response in breast cancer patients. *Cancer Res., 57,* 2598-25601.

Müller, M., Bockenheimer, J., Zellenger, U., Klein, N., Steger, G.G., Eichler, H.G., and Mader, R.M. (2000) Relationship between *in vivo* drug exposure of the tumor interstitium and inhibition of tumor cell growth *in vitro*: a study in breast cancer patients. *Breast Cancer Res. Treat., 60,* 211-217.

Nordbring, F. (1978) Tissue penetration of antibiotics. Introduction. Focus on some problems involved in the treatment of infectious disease. *Scand. J. Infect. Dis. Suppl., 14,* 21-22.

Nutt, R. (2002) 1999 ICP Distinguished Scientist Award. The history of positron emission tomography. *Mol. Imaging Biol., 4,* 11-26.

Palsmeier, R.K., and Lunte, C.E. (1994) Microdialysis sampling in tumor and muscle: study of the disposition of 3-amino-1,2,4-benzotriazine-1,4-di-Noxide (SR 4233). *Life Sci., 55,* 815-825.

Perez-Urizar, J., Granados-Soto, V., Flores-Murrieta, F.J., and Castañeda-Hernandez, G. (2000) Pharmacokinetic-pharmacodynamic modeling: why? *Arch. Med. Res., 31,* 539-545.

Plock, N., and Kloft, C. (2005) Microdialysis—theoretical background and recent implementation in applied life-sciences. *Eur. J. Pharm. Sci., 25,* 1-24.

Prentis, R.A., Lis, Y., and Walker, S.R. (1988) Pharmaceutical innovation by the seven UK-owned pharmaceutical companies (1964-1985). *Br. J. Clin. Pharmacol., 25,* 387-396.

Presant, C.A., Wolff, W., Walush, V., Wiseman, C., Kennedy, P., Blayney, D., and Brechner, R.R. (1994) Association of intratumoral pharmacokinetics of fluorouracil with clinical response. *Lancet, 343,* 1184-1187.

Prokai, L., Kim, H.S., Zharikova, A., Roboz, J., Ma, L., Deng, L., and Simonsick Jr., W.K. (1998) Electrospray ionization mass spectrometric and liquid chromatographic-mass spectrometric studies on the metabolism of synthetic dynorphin A peptides in brain tissue *in vitro* and *in vivo*. *J. Chromatogr. A, 800,* 59-68.

Raje, S., Cornish, J., Newman, A.H., Cao, J., Katz, J., and Eddington, N.D. (2005) Pharmacodynamic assessment of the benztropine analogues AHN-2055 and AHN-2005 using intracerebral microdialysis to evaluate brain dopamine levels and pharmacokinetic/pharmacodynamic modeling. *Pharm. Res., 22,* 603-612.

Regesta, G., and Tanganelli, P. (1999) Clinical aspects and biological bases of drug-resistant epilepsies. *Epilepsy Res., 34,* 109-122.

Remy, S., and Beck, H. (2006) Molecular and cellular mechanism of pharmacoresistance in epilepsy. *Brain, 129,* 18-35.

Rice, M.E., Gerhardt, G.A., Hierl, P.M., Nagy, G., and Adams, R.N. (1985). Diffusion coefficients of neurotransmitters and their metabolites in brain extracellular fluid space. *Neuroscience, 15,* 891-902.

Rizzi, M., Caccia, S., Guiso, G., Richichi, C., Porter, J.A., Aronica, E., Aliprandi, M., Bagnati, R., Fanelli, R., D'Incalci, M., Maman, R., and Vezzani, A. (2002) Limbic seizures induce P-glycoprotein in rodent Varin: functional implications for pharmacoresistance. *J. Neurosci., 22,* 5833-5839.

Rocha, L.L., Evans, C.J., and Maidment, NT. (1997) Amygdala kindling modifies extracellular opioid peptide content in rat hippocampus measured by microdialysis. *J. Neurochem., 68,* 616-624.

Rosdahl, H., Ungerstedt, U., and Henriksson, J. (1997) Microdialysis in human skeletal muscle and adipose tissue at flow rates is possible if dextran-70 is added to prevent loss of perfusion fluid. *Acta Physiol. Scand., 159,* 261–2.

Sandberg, M., and Lindström, S. (1983) Amino acids in the dorsal lateral geniculate nucleus of the cat-collection *in vivo*. *J. Neurosci. Methods, 9,* 65-74.

Scheyer, R.D., During, M.J., Hochholzer, J.M., Spencer, D.D., Cramer, J.A., and Mattson, R.H. (1994a) Phenytoin concentrations in the human brain: an *in vivo* microdialysis study. *Epilepsy Res., 18,* 227-32.

Scheyer, R.D., During, M.J., Spencer, D.D., Cramer, J.A., and Mattson, R.H. (1994b) Measurement of carbamazepine and carbamazepine epoxide in the human brain using *in vivo* microdialysis. *Neurology, 44,* 1469-72.

Schoemaker, R.C., van Gerven, J.M.A., and Cohen, A.F. (1998) Estimating potency for the Emax-model without attaining maximal effects. *J. Pharmacokinet. Biopharm., 26,* 581-593.

Schuck, E.L., and Derendorf, H. (2005) Pharmacokinetic/pharmacodynamic evaluation of anti-infective agents. *Expert Rev. Antiinfect. Ther., 3,* 361-373.

Schumacher, G.E., and Barr, JT. (2002) Therapeutic drug monitoring: is it cost-effective? *Expert Rev. Pharmacoeconomics Outcomes Res., 2,* 89-94.

Sheiner, L.B., Stanski, D.R., Vozeh, S., Miller, R.D., and Ham, J. (1979) Simultaneous modeling of pharmacokinetics and pharmacodynamics: application to D-tubocurarine. *Clin. Pharmacol. Ther., 25,* 358-371.

Sjögren, F., Svensson, C., and Anderson, C. (2002) Technical prerequisites for *in vivo* microdialysis determination of interleukin-6 in human dermis. *Br. J. Dermatol., 146,* 375–82.

Ståhle, L. (1992) Pharmacokinetic estimations from microdialysis data. *Eur. J. Clin. Pharmacol , 43*, 289-294.

Ståhle, L., Segersvärd, S., and Ungerstedt, U. (1990) A comparison between three methods for estimation of extracellular concentrations of exogenous and endogenous compounds by microdialysis. *J. Pharmacol. Meth., 25*, 41-52.

Steele, K.M., and Lunte, C.E. (1995) Microdialysis sampling coupled to on-line microbore liquid chromatography for pharmacokinetic studies. *J. Pharm. Biomed. Anal.*, *13*, 149.

Thompson, J.F., Siebert, G.A., Anissimov, Y.G., Smithers, B.M., Doubrovsky, A., Anderson, C.D., and Roberts, M.S. (2001) Microdialysis and response during regional chemotherapy by isolated limb infusion of melphalan for limb malignancies. *Br. J. Cancer, 85,* 157-165.

Tomaselli, F., Maier, A., Matzi, V., Smolle-Jüttner, F.M., and Dittrich, P. (2004) Penetration of Meropenem into Pneumonic Human Lung Tissue as Measured by *In vivo* Microdialysis. *Antimicrob. Agents Chemoter., 48,* 2228-2232.

Torto, N., Mikeladze, E., Gorton, L., Csoregi, E., and Laurell, T. (1999) Maximising microdialysis sampling by optimising the internal probe geometry. *Anal. Commun., 36,* 171–174.

Toutain, P.L. (2002) Pharmacokinetic/pharmacodynamic integration in drug development and dosage–regimen optimization for veterinary medicine. *AAPS J., 4,* 1-29.

Treiman, D.M., and Woodbury, D.M. (1995) Phenytoin: absorption, distribution, and excretion, In: R.H. Levy, R.H. Mattson, Meldrum B.S. (Eds.), Antiepileptic Drugs, Raven Press, New York, 4th edn., pp 301-314.

Ungerstedt, U. (1991) Introduction to intracerebral microdialysis. In: Robinson, T.E., Justice Jr., J.B. (Eds.), *Microdialysis in the Neurosciences.* (pp. 3-22). Amsterdam: Elsevier.

Vogt, G.L., Janle, E., Ash, S.R., Venema, K., and Korf, J. (2003) Utilization of *in vivo* ultrafiltration in biomedical research and clinical applications. *Life Sci., 73,* 2005-2018.

Wang, Y., Wong, S.L., and Sawchuk, R.J. (1991) Comparison of *in vitro* and *in vivo* calibration of microdialysis probes using retrodialysis *Curr. Sep., 10,* 87.

Weikop, P., Egestad, B., and Kehr, J. (2004) Application of triple-probe microdialysis for fast pharmacokinetic/pharmacodynamic evaluation of dopamimetic activity of drug candidates in rat brain. *J. Neurosci. Meth., 140,* 59-65.

Wise, R. (1986) The clinical relevance of protein binding and tissue concentrations in antimicrobial therapy. *Clin. Pharmacokinet., 11*, 470–82.

Wisniewski, N., and Torto, N. (2002) Optimisation of microdialysis sampling recovery by varying inner cannula geometry. *Analyst, 127,* 1129–1134.

Zeitlinger, M.A., Traunmuller, F., Abrahim, A., Muller, M.R., Erdogan, Z., Muller, M., and Joukhadar, C. (2007) A pilot study testing whether concentrations of levofloxacin in interstitial space fluid of soft tissues may serve as a surrogate for predicting its pharmacokinetics in lung. *Int. J. Antimicrob. Agents., 29,* 44-50.

Zeitlinger, M.A., Marsik, C., Georgopoulos, A., Müller, M., Heinz, G., and Joukhadar, C. (2003) Target site bacterial killing of cefpirome and fosfomycin in critically ill patients. *Int. J. Antimicrob. Agents, 21,* 562-567.

Zhouh, Q., and Gallo, J.M. (2005) *In vivo* microdialysis for PK and PD studies of anticancer drugs. *AAPS J., 7,* E659-E667.

Zhuang, Y., Fraga, C.H., Hubbard, K.E., Hagedorn, N., Panetta, J.C., Waters, C.M., and Stewart, C.F. (2006) Topotecan central nervous system penetration is altered by a tyrosine kinase inhibitor. *Cancer Res., 66,* 11305-11313.

In: Drug Design Research Perspectives
Editor: Stanley P. Kaplan, pp. 171-211

ISBN: 978-1-60021-823-1

Chapter VII

Magnetic Resonance Imaging in Drug Development for Ischemic Stroke: A Translational Approach

Aysan Durukan and Turgut Tatlisumak*
Department of Neurology, Helsinki University
Central Hospital, Helsinki, Finland

1. Abstract

Stroke is the third most common cause of death worldwide and the most common cause of adult disability in industrialized countries. Stroke alone absorbs 6% of all heath costs. To date the only approved therapy for acute ischemic stroke is thrombolysis within 3 hours of stoke onset in well-selected cases with only 5% of all ischemic stroke patients currently receiving thrombolytic therapy. Therefore, new therapies must be developed urgently. One major obstacle in the development of effective therapies for ischemic stroke has been the lack of versatile imaging techniques. Magnetic resonance imaging (MRI) in stroke permitted major pathophysiological insights into the understanding of acute cerebral ischemia, but now is also considered of major importance in patient management and can serve as a surrogate end point. Because of its noninvasive nature, versatility, patient compatibility with few contraindications and its potential for imaging with relatively high spatial and temporal resolution, MRI serves as an indispensible tool in stroke evaluation for two major purposes: to ensure a definite diagnosis and to follow-up the ischemic lesion size over time. Besides, its availability both in clinical and preclinical settings allows direct bridging of findings in humans and animals. More than a decade, new MRI modalities, especially diffusion- and perfusion-weighted MRI (DWI and PWI, respectively) have been used in experimental studies with great success. DWI first has been performed and developed in animals and rapidly applied to clinical practice. Multimodal stroke MRI provides numerous advantages: DWI enables detecting

* Correspondence to: Dr. Turgut Tatlisumak, Department of Neurology, Helsinki University Central Hospital, Haartmaninkatu 4, 00290 Helsinki, FINLAND; Tel: + 358 9 47171762, + 358 50 4270309, Fax: + 358 9 47174089, E-mail: turgut.tatlisumak@hus.fi

very early ischemic lesion(s), allows for observation of lesion evolution with repeated measurements and gives clues for etiology; PWI shows where and to what degree cerebral blood flow is compromised; PWI and DWI can together approximate penumbra (the salvageable brain tissue); MR angiography discloses the occluded artery and other vascular patholcgies; T2*-weighted images helps to exclude new and old hemorrhagic lesions; and conventional images show the final infarct volume. Thus, MRI improves patient triage to clinical therapy trials. We propose that implementation of a similar MRI protocol we used in animal models to human therapy trials will deliver quick answers to whether a novel therapy is effective or not, in fairly small and homogenous patient groups unnecessitating large, expensive, and time-consuming old-fashioned clinical trials.

Keywords: Cerebral ischemia, drug development, experimental stroke, magnetic resonance imaging, diffusion, perfusion, neuroprotection, thrombolysis, clinical trials.

2. Abbreviations

ADC	Apparent diffusion coefficient
ASL	Arterial spin labeling
BBB	Blood-brain barrier
CBF	Cerebral blood flow
CBV	Cerebral blood volume
CT	Computed tomography
DIAS	The Desmoteplase in Acute Ischemic Stroke Trial
DSC	Dynamic susceptibility contrast
DWI	Diffusion-weighted magnetic resonance imaging
DEDAS	Dose Escalation of Desmoteplase for Acute Ischemic Stroke Trial
DEFUSE	The Diffusion and Perfusion Imaging Evaluation for Understanding Stroke Evaluation Trial
EPITHET	The Echoplanar Imaging Thrombolysis Evaluation Trial
GAIN	Glycine Antagonist in Neuroprotection
Gd-DTPA	Gadolinium diethyltriaminepentacetic acid
ICH	Intracerebral hemorrhage
MRA	Magnetic resonance angiography
MRI	Magnetic resonance imaging
MTT	Mean transit time
PET	Positron emission tomography
PWI	Perfusion-weighted magnetic resonance imaging
rt-PA	Recombinant tissue plasminogen activator

3. Introduction

Stroke is the second most common cause of death globally (5.7 million deaths per year) and the most common cause of adult disability in industrialized countries [1]. From the early 1970s to early 1990s, the estimated number of noninstitutionalized stroke survivors increased for 1.6 fold [2] and according to the WHO estimates the number of stroke events in European countries is likely to increase from 1.1 million per year in 2000 to more than 1.5 million per year in 2025 [3]. In addition, because of its huge socioeconomic burden absorbing 6% of all health care budgets, stroke is already, and will continue to be, the most challenging disease. Majority (approximately 80%) of all strokes are caused by brain ischemia (infarction).

Therapeutic approaches for acute ischemic stroke fall into two categories: recanalization and neuroprotection. Although it has been a revolutionary step in stroke therapy, to date, intravenous thrombolysis with recombinant-tissue plasminogen activator (rt-PA) is still the only approved therapy of acute stroke within 3 hours of symptom onset. Discovery of novel therapies, expanding time window of thrombolysis and translation of neuroprotectants from laboratory to clinics are major investigation fields of contemporary stroke research. Drug development studies inevitably use animal stroke models and a novel compound is tested in humans after have been found safe and effective first in animals with a multistep research programme [4]. New therapies are urgently needed and to our opinion magnetic resonance imaging (MRI) should be part of drug development process as MRI offers an in vivo versatile approach, which could directly be transferred to human trials.

Ischemic stroke is a highly variable clinical state. Many variables, mainly in terms of the causes, duration, localization, and severity of ischemia and co-existing systemic diseases, can all contribute to progress of the infarction and patient outcome. Because of these uncontrollable aspects in clinical research, very large patient group sizes are required to enable appropriate statistical evaluation. Consequently, clinical stroke investigations in the human are difficult, expensive, and time consuming. In contrast with human studies, experimental animal models of brain ischemia can be physiologically controlled and may eliminate many of the variables. Therefore, relevant questions can be easily and quickly answered in experimental models. For decades, neuroscientists have made significant efforts to develop animal models that mimic closely human ischemic stroke. To date, the most common model used is intraluminal middle cerebral artery occlusion in rodents. Embolic models offer researchers a laboratory tool when the aim is to evaluate thrombolytics.

With the use of MRI, animal stroke models enable to examine very early physiologic and biochemical ischemic changes in a temporal manner. Because of its noninvasive nature, versatility, patient compatibility with few contraindications and its potential for imaging with relatively high spatial and temporal resolution, MRI serves as an indispensable tool in stroke research for two major purposes: evaluation of ischemic lesion and assessment of drug efficacy. Functional diagnostic versatility of newer MRI modalities, namely diffusion-weighted imaging (DWI) and perfusion-weighted imaging (PWI), opened a new era first in the acute stroke evaluation, but also in patient management. Detecting an ischemic lesion is just the very first interest of today's neurologist, the following aim is to detect penumbra and predict its fate. Imaging of penumbra, in other words differentiating infarct core from surrounding potentially salvageable tissue-at-risk with MRI is already in clinical use.

Penumbra once detected approximately with DWI and PWI, is the target of therapeutic approaches (figure 1,2). Time seems to be a reliable marker for the presence of penumbra in the first 90 minutes, but afterwards heterogeneity of stroke pose again a problem [5] which affects tissue viability, thus penumbra's and patient's fate. Several clinical trials such as ongoing DIAS-2, EPITHET, and some other terminated trials [6, 7] already included MRI-based criteria for patient selection with the hope to prolong 3 hours limitation for the use of thrombolysis and to evaluate whether the presence of a PWI-DWI mismatch influences patient outcome. Some measurements obtained by PWI may be used to define thresholds that discriminate between normal and ischemic tissue, and penumbra and infarct core within ischemic tissue. However, these thresholds are not systematically studied in humans (Bandera 2006) and there is need for further research in perfusion imaging to refine threshold measurements and to generate automated measurements of reversible and irreversible ischemia volumes [5]. Obtaining information about the site of arterial occlusion with magnetic resonance angiography (MRA) might help in taking therapy decision between mechanical or pharmacological recanalization therapies (ongoing Multi-MERCI trial). In acute stroke MR imaging, exclusion of brain hemorrhage and detection of old bleeds is now possible with T2*-weighted images. Detecting patients with higher risk of hemorrhagic transformation and prediction of hemorrhagic transformation are other important topics to address in stroke research. Contrast-enhanced MRI seems to be a promising tool in this field.

In preclinical drug efficacy experiments, the lesion volume is usually used as an important index [4], which traditionally quantified by postmortem histological staining techniques accepted as the gold standard [8] at a first and last single time point. MRI enables not only in vivo initial and final detection of lesion volumes, overcoming the potential pretreatment bias, but also to monitorize lesion growth and microvascular perfusion status in a real time manner, giving clues about mechanism of action of the drug used. Since MRI is a noninvasive and completely safe tool, it allows repetitive measurements in the same animal that eliminates interanimal variability. Thus in pharmaceutical research, the number of animals required to reach sufficient statistical power, is much smaller when MRI is used. Thrombolysis with rt-PA was first found to be effective in experimental models [9] and then successfully applied to humans [10]. Neuroprotective therapy approach has been found effective in animal models with numerous drugs, as reviewed elsewhere [11], but failed in clinical trials [12]. One of the leading reasons for the failure of the data translation from laboratory to the clinic has been the lack of versatile imaging techniques; other possible reasons will be discussed briefly. Thus, our main purpose is to point out the role of multimodal MRI in antiischemic drug research and its bridging function between animal and human drug efficacy studies. After a brief pathophysioplogical overview of stroke, we will look at the basic principles of major MRI modalities available both in experimental stroke and clinics. The utility of multimodal MRI approach in drug developmental studies and clinical trials (figure 1) for data translation from laboratory to clinics will be the focuses of the second part of this chapter.

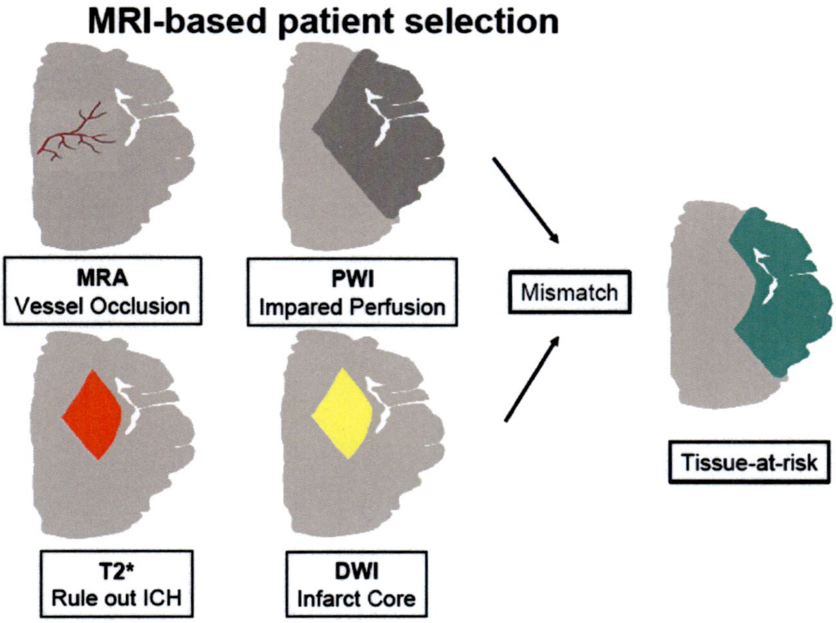

Figure 1. MRI-based patient enrolment into clinical trials. Baseline multimodal imaging may detect vessel occlusion site (by MR angiography,MRA), rule out intracranial hemorrhage (ICH, by T2*-weighted image), show hypoperfusion (by PWI), ischemic lesion (by DWI), and mismatch (PWI-DWI). A substantion volume of mismatch (tissue-at-risk), is the main target of the therapy.

4. Pathophysiology of Brain Ischemia and the Concept of Penumbra

Survival of brain cells is dependent on adequate brain perfusion. Within minutes of vascular occlusion, the cessation of blood flow to a region in the brain leads to a rapid decrease of oxygen and glucose within the tissue, triggering a complex sequence of pathophysiological spatial and temporal events (ischemic cascade). Leading pathogenic mechanisms of ischemic cascade include energy failure, elevation of intracellular Ca^{2+} level, excitotoxicity, waves of spreading depression, generation of free radicals, blood-brain barrier (BBB) disruption, inflammation, and apoptosis.

The key pathophysiological concept is the distinction of hypoperfused tissue into tree operational compartments (figure 2): 1) tissue that will inevitably die (core); 2) tissue that may either die or survive (the penumbra); and 3) tissue that will in principle survive (oligemia) [13]. The cerebral blood flow (CBF) in the core region is severely depressed (0 to 20% of normal) and gradually improves through the periphery of the lesion. The CBF thresholds for the energy failure was reported to be around 18-20 mL / 100 g per minute in different ischemia models and species. This threshold is time dependent and increases approximately 40 mL / 100 g per minute over time. Reduction of CBF to less than 20 mL / 100 g per minute results in impaired neural function but preserved tissue integrity. This region with a silent function but intact structure has been termed by Astrup et al. as penumbra, a word etymologically originating from Latin (paene, almost + umbra, shadow), with an analogy to the partly illuminated area around the complete shadow of the moon in full eclipse [14]. Unless early reperfusion occurs, the penumbra will gradually undergo irreversible changes over time [15]. Thus, penumbra represents the tissue-at-risk, which is the optimal target for stroke therapy, tissue destined for infarction if untreated but potentially salvageable if effectively treated [16].

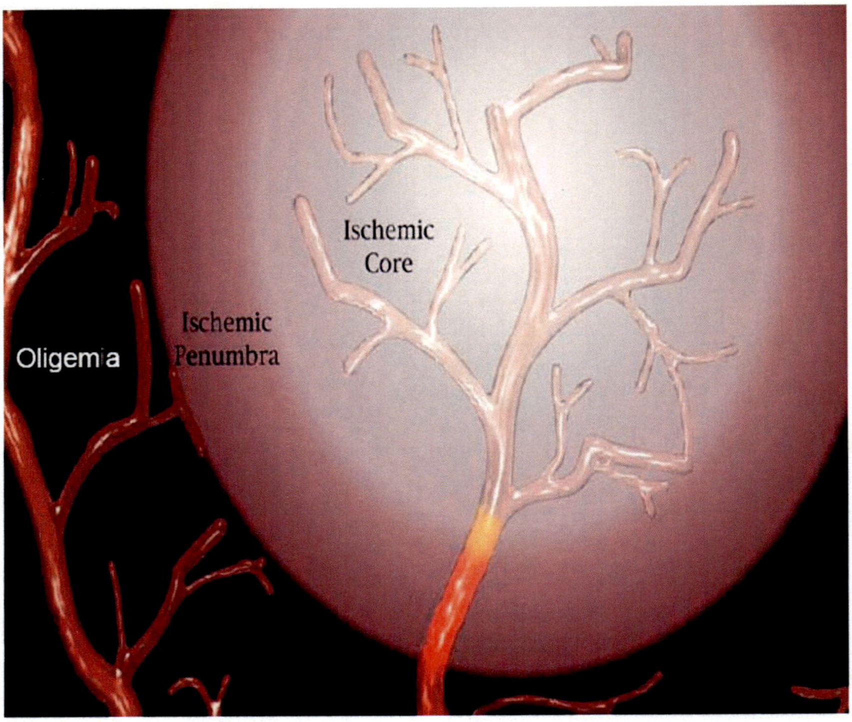

Figure 2. Schema of tree operational compartments in hypoperfused brain region. Ischemic core represents the tissue destined to die, ischemic penumbra may either die or survive, and oligemia will in principle survive.

Therapies in the acute phase may salvage a large portion of the penumbral tissue and may quickly result in rapid relief of symptoms with less severe disability, shorter hospital stay, and lesser mortality in stroke patients. The perfusion-diffusion mismatch in MRI is widely accepted to approximate the penumbra. The MRI characteristics of penumbra have been well studied in both permanent and transient focal cerebral ischemia models [17, 18], with different methods to induce ischemia [19] and in different rat strains [20]. In experimental studies penumbra exists for 120-180 minutes or as long as 22 hours [21]. Substantial penumbra is present in up to 90% of patients within 6 h of symptom onset; this falls about 50% within 9 h but still about 30% 18 h after symptom onset [22]. It is estimated that up to 6 h after stroke onset greater than 50% of stroke patients will have at least 20% volume mismatch between PWI and DWI [23]. Although two critical thresholds are widely accepted (20 and 10 mL / 100 g per minute; first to define penumbra where CBF is 20-10 mL / 100 g per minute and second is to define core where CBF is <10 mL / 100 g per minute)

[24], a systematic review of perfusion studies in stroke patients to optimize CBF thresholds for differentiating between penumbra and infarct core showed insufficiency of present knowledge [25]. Recently operational criteria for the definition of the penumbra, presumably for a clinically more relevant definition, have been proposed by Muir et al. [13]. According to the authors penumbral tissue characteristics are: 1) hypoperfusion <20 ml/100 g per minute, 2) abnormal neuronal function documented by a correlation with neurological deficit, 3) physiological and/or biochemical characteristics consistent with cellular dysfunction but not death, 4) uncertain fate, and 5) salvage of tissue is correlated with better clinical recovery.

Many laboratory-based and clinical functional studies, using different methods, including positron emission tomography (PET), single photon emission computed tomography, and xenon-computed tomography analyzed different aspects of penumbra. However, PET seems superior to other modalities in penumbra imaging [26], applying PET routinely in clinical environments, especially in the case of ischemic stroke within the few hours of symptom onset, is nearly impossible. Advantages and disadvantages of available perfusion and penumbra imaging methods have been discussed recently [13, 21, 24]. Multimodal MRI including DWI and PWI is the most convenient modality in stroke imaging [27], that provides information on the morphological and physiological status of the ischemic brain tissues.

5. Basic Principles of DWI and PWI and the Concept of Mismatch

With the rapid technological development since the late 80s, MRI has offered researchers the benefits of multipurpose/versatile imaging. Echo planar imaging, which uses strong, rapidly switching magnetic field gradients, further, enabled the researchers to collect imaging data in seconds only, virtually without motion artifacts [28]. Echo planar DWI provides information that is not available on conventional MRI sequences. DWI has revolutionized stroke research since this technique is extremely sensitive to the hyperacute phase of brain ischemia. Patients with acute stroke, even those with depressed consciousness or restlessness, could be safely imaged with MRI scanners capable of performing echo-planar imaging, thus DWI and PWI, with only few and rare contraindications [29].

Diffusion is referred to as random translational motion of water molecules due to thermal energy, so-called Brownian motion. In biological tissues evaluated with DWI, the apparent diffusion coefficient (ADC) of water is used to obtain tissue contrast. In the presence of a magnetic field gradient, moving protons diphase in their transverse plane, resulting in signal attenuation as it happens in conventional MRI sequences. DWI is based on the additional signal intensity loss of phase coherence accumulated by randomly moving spins, stimulated by adding a pair of strong and short diffusion-sensitizing gradients to a conventional spin-echo sequence. Because DWI measures all incoherent motions inside a voxel and motion in tissue is influenced by several components, such as membranes, fibers, and macromolecules [30], in vivo DWI does not actually measure the intrinsic self-diffusion of water molecules. Pathophysiologically, the ADC decline (by about 30 to 50%) starting within few minutes following onset of ischemia, occurs at the same time as anoxic depolarization. Energy failure which stems from blood flow cessation to a region of the brain, leads to perturbation of the ion pumps [31]. Subsequent ion dyshomeostasis (elevation of intracellular Na^+, Ca^{2+}, and Cl^- with simultaneous elevation of extracellular K^+ levels) leads to cytotoxic edema. Despite the evident association of tissue water ADC reductions with the development of cytotoxic edema, there seems to be other factors to contribute to the tissue water ADC decrease: increased intracellular viscosity due to fragmentation of microtubules and other intracellular components, increased tortuosity of the intracellular space, decreased cytoplasmic mobility, shrinkage of extracellular space, and increase in extracellular tortuosity [32, 33].

In DWI, tissue with faster diffusion is darker because of more signal loss, while tissue with slower diffusion is brighter because of lesser signal loss. At the early stages of ischemia, regions with decreased diffusion appear hyperintense on DWI and hypointense on ADC images. Presumably due to the cell lysis of necrotic neurons and astrocytes leading to an increase in extracellular space, the ADC begins to rise. The ADC returns to baseline or pseudonormalizes at approximately 2 days in animals and is elevated afterwards. In humans, the ADC pseudonormalizes at 1 to 4 weeks, while DWI images usually show mild hyperintensity due to T2 component [32, 34]. Based on a formula, the ADC values of water can be calculated on a pixel-by-pixel basis and ADC image can then be reconstructed. Typical ADC values in human brain are age-dependent [34] and differ between grey and white matter [35]. Figure 3 illustrates serial postreperfusional DWI images and ADC maps obtained from a single rat brain underwent 90 minutes of transient ischemia with the intraluminal suture middle cerebral artery occlusion technique.

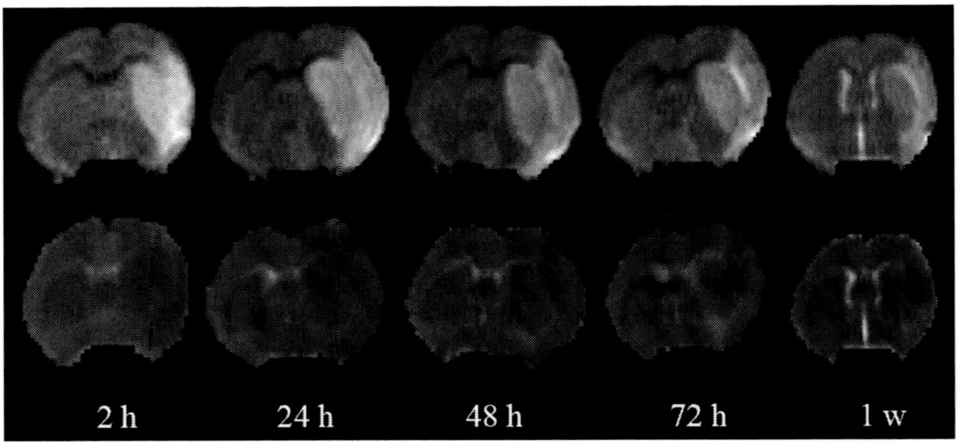

Figure 3. DWI images (except T2-weighted image at 1 week, upper row) and ADC maps (lower row) obtained in a rat after reperfusion, underwent 90 minutes of transient ischemia with the intraluminal suture MCAO technique. Note that ischemic lesion is hyperintense in every time points, ADC shows hypointensity in the acute period, then pseudonormalizes at 48 hours, afterwards shows heterogeous hypointense-hyperintense areas.

PWI is a semiquantitative method of evaluating brain perfusion, which provides imaging and measuring blood flow at the capillary level. PWI may serve to differentiate and characterize the pathological tissues by their capillary vasculature, follow treatment effects by monitoring perfusion changes (figure 4), and allow brain functional mapping associated with functional activity. Three important parameters are cerebral blood volume (CBV), CBF, and mean transit time (MTT). CBV is the total volume of blood in a given region of brain, usually expressed as mL/100 g of brain tissue. Average CBV values of brain are reported to be 4-5 mL/100 g of brain tissue, gray matter having much higher values than white matter. CBF is the amount of blood that moves through a given volume of brain tissue per unit time, usually expressed as mL/100 g of brain tissue per minute. Typical CBF values for adult human beings are 50-80 mL/100 g of brain tissue per minute with gray matter receiving 3- to 4-fold higher blood circulation than the white matter. MTT is the time required for a given substance or blood to pass through the brain tissue, usually measured in seconds. Typical MTT values reported in the literature are 3 to 5 seconds in a healthy human brain. Although CBF is defined as CBF = CBV/MTT, it has been shown that MTT is contaminated by the bolus dispersion on its way between injection site and brain [36]. However, MTT yields indirect information about the change of blood flow: An increase in MTT indicates that CBF has decreased [37]. Figure 5 shows various perfusion maps obtained with most often used parameters.

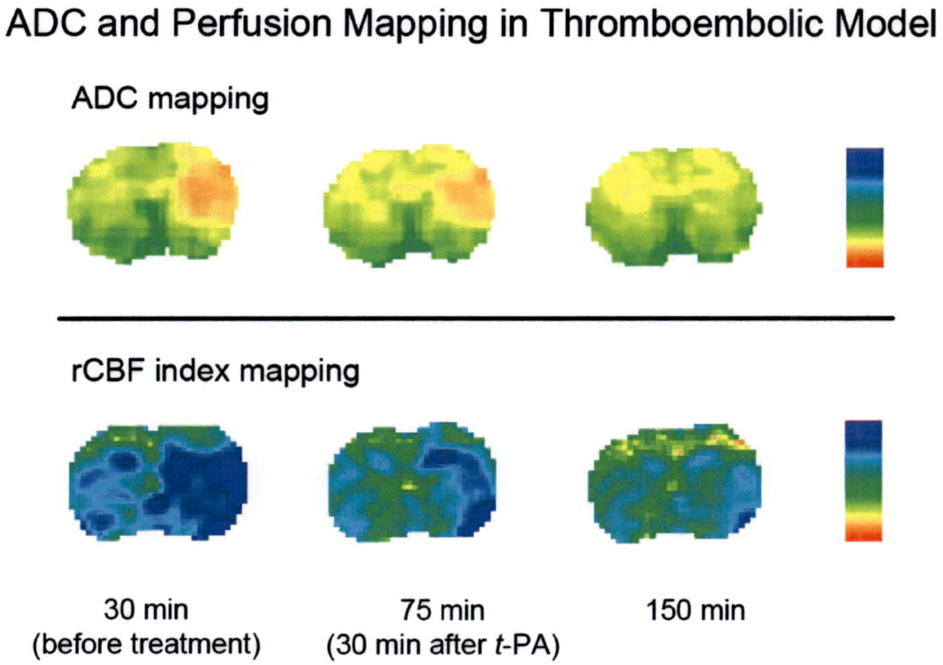

Figure 4. ADC and perfusion mapping during thrombolysis in a thromboembolic MCAO rat model. Large ischemic lesion with ADC decrease and larger perfusion defisit recovers almost completely with therapy in a stepwise manner.

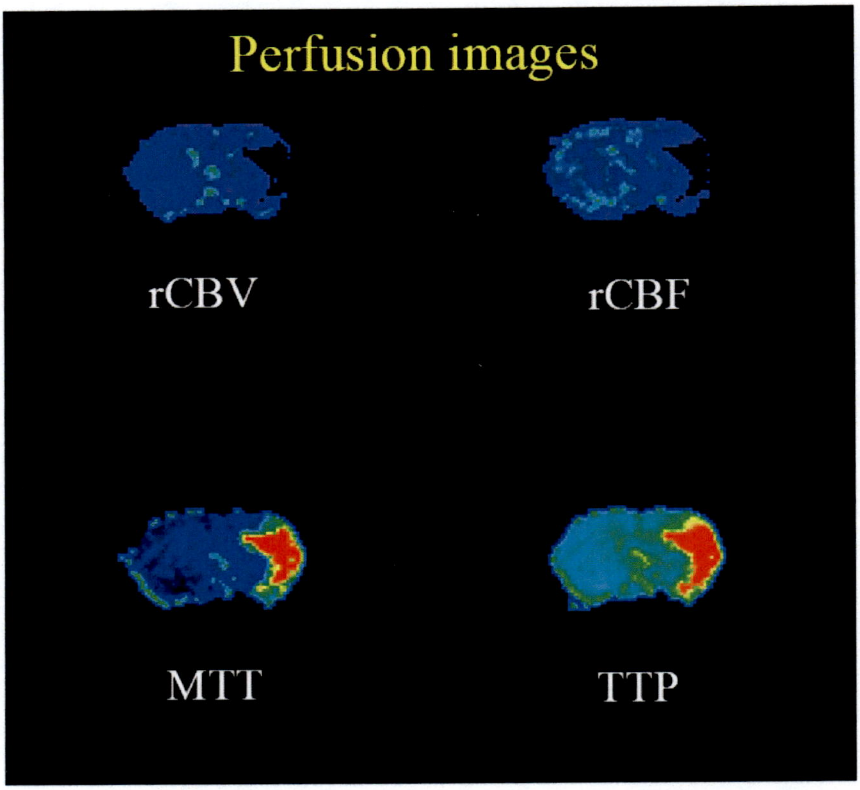

Figure 5. Perfusion images in a rat middle cerebral artery occlusion model. Various parameters such as regional cerebral blood volume (rCBV), regional cerebral blood flow (rCBF), mean transit time (MTT), and time-to-peak (TTP) can be used to assess the extent and severity of hypoperfusion.

The parameters mentioned above are only semiquantitative values, unless precise information about arterial input function is combined. Among several methods used for this aim, each has its own limitations and the method used by Ostergaard et al. [38] seems to be the most reliable [37]. Quantitative CBF mapping is still far to apply into clinical practice. Flow heterogeneity may develop a useful parameter to predict infarct size but there is yet scarce data [39, 40].

There are two major modalities for measurement of PWI (for reviews, see [41, 42]). The most widely used method in both research and clinical practice is dynamic susceptibility contrast (DSC) technique, which is based on the measurement of the first passage of a bolus of an exogenous MR contrast agent, typically gadolinium diethyltriaminepentacetic acid (Gd-DTPA). Nowadays echo planar imaging sequences (spin echo or gradient echo sequence) are preferred because of their ability to acquire multi-slice images, which can encompass the whole brain in only seconds. Gradient-echo sequences are strongly T2*-weighted resulting a highest signal drop for given Gd-DTPA and better signal-to-noise ratio [43] with the disadvantage of being sensitive to contrast agent both in capillary and larger capacitance vessels. In contrast, spin-echo sequences are predominantly T2-weighted, signal drop is less and largely related to contrast agent molecules in capillaries [44]. The bolus of Gd-DTPA (0.1-0.2 mmol/kg), during its passage through the microvascular bed of the target tissue, transiently alters the magnetic susceptibility of the tissue and leads to shortening of T2 and

T2*. Although the volume fraction of the vascular space in the brain is small (3-5%), the compartmentalization of contrast agent causes to a highly significant (>50%) transient signal drop. The signal drop depends on the contrast agent concentration within a voxel, that is influenced mainly by contrast agent dosage [45]. A double dose (0.2 mmol/kg) application of routine clinical dose of Gd-DTPA, has been found to be the optimum dosage [46]. PWI takes altogether approximately one minute. With this method, hypointense areas are observed in normally perfused brain tissues due to paramagnetic effect of the contrast agent passage, whereas relative hyperintense areas are seen in poorly perfused brain tissues because of a relatively lesser penetrance of the contrast agent into this region.

An alternative method to obtain PWI is arterial spin labeling (ASL) technique in which the blood is used as an intrinsic contrast agent, which does not require the injection of an exogenous contrast agent. This technique is more amenable to the quantification of tissue perfusion than the DSC method. However, as compared to DSC, ASL is currently hampered by low signal-to-noise ratios and relatively long scan times, and it does not allow measurements of multiple hemodynamic parameters [47]. The use of high-field MRI systems may help to overcome this problem [37]. Because ASL does not need any contrast agent, ASL measurements can be repeated several times. This makes ASL a useful method for measuring CBF changes in experimental stroke studies, including pharmacological research [37].

6. The Mismatch Concept

Survival of penumbra is suggested to be the main determinant of clinical recovery from stroke [48]. Therefore, mapping the penumbra is of great interest in stroke research. It is now generally accepted that the difference between PWI-based hypoperfusion volume and the DWI-based lesion volume (the so-called mismatch) is operationally approximate the ischemic penumbra. [49, 50] The transition from reversible to irreversible injury is complex and highly dependent on the duration and severity of ischemia, and additionally different areas of the penumbra can have variable outcomes [51]. Generally speaking, when the PWI CBV mapping-based hypoperfusion volume is bigger than the DWI-based lesion volume, the lesion frequently grows to the size of the CBV lesion, unless early reperfusion occurs. Recent findings on DWI lesion characteristics - beside DWI lesion reversibility and delayed reappearance of DWI lesion that we discussed below- are assuming that DWI lesion contains not only core but also some of the penumbra [52], and PWI-based hypoperfusion volume contains also part of oligemia [53]. A modified MRI based definition has been proposed by Maulaz et al. [21] as following: Penumbra = (mismatch area - oligemic rim) + (DWI core – severely depressed ADC values). PWI-DWI mismatch concept needs to be evaluated and clarified in experimental studies and large clinical trials.

Parsons et al. reported the first prospective clinical study systematically performed PWI and DWI before thrombolysis in acute stroke patients, within 6 hours of symptom onset [54]. They showed that patients with PWI-DWI mismatch up to 6 hours after stroke onset had a better clinical improvement than without mismatch patients. In addition, some of the patients had a significant reduction of lesion volume between acute DWI and final T2-weighted image. Two dose-escalating trials with desmoteplase showed that patients with PWI-DWI mismatch thrombolyzed within 3-9 hours of stroke onset responded to therapy (Hacke 2005, DIAS)(Furlan 2006 DEDAS). These studies provided evidence to design randomized controlled clinical trials to test the utility of PWI and DWI to prolong time window of thrombolysis to 6-9 hours. ARTIST-MRI is another study including patients with PWI-DWI mismatch, which currently tests the safety and the possible neuroprotective efficacy of a glutamate blocker YM872. The clinical trial MR-RESCUE aims to determine whether combined PWI and DWI can identify patients who might benefit from mechanical embolectomy.

Apart from the mismatch pattern, DWI-based lesion may be equal to PWI hypoperfusion volume in some acute stroke patients. It is reasonable to think that therapies will not be effective in these patients without mismatch. However, in the absence of data from prospective randomized controlled trials, no mismatch lesion pattern will continue to be a management dilemma, because it is unclear if the entire DWI lesion is already irreversibly damaged or whether it still contains a sizeable penumbra [13].

Studies comparing PWI-DWI to quantitative imaging of flow and oxygen consumption by PET indicate that: 1) PWI-DWI mismatch overestimates penumbra [53], 2) the degree of restricted water diffusion reliably reflects the severity of oxygen deprivation below the penumbral threshold but is less strongly related to metabolic disruption and severe ADC reductions may present in ischemic penumbra as well as in the core, that may explain why the ADC alone does not reliably predict tissue outcome [55, 56], 3) within DWI lesion, the degree of metabolic disturbance and flow rates are variable, in other words DWI lesion contains substantial proportions of penumbra, which may explain fate variability of DWI lesion and why sometimes DWI lesion is reversible [55-57]. One therapeutic implication is that, a matched PWI-DWI lesion may still represent, at least in part, salvageable tissue [58].

7. Lesion Evaluation: DWI and PWI

In humans, in addition to the severity and duration of ischemia, many other factors may influence lesion growth to a certain extent, such as intrinsic tissue factors (preexisting determinants of susceptibility), induced ischemic tissue tolerance induced by preceding vascular events, age [59], gender, serum glucose, temperature, and blood pressure [60]. Imaging has documented substantial individual differences in both the time course of infarct growth and the occurrence of spontaneous recanalization supporting the idea that management decisions should be based on the individual patient's pathophysiological diagnosis [13] mainly based on information achieved by MRI.

In the hyperacute phase of ischemic stroke, computed tomography (CT) images are often normal or show subtle changes only [61]. DWI is superior to both conventional MRI and CT imaging in identifying early ischemic changes and delineating the extent of ischemia [62-64]. Studies have shown that DWI can identify ischemic lesions as early as three minutes after the onset of ischemia in experimental models [65] and only 11 minutes after symptom onset in humans [66]. Within 6 hours of onset DWI has reported sensitivity of 95% and specificity of nearly 100% [67].

In animal stroke models, DWI signal abnormalities correlates highly with the location and volume of pathologically confirmed cerebral infarction [68] and recently same correlation has been shown in humans [69]. In an MRI follow-up study of lesion progression in acute stoke patients not thrombolyzed, initial DWI lesion volume was in correlation with maximum infarct size detected with T2-weighted image and clinical severity [70]. Capability of DWI to monitor infarct progression has been shown in several studies. Infarct expansion mainly occurs in the first 2-3 hours of focal ischemia in rats [71]. Without therapeutic intervention, as within permanent focal ischemia models, areas with severely decreased ADC values invariably become infarcted [72] and when cerebral artery occlusion times exceed 90 minutes, no recovery in ADC is expected until approximately 2 days later when occurs pseudonormalization [73]. In the permanent intraluminal MCAO model, ADC thresholds that can be applied for in vivo quantification of the final infarct (histologically derived lesion volumes evaluated at 24 hours) are well defined for different time points after induction of the ischemia [74, 75]. Initially, it was believed that DWI abnormalities are irreversible and depict the ischemic core. However, recent experimental as well as clinical studies have shown that initial DWI abnormalities and decreased ADC values may normalize if the hypoperfusion is moderate or transient (figure 5). Early reperfusion (less than 30 minutes of transient ischemia) results in acute normalization of brain ADC values transiently and secondary ADC declines occur [76]. Partial recovery and secondary decline in initial ADC values has been reported by other investigators, using a model with longer ischemia duration [77]. In rats with shorter ischemia duration (10 minutes), although initial DWI lesion disappears permanently, delayed pathological changes may still be found in the affected regions [78]. Findings from DWI studies in acute stroke patients with reperfusion, well overlap with experimental data. Normalization of initially decreased ADC values [79-81] and secondary delayed reappearance of DWI lesion [82] have been reported also in humans. The pathophysiology behind the secondary ADC decline is not fully understood. The ischemic tissue changes in the ADC normalization period that later are recognized as delayed infarction, cannot be detected with current imaging methods. In the future, late secondary tissue injury may become an important therapeutic target for postreperfusional neuroprotective therapies, with treatment efficacy monitored by serial DWI.

An important question is whether DWI can provide prognostic information on the ischemic tissue viability and whether it can differentiate between reversibly injured regions and irreversibly damaged tissue. DWI lesion volume of >89 cm^3 detected within 6 hours of symptom onset has been shown as a predictor of early neurological deterioration [83]. Davalos et al. showed that initial DWI lesion of $\leq$ 25 mL had a higher probability to grow [84]. In their study, mismatch between stroke severity and the volume of the DWI lesion (so-called clinical-DWI mismatch) predicted infarct expansion and response to thrombolysis. It has been shown, both in experimental and clinical studies, that a relatively greater ADC decrease is associated with greater risk of irreversibility of the lesion. In animal stroke models, because variables are much under control (especially in terms of ischemia duration and severity) initial ADC values can better predict the reversibility of ischemic injury than in humans. Thus, in a stroke experiment a threshold for the recovery of initially reduced ADC values which reflects the tissue viability can be defined [85, 86] presumably specific to the model used. However, the change in ADC does not reliably predict tissue outcome in human stroke and there is no ADC threshold below which tissue will progress to infarction with certainty [87]. Combining ADC values on a voxel basis with perfusion parameters may be better at predicting tissue viability [88]. Predicting the fate of the tissue and clinical outcome from early diffusion and perfusion measurements remains a topic of great interest and little consensus [89].

Other advantages of DWI in stroke imaging is high sensitivity in detection of multiple acute ischemic lesions [90-92] pointing an embolic source (figure 6) and when DWI and ADC evaluated together, ability to approximate the age of ischemic lesion (figure 3). The advantage of differentiating between acute and chronic lesions, suggest DWI for a surrogate measure for recurrent stroke in a clinical trial [93]. Thus, DWI does not only deliver definite diagnosis of acute ischemic stroke, gives also clues for etiology and guide the decision of etiologic investigations.

Much debate revolves around which PWI parameter can better assess the severity and the extent of ischemia (figure 4). A recently published systematic review on CBF thresholds for the diagnosis of penumbra and infarct core in acute ischemic stroke patients showed the large variability of CBF thresholds reported for penumbra and infarct core [25]. Initial CBV volumes seem to correlate more closely with the final stroke volume and with initial DWI. [94] Also, time to peak more than 5.2 [95] or 6 seconds [96] or MTT more than 4-6 seconds was shown to well correlate with final infarct volume in acute stroke. [54] Simonsen et al. [97] and Perkiö et al. [40] showed that flow heterogeneity maps were more accurate predictors of final infarct than MTT maps.

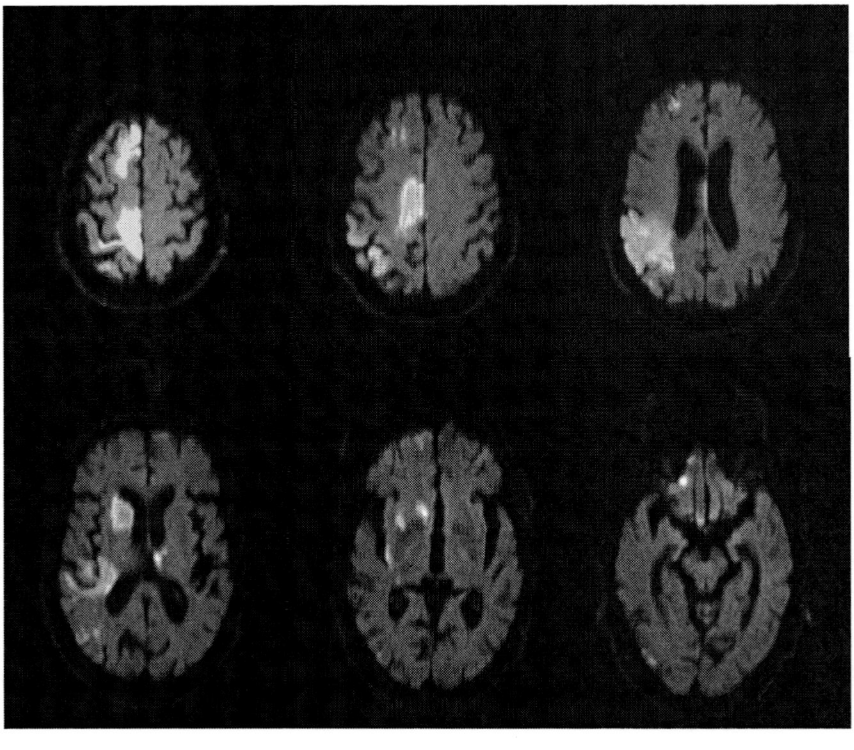

Figure 6. Diffusion-weighted imaging showing multiple acute ischemic lesions. Detection of multiple multiterritorial infarcts raises the suspicion of embolic source for etiology.

8. Detection of Vessel Occlusion: MRA

MRI can provide high-quality assessment of intra- and extracranial macrovascular status in stroke. MRA techniques are based upon the differentiation between stationary and moving spins. Time-of-flight MRA, a gradient echo technique, visualizes the arteries by the change in flow (figure 7). Contrast-enhanced MRA visualizes the vascular lumen.

Compared with intra-arterial angiography, 3D time-of-flight angiography appears adequate for the diagnosis of carotid artery steno-occlusions [94] with a high sensitivity and specificity comparable to digital subtraction angiography [50]. Because of slow and disturbed flow, conventional 3D time-of-flight MRA tends to overestimate stenotic lesions and contrast agent application may overcome this problem [98]. Limitation of MRA appears in the evaluation of leptomeningeal collateral circulation.

High resolution MRA has been successfully obtained in rats [99] and mice [100] (figure 8). In experimental and clinical studies, by obtaining MRA before and after thrombolysis, success of the therapy in terms of lysis of the blood clot can be evaluated. Additionally, MRA is useful to disclose other vascular pathologies such as carotid artery dissection and may improve patient enrolment into clinical stroke trials.

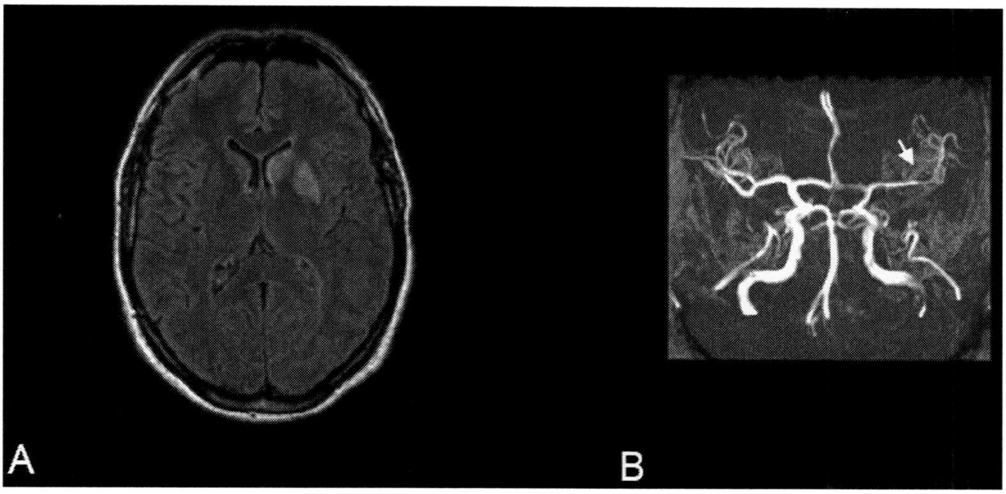

Figure 7. A. FLAIR image of a patient showing acute ischemic lesion involving left putamen and caudate nucleus. B. Intracranial MR angiography of the same patient. Note the occlusion of distal M1 (arrow).

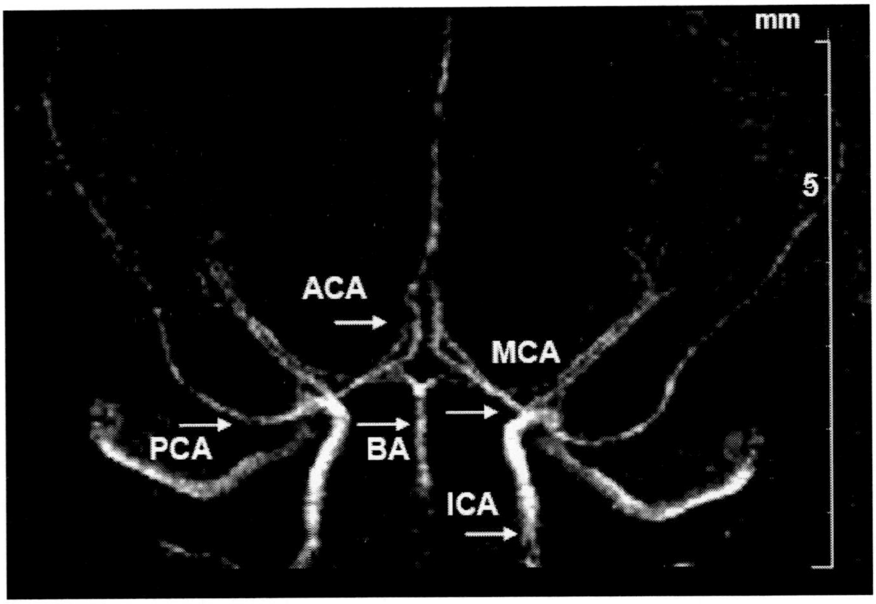

Figure 8. 3D Time-of-flight MR-angiogram of cerebral arteries of a mouse. ACA, anterior cerebral artery; BA, basilar artery; ICA, internal carotid artery; MCA, middle cerebral artery; PCA, posterior cerebral artery.

9. Detection of Old and New Brain Hemorrhages, T2*-Weighted Image

MRI signatures of intracerebral hemorrhage (ICH) depends on factors both related to MRI (sequence used, field strength) and blood's constitution (mainly oxygen saturation of hemoglobin and its degradation) that causes the complexity of ICH appearance in different sequences at different stages. As a hematoma ages, the hemoglobin passes through several forms (oxyhemoglobin, deoxyhemoglobin, and methemoglobin) followed by breakdown into ferritin and hemosiderin after the red cell lysis. Acute hematomas are characterized by central hyperintensity on T2-weighted image as opposed to hypointensity on T1-weighted image due to T2 relaxation effect of deoxyhemoglobin [101]. The appearance of methemoglobin formation marks the transition from the acute to the subacute phase. Conformational change provided by the presence of methemoglobin in the hemoglobin molecule causes shortening of both T1 and T2 but predominantly T1 [102] and, hence to a centripetal hyperintensity on T1-weighted image, while also affecting T2 relaxation and causing a hyperintense signal. This may allow to classify approximately the hematoma between acute and early subacute stages [101]. The T2 shortening effects (the susceptibility effect) are directly proportional to the field strength that at low field (less than 0.5 T) acute hematomas may appear isointense to brain parenchyma [103]. In contrast, paramagnetic effect of methemoglobin is not significantly affected by field strength [102]. Late subacute and chronic ICH is characterized by hypointense rim caused by paramagnetic effect of ferritin and hemosiderin, which demarcates the border zone of hematoma (less conspicuous on T1-weighted sequences and more conspicuous on T2- and T2*-weighted sequences) completely surrounding a central area hyperintense in all sequences due to the persistence of extracellular methemoglobin [104].

It was initially believed that MRI is not sensitive for reliable detection of hyperacute ICH (less than 24 hours duration) because of an unspecific pattern in T1- and T2-weighted images imparted by the predominance of oxyhemoglobin, which is not a paramagnetic molecule. The introduction of gradient-recalled echo sequences, which are T2*-sensitive, into MRI, changed this belief. The hyperacute lesion on T2*-weighted image typically consists of a core of heterogeneous signal intensity, reflecting the most recently extravasated blood, that mainly contains significant amounts of oxyhemoglobin, surrounded by a rim of hypointensity, signifying parenchymal blood that became deoxygenated and paramagnetic (figure 9). The need for CT to rule out ICH in acute stroke patients is not anymore valid, since it has been proven that hyperacute ICH can be detected by MRI with 100% sensitivity [105] and MRI is equivalent to CT [106]. In the case of subarachnoid hemorrhage suspicion, CT is still the first choice of imaging [50, 106]. In experimental ICH models, MRI was successfully used to evaluate signal characteristics of the hematoma [107] and in drug development studies for ICH to monitorize changes in hematoma volume (figure 10) [108, 109].

Besides hyperacute stage, MRI is also powerful to detect ICH in later stages of bleeding. Moreover, MRI was found to be more accurate than CT for the detection of chronic ICH [106]. Detection of bleedings is crucially important especially in patients candidate for thrombolysis. Cerebral microbleeds are now better recognized with the increasing use of T2*-weighted sequence as small sharply defined signal loss (figure 11). The presence of cerebral microbleeds was suggested to increase the risk of hemorrhagic transformation after thrombolysis [82]. However, another study failed to show a correlation [110, 111]. Additional studies with large groups of patients are needed to clarify this issue. Another utility of T2*-weighted sequence in acute stroke patients is in detection of hemorrhagic transformation, in which it is at least as efficient as CT scan [112] (figure 9).

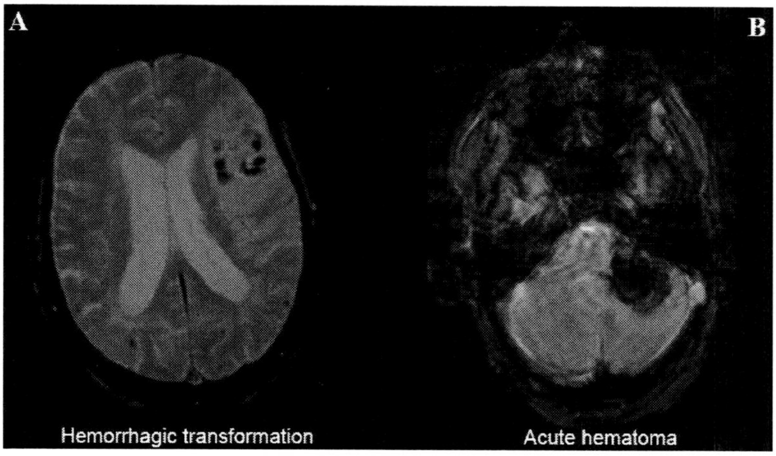

Figure 9. T2*-weighted images from two separate patients. A, large infarction with hemorrhagic transformation (black spots within the infarct) in the left middle cerebral artery territory; B, acute left cerebellar hematoma.

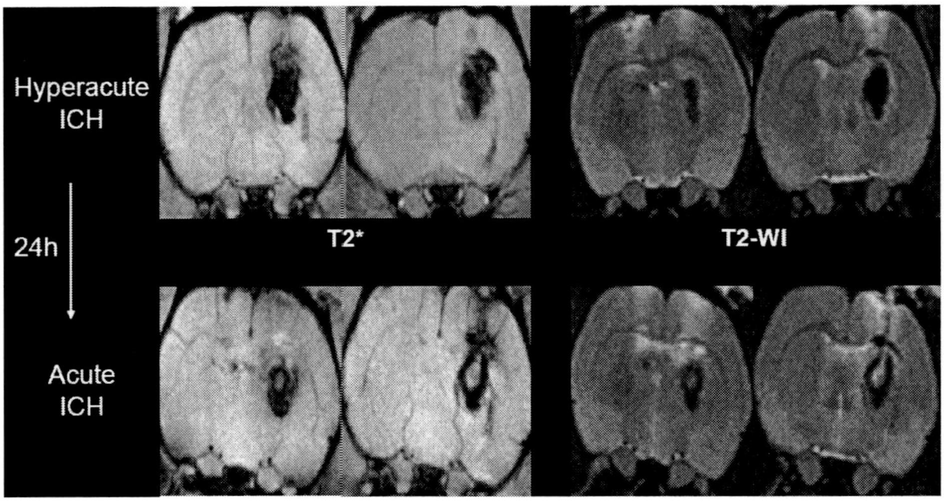

Figure 10. Experimental intracerebral hemorrhage. MRI of hyperacute ICH (upper row, 40 min after intraparenchymal injection of blood) and follow up MRI after 24 hours (below row, acute ICH) of the same rat are represented. Two consecutive slices from T2-weighted and T2-weighted images are given.

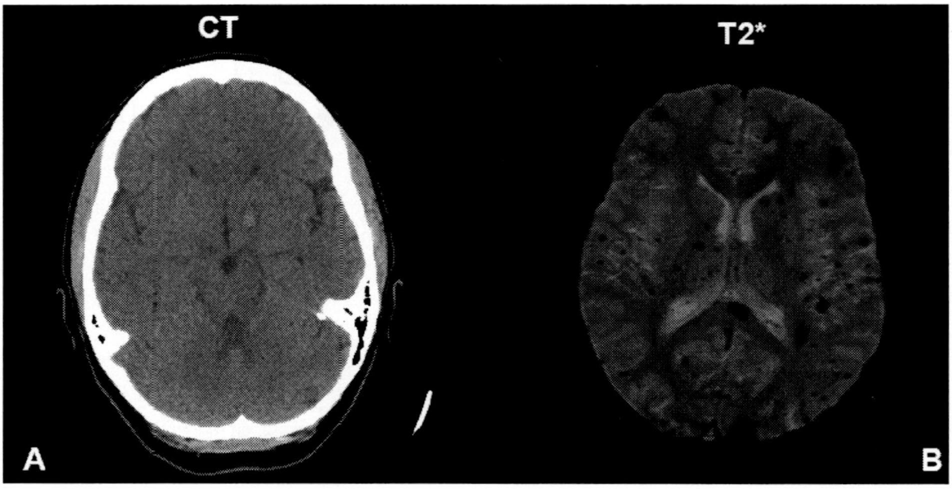

Figure 11. A. CT of a patient showing acute-subacute small thalamic hematoma. B. T2*-weighted image displays over 50 cerebral microbleeds not visible on CT of the same patient.

10. BBB Disruption: Contrast-Enhanced MRI

Intact BBB keeps a strict milieu around neurons. After ischemic injury BBB is disturbed and allows passage of contrast agents from blood to brain. Contrast-enhanced MRI may show the BBB disturbance in the ischemic area (figure 12). In agreement with animal studies suggesting early gadolinium enhancement in the ischemic area after reperfusion as a predictive factor of subsequent hemorrhagic transformation [113-116] early BBB disruption detected by parenchymal enhancement in human ischemic stroke was found highly specific for hemorrhagic transformation [117-119]. Quantitative evaluation of BBB with several methods is available in animal models of stroke [120-122], but has not yet been applied to humans.

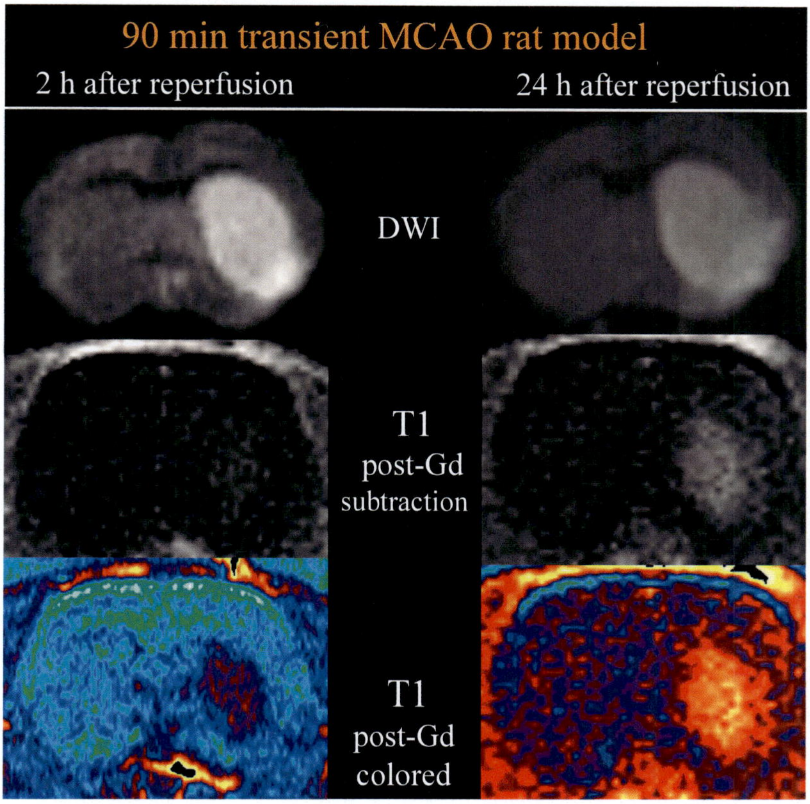

Figure 12. BBB damage after transient focal cerebral ischemia. MRI of a rat brain 2 h (on the left) and 24 h after reperfusion (on the right). 90 min of transient ischemia was induced by the suture MCAO method. Note post-contrast T1-weighted image shows enhancement in the ischemic region already at 2 h and at 24 h ischemic lesion is bigger and contrast enhancement is more pronounced.

11. Drug Development for Ischemic Stroke

In ischemic stroke, most of the drug development efforts have focused on the very early phase (within hours of stroke onset) because it is the most appropriate time point to salvage a large portion of the brain tissue-at-risk (or penumbra) before it evolves into irreversible injury and later to pannecrosis. Salvaging large portions of the tissue-at-risk directly reduces neurological deficits and disability, leading to a lower mortality, better outcome, shorter hospital stay, reduced need for assistance, and better preservation of ability to work after stroke.

The therapeutic approaches in reducing brain tissue injury include recanalization and neuroprotection. Recanalization, which aims to reopen the occluded brain artery and restore the blood flow, is available with thrombolysis and mechanical thrombus retraction. Neuroprotection aims at preserving neurons and brain function by reducing the extent of cellular damage. More than 1,000 drugs and various nonpharmacological strategies have been tested for their neuroprotective efficacy in animal models and some degree of neuroprotective efficacy was found in more than half of experiments [11]. No particular neuroprotectant drug mechanism distinguished itself on the basis of superior efficacy in animal models of focal ischemia, perhaps because of the pathophysiological complexity of stroke [123]. Despite these successes in the laboratory, all of the hundreds of clinical trials aiming at developing a panacea with neuroprotectans for ischemic stroke have failed during the past two decades, except one positive phase III trial of a free radical scavanger (NXY-059)[124], one pilot study of erythropoietin [125] and few trials which can be considered as partially positive [126, 127]. The use of intravenous rt-PA is still the only therapy in acute ischemic stroke since its approval in the United States in 1996. Intravenous rt-PA is used in only a fraction of ischemic stroke patients mainly because of the restrictive 3-hour time window [12]. Several reasons for failure of translation from laboratory to clinics were previously well summarized reflecting problems both in experimental studies and clinical trials [11, 15, 128, 129]. A comprehensive overview of these contentious issues has been provided recently by Fisher et al. [130]. Conditions that should be met for standardization of preclinical antiischemic drug studies before a compound progresses to clinical trials were published by Stroke Therapy Academic Industry Roundtable [4], later additional recommendations have been proposed by several authors [15, 129, 131]. Only neuroprotective drug found effective in human, NXY-059, a free radical scavanger, underwent an extensive preclinical testing program that largely fulfilled the Stroke Therapy Academic Industry Roundtable's first meeting's recommendations [4], suggesting that a well-designed neuroprotective drug development program is essential for the success in clinic [132]. However, the failure in the second multicenter large-scale NXY-059 study (SAINT-II) [133] was disappointing and calls for more innovative approaches, such as incorporation of MRI into clinical trials. We feel that, in order to translate positive findings in animal stroke models into beneficial and safe treatments in stroke patients, in addition to proper design of both preclinical and clinical studies, the use of versatile imaging modalities, especially MRI may build a bridge to step from bench to bedside. MRI is now incorporated in the design of many randomized controlled trials (table 2), particularly at the dose-finding stage, and is the basis of current shift towards smaller trials but more homogenous samples, with an expected increase in trial efficiency [13].

In clinical trials, by a multimodal baseline imaging, DWI-based ischemic lesion, PWI-based hypoperfusion regions, and mismatch between PWI and DWI can easily and rapidly be calculated volumetrically. After treatment, MR imaging can be repeated at several time points and lesion (or infarct at later time points) volumes can be calculated. At later time points (week(s) or months), T2-weighted images replace DWI in final infarct volume measurements. Volumetric changes in lesion size in the drug-treated and placebo-treated patient groups can be compared to obtain an objective measure of drug efficacy. Treatment effect can be determined as "infarct growth" or "reduction in infarct size". The Food and Drug Administration has already approved this approach in certain circumstances. Although extremely appealing, this approach cannot yet completely replace neurological examination and scoring.

12. The Role of Multimodal MRI in Drug Development for Ischemic Stroke

Table 1 summarizes the advantages of incorporating MRI into studies.

Table 1. The utility of MRI in stroke therapy research

	MRI sequence to be used
Study quality	DWI
By baseline lesion volume determination overcoming a potential pretreatment bias	
By elimination of variables achieving a more homogeneous stroke patient group reducing the number of animals or patients required	
Due to ease to perform repeated measurements ease to conduct longitudinal studies	
By detecting lesion or vascular occlusion location obtaining a more refined group of patients†	DWI, MRA
Diagnosis	
Diagnosis of stroke with high sensitivity at every phase of stroke starting from hyperacute stage	DWI, ADC, PWI
Exclusion of non-stroke patients or patients with other vascular diseases	DWI, MRA, T2*
Monitoring recurrent strokes	DWI, ADC
Outcome††	
Surrogate marker in drug studies by monitoring lesion volume, hypoperfusion or recanalization	DWI, PWI, T2, MRI
Predicting hemorrhagic transformation	Contrast-enhanced MRI, T2*

† e.g. excluding patients with only white matter ischemia when evaluating a drug known to be ineffective to white matter ischemia

†† This issue needs further evaluation

ADC, apparent diffusion coefficient; DWI, diffusion-weighted image; MRI, magnetic resonance imaging; MRA; magnetic resonance angiography; PWI, perfusion-weighted image.

13. Diagnosis of Stroke

High sensitivity and the high predictive value of MRI to ischemia enable an early, rapid (takes 20 seconds to image whole brain), and definite diagnosis of ischemic stroke and can exclude ischemic stroke in patients with stroke mimics (avoiding unnecessary and potentially dangerous therapeutic interventions in nonstroke patients). Because DWI can identify the localization and number of the infarcts, gives hints regarding the etiology). When DWI and ADC evaluated together, ischemic lesion can be depicted in its full extent, the age of ischemic lesion can be determined, and recurrent strokes can be monitorized. Before starting the treatment ischemic lesion volume can be determined from baseline DWI images. PWI is able to detect ischemic hypoperfusion immediately following ischemia, thus enhancing the ability to diagnose ischemic stroke in a rapid and effective way. PWI also delivers versatile information regarding occlusion status of the artery. MRA can identify extracranial carotid artery steno-occlusions adequately and intracranial steno-occlusive lesions mainly in the proximal segments. MRA also discloses other vascular pathologies. T2*-weighted image excludes hemorrhagic stroke, shows subacute or chronic bleedings, as well as cerebral microbleeds. Conrast-enhanced MRI may disclose BBB disruption. Conventional MR modalities exclude nonstroke etiologies and demonstrate old lesions of various reasons.

14. Study Quality

In preclinical drug efficacy experiments, the ischemic lesion volume is usually used to determine the therapeutic efficacy of drugs and it is traditionally quantified by postmortem histological staining techniques after the animals are sacrificed at an appropriate time point after the induction of ischemia (usually after 1-3 days). The interpretation of a positive result determined by such single end point needs to be cautious, since it requires the assumption that the lesion size or the severity of ischemic damage is equal in the drug-treated and placebo-treated groups before the therapeutic intervention is started. Without a similar initial lesion volume in the two groups, a conclusion cannot be drawn that the difference of the final lesion volume in the two groups is attributed to the therapeutic intervention. This pretreatment bias can be overcome by utilizing DWI as it allows for precise calculation of infarct volumes before initiation of treatment (figure 13,14). An unequal initial lesion size before treatment in two groups of animals (or as well in patients) can occur especially with a small sample size and variability of the initial lesion can be much greater than any effect of the treatment, that may lead to a false positive result. Standardizing stroke severity in drug developmental research is required to optimize the ability to see a treatment effect between the drug and placebo [128]. If stroke severity is too pronounced, then animals will die whether or not they receive treatment. If severity is too mild, any significant differences cannot be detected. Including too many mild patients in a clinical trial will increase the placebo responder rate and reduce the power of the trial to detect a treatment effect [134]. Thus, in both cases the experiment can fail by a false negative result. In clinical trials, stroke severity is defined by stroke scales, usually by the National Institutes of Health Stroke Scale, and pretreatment baseline severity is compared to end-point severity. DWI-based stroke severity, namely ischemic lesion volume, as an additional information before treatment is started, may refine the definition of stroke severity. Infarct volume has been hypothesized to be a more sensitive and more reliable measure of relevant drug effects than outcomes measured by clinical disability scales [135]. Ongoing clinical trials using MRI in detecting outcome measures (table 2) perhaps will clarify the use of MRI as a surrogate marker.

Table 2. Clinical acute stroke trials that included MRI in study design

Study name	Drug tested	MRI in inclusion or exclusion criteria	MRI in outcome determination
Completed trials			
Citicoline 010	citicoline	DWI lesion:1-120cc, within the territory of MCA, including gray matter	prim. LV at day 90
Citicoline ECCO 2000	citicoline		sec. LV (T2) at day 90
COOL AID I	hypothermi a		sec. LV (MRI)
DIAS	desmotepla se	M= 20%; perfusion deficit>2cm involving the hemispheric grey matter exclusion:DWI>1/3 MCA; no PWI deficit; ICA occlusion	prim. reperfusion by MRA
DEDAS	desmotepla se	M= 20%; perfusion deficit>2cm involving the cerebral cortex	prim. DWI, PWI deficits, MRA at 4-8 h
EPITHET-pilot	rt-PA		prim. chage in DWI lesion size at day 3-5 and day 10; sec. perfusion change (PWI), HT (MRI), LV (T2) , recanal. (MRA)
GAIN MRI Substudy	gavestinel	cortical acute DWI =1,5cm of diameter	prim. LV (T2) at day 90; sec. reduction in LV growth (DWI)
MR Images	magnesium sulfate	DWI= 5cc	prim. frequency of lesion growth at day 90
MRI-guided abciximab for ischemic str	abciximab	acute supratentorial DWI lesion	prim. percent change in lesion size of DWI at 24 h
SIS	sipatrigine	cortical LV= 20cc in the MCA territory; M= 30%	prim. LV at day 7
Ongoing trials			
ARTIST MRI	YM872, AMPA receptor antagonist	DWI: 5-120cc; M= 20%	prim. LV (FLAIR) at day 28
ASP I	ancrod		prim. LV (MRI) at day 5
AXIS	G-CSF	M existence	prim. lesion growth (MRI) at day 90
DEFUSE	rt-PA	pretreatement MRI in all patients	prim. MRI at 3-6 h

Table 2. (Continued)

Study name	Drug tested	MRI in inclusion or exclusion criteria	MRI in outcome determination
Ongoing trials			
DIAS-2	desmoteplase	M=20%	prim. LV (MRI) at day 90; sec. change in LV (MRI or perfusion CT) at day 30
Ebselen Trial - Phase III	ebselen	cortical/boundary lesion (CT or MRI) exclusion: lacunar (CT or MRI); extensive infarct (CT or MRI)	
EPIC	piclozotan	M existance exclusion: subacute stroke (FLAIR); DWI> 2/3 MCA	prim. change in LV (MRI)
Study name	Drug tested	MRI in inclusion or exclusion criteria	MRI in outcome determination
EPITHET	rt-PA	pretreatment MRI	prim. change in lesion size (DWI) at days 3-5 and 10; sec. perfusion change (PWI), HT (MRI), LV (T2), recanal. (MRA) at day 3-5
IST-3	rt-PA	acute ischemic lesion (MRI or CT)	
ITAIS	rt-PA ia/iv	pretreatment MRI in all patients	prim. reperfusion (PWI and MRA) at 24 and 48 h
MITI-IV	microplasmin	pretreatment MRI in all patients exclusion: DWI>1/3MCA	sec. LV (DWI), rep (PWI), recanal. (MRA) at 6h and day 7
MRI Rescue	mechanical thrombectomy	pretreatment MRI in all patients	sec. MRI at day 7
ROSIE	abciximab+ reteplase	perfusion deficit (PWI or MRA) exclusion: uninterpretable DWI or PWI; DWI lesion>1/3 MCA	prim. reperfusion (MRI)
ROSIE-2	aspirin+tinzaparin+eptifibatide+rt-PA	pretreatment MRI in one arm of the trial	prim. reperfusion (MRI) at 2h and 24h
RREAC	ONO-2506	pretreatment MRI in some patients	sec. MRI

CT, computed tomography; DWI, diffusion-weighted image; FLAIR, fluid attenuated inversion recovery; HT, hemorrhagic transformation; LV, lesion volume; M, PWI/DWI mismatch; MRI, magnetic resonance imaging; MRA, magnetic resonance angiography; PWI, perfusion-weighted image; prim. primary outcome; recanal. recanalization; rt-PA, recombinant.

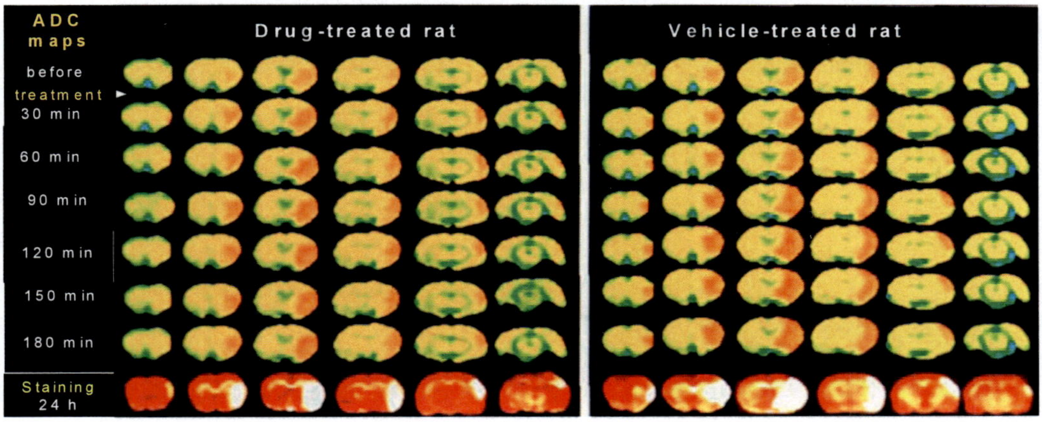

Figure 13. In vivo demonstration of efficacy by DWI. ADC maps obtained (six 2-mm-thick coronal slices) before treatment and every 30 min up to 3 h are given. The lowest row represents postmortem brain staining at 24 h. Note the ischemic lesion growth in the vehicle-treated rat brain, while drug-treated rat brain shows smaller lesion.

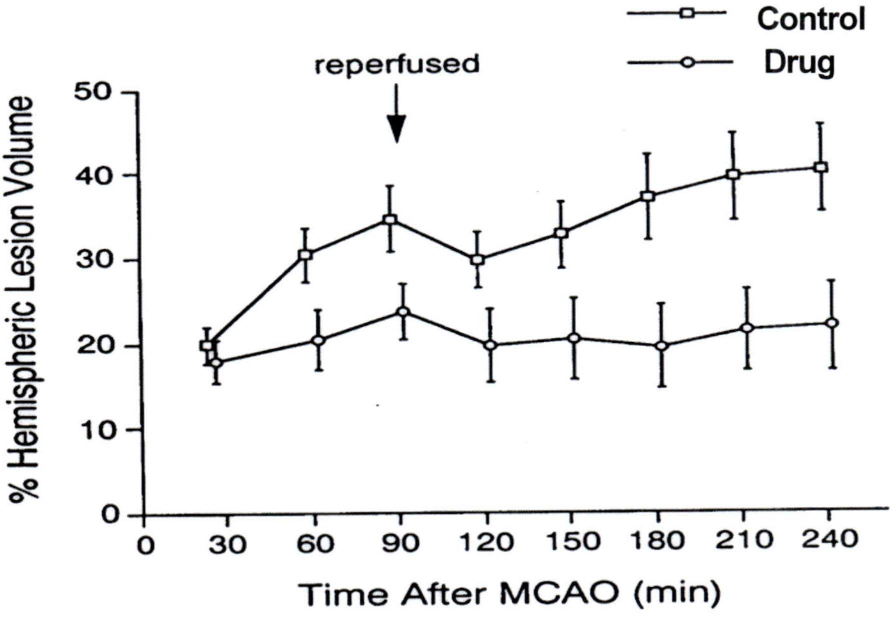

Figure 14. Ischemic lesion volume evaluation in rats underwent focal cerebral ischemia and treated with drug (empty circles) and vehicle (empty squares). Initial lesion volumes were equal between the groups. Although lesion volume decreased in both groups after reperfusion, later it freezed in the drug-treated group, but continued to grow in the control group. 4 h after ischemia the lesion volume in the treated group was significantly smaller.

Another concern in clinical trial design is the inclusion of patients with a stroke subtype unlikely to respond to the drug being tested because of its mechanism of action [134]. For instance, white matter ischemic injury has some pathophysiological mechanisms that differ from gray matter and it would be unlikely that drugs which act primarily on mechanisms of cell injury unrelated to white matter injury (such as N-methyl-D-aspartate or α-amino-3-

hydroxy-5-methylisoxazole-4-propionic acid antagonists) would be effective in this stroke subtype [134]. Similar to animal models in which stroke is less variable, elimination of some variables in clinical trials in terms of lesion location (gray or white matter) and vessel occlusion site is feasible by the use of DWI and MRA. To our knowledge, among ongoing acute stroke trials, only Ebselen Trial-Phase III is including patients with only cortical or boundary ischemic lesions while excluding patients with lacunar infarction by using imaging modalities (CT or MRI). Thus, in a clinical trial using MRI, several variables can be standardized; consequently a more homogenous patient population can be achieved. According to Warach et al. [135] homogeneity achieved by image-based selection will limit the proportion of eligible patients and thereby recruitment rate. Same authors add that an imaging outcome prior to hospital discharge may reduce loss to follow-up problem.

Another advantage of incorporating MRI into preclinical experiments and clinical trials is that since MRI is a noninvasive and safe method (lacking ionizating radiation), it allows repetitive measurements in the same animal or individual. This is of relevance in longitudinal studies since the inter-individual variance is eliminated and the number of subjects required to reach the statistical significance is much smaller [136].

15. In Vivo Demonstration of Efficacy

With the monitoring of DWI, many research groups demonstrated early in vivo beneficial effects or delayed therapeutic effects of several neuroprotective agents in experimental stroke models (figure 13). Drug-related regression of the DWI lesion has been observed with various neuroprotective agents (for a review see [136], such as N-methyl-D-aspartate antagonists MK-801 [137] and CNS-1102 [138], calcium channel blocker [139], sodium-calcium channel blocker [140], free radical scavenger [141], cation channel blocker LOE 908 MS [142], basic fibroblast growth factor [143], endothelin antagonist [144], immunosuppressant FK 506 [145], estrogens [146], that interfere with various steps of the ischemic cascade. Although the results of preclinical trials from each laboratory were promising, there has been only few published report about in vivo neuroprotective benefit monitored by DWI in stroke patients. An acute stroke trial of citicoline used MRI to select [147]. The mean increase in lesion volume from the baseline DWI scan to the day 90 T2-weighted scan was 180% in the placebo group and only 34% in the citicoline group. This trend toward lesion growth reduction with citicoline therapy did not reach statistical significance perhaps because of the large variability in percent lesion growth and the small sample size [51]. This study has validated the use of DWI in designing the clinical trial of neuroprotective agents. Glycine Antagonist in Neuroprotection (GAIN) MRI substudy [135] was conducted although both GAIN Americas [148] and GAIN International [149] trials failed to show a clinical benefit of the drug tested (gavestinel, GV150526, selective antagonist at the glycine site of the N-methyl-D-aspartate receptor). The study was designed to use MRI as a surrogate marker in terms of both primary and secondary outcomes. Baseline imaging included DWI, PWI, T2*-weighted image, and MRA. Patients with acute cortical pretreatment lesion on DWI scan at least 5cc of volume have been investigated. No effect of gavestinel on infarct volume at day 90 were observed in the primary or other analysis. On the other hand, a significant correlation has been found between ischemic lesion volume decrease and clinical improvement over the 3 month period of observation. This study further confirmed the previously known phenomena that ischemic lesion growth is greater in patients with PWI/DWI mismatch. ARTIST-MRI is an ongoing clinical trial including patients with PWI-DWI mismatch to test safety and possible neuroprotective efficacy of a glutamate antagonist.

According to the preclinical studies, early (within 3 hours after stroke) reperfusion of occluded brain vessels reduces infarct size and improves outcome in several species [150]. Using an embolic animal model combined with DWI and PWI, several experimental studies have shown that both prourokinase and rt-PA when given intraarterially or intravenously shortly after induction of focal brain ischemia could significantly decrease the final infarct size and ameliorate neurological outcome [151-153]. Both DWI and PWI have been recently used to evaluate the efficacy of rt-PA treatment in ischemic stroke patients (for a review see [154]). Kidwell et al. [79] demonstrated that thrombolytic therapy decrease the DWI-based ischemic lesion size by about 50% at 2.5 to 9.5 h after treatment. The Desmoteplase In Acute Ischemic Stroke Trial (DIAS), was the first prospective, placebo-controlled randomized acute stroke thrombolysis trial to use MRI both for patient selection and as a primary efficacy endpoint [6]. The hypothesis tested was that better patient selection by using MRI, may extend the 3-hour window for intravenous thrombolysis. Patients with a PWI/DWI mismatch of $\geq$ 20% and perfusion abnormality of >2 cm in diameter involving hemispheric gray matter up to 9 h after stroke received thrombolysis. Mismatch was defined by each investigator's preferred technique, later confirmed by a single method in the core laboratory as 95% accurate [155]. Patients with lesions larger than 1/3 of MCA territory on baseline DWI scan or those without perfusion deficit were excluded from the study. DIAS researchers observed a significant benefit on both the target imaging outcome, early reperfusion, and 90-day clinical outcome. The results of DIAS were later supported by Dose Escalation of Desmoteplase for Acute Ischemic Stroke trial (DEDAS). In DEDAS [7], the MRI PWI-DWI mismatched identified patients in the 3 to 9 hour time window for thrombolysis with desmoteplase, with apparent benefit especially in patients who fulfilled all MRI criteria. As in DIAS, there was a strong positive correlation between reperfusion and good clinical outcome. These studies provide optimism about expanding time window of thrombolytic therapy by using MRI technology to detect a "tissue-clock". Which parameters obtained by MRI may better predict ischemic tissue fate will be found out shortly. The Echoplanar Imaging Thrombolysis Evaluation Trial (EPITHET) was designed to test the hypothesis that PWI-DWI mismatch might be predictive of the response to thrombolysis 3 to 6 hours after stroke. A recently published preliminary analysis of the initial 40 EPITHET patients compared systematically PWI parameters and their effect on mismatch calculations [156]. According to the authors, the presence and degree of mismatch varies considerably with the definition, particularly the PWI threshold used. Among the parameters evaluated (CBF, MTT, TTP, Tmax and first moment transit time, FMT), threshold defined by Tmax +2s has been found to better estimate the PWI-DWI mismatch. However, mismatch volume calculated with all parameters and thresholds did not correlate with DWI expansion. Tissue reperfusion was more predictive of lesion growth. The Diffusion and Perfusion Imaging Evaluation for Understanding Stroke Evaluation Trial (DEFUSE) was designed as an observational study in which acute stroke patients within 3 to 6 hours of symptom onset underwent pretreatment MRI and regardless of the MRI results received rt-PA therapy [157]. MRI was repeated 3 to 6 hours after treatment. Patients with PWI-DWI mismatch and early reperfusion had more favorable clinical response than patients without mismatch. Investigators identified two distinct profile of mismatch: "malignant and target mismatch". Large baseline DWI volume and/or large PWI volume with severe hypoperfusion (long delays on the Tmax maps) termed as malignant mismatch because patients having this type of imaging were at high risk for symptomatic ICH and a poor outcome. In contrast, target mismatch profile (simply the profile not having malignant mismatch characteristics) has been found to identify patients who have robust clinical response rate after early reperfusion. One limitation of the study is the calculation of PWI lesion by only one parameter, because there is not yet consensus about which PWI parameter estimates better the penumbra. The number of patients without PWI-DWI mismatch did not suffice to provide reliable data about management of such patients. On the other hand, DEFUSE supports the view that MRI may serve as a "brain clock" [154] and should be involved into clinical trials to refine selection of candidates for thrombolytic treatment.

16. Conclusion

Availability of MRI in both clinical and preclinical acute ischemic stroke settings is a key bridge in the translation of positive findings of drug development studies from animals to humans. Incorporation of MRI, especially DWI and PWI, in the design of acute stroke trials may lead to rapid, reliable, safe, and cost-effective testing of novel drugs. We believe that the era started with the revolutionary thrombolytic therapy, will continue with the neuroprotectant drugs effective in humans due to a better drug developmental program and a better clinical trial design, with MRI technology playing a key role in both steps. The use of multimodal MRI approach in stroke research will improve the study quality both in preclinical and clinical research. Advantages of this approach are: 1) ascertaining diagnosis of ischemic stroke, excluding other etiologies, 2) demonstrating whether a major artery occlusion is present, 3) predicting lesion's age, 4) obtaining clues for etiology, 5) pretreatment lesion size determination and lesion size standardization enables to avoid potential pretreatment bias, eliminates inter-individual high variability of lesion sizes and avoid false positive or negative results, 6) longitudinal studies can be easily performed, leading the need of less number of subjects to reach the statistical power, 7) patient inclusion into the trial may improve by selecting with high sensitivity patients with acute stroke who will likely benefit from the drug under testing, by excluding non-stroke patients, or without the target ischemic tissue-at-risk, and 8) monitoring ischemic lesion volume as an objective outcome measure is easily available with repeated MRI in search of *in vivo* therapeutic benefit of the drug under evaluation. DWI as the most sensitive MRI sequence to diagnose acute stroke, T2*-weighted image, as sensitive as CT to exclude brain hemorrhage, PWI, as giving information about perfusional status of the brain, DWI and PWI combination to evaluate penumbra, and finally MRA to detect vascular occlusion site are the MRI sequences we propose to include in the multimodal MRI evaluation of acute stroke. A "stroke MRI package" including positioning (scout) images, DWI, PWI, T2*-weighted images, MRA can be completed in 15-20 minutes.

17. Acknowledgments

This work was supported in part by the University of Helsinki, the Helsinki University Central Hospital, the Sigrid Juselius Foundation, the Finnish Medical Foundation, the Center for International Mobility, and the Finnish Academy of Sciences.

18. References

[1] Murray CJ; Lopez AD. Mortality by cause for eight regions of the world: Global Burden of Disease Study. *Lancet,* 1997, 349, 1269-76.

[2] Muntner P; Garrett E; Klag MJ; Coresh J. Trends in stroke prevalence between 1973 and 1991 in the US population 25 to 74 years of age. *Stroke,* 2002, 33, 1209-13.

[3] Truelsen T; Piechowski-Jozwiak B; Bonita R; Mathers C; Bogousslavsky J; Boysen G. Stroke incidence and prevalence in Europe: a review of available data. *Eur. J. Neurol,* 2006, 13, 581-98.

[4] STAIR. Recommendations for standards regarding preclinical neuroprotective and restorative drug development. *Stroke,* 1999, 30, 2752-8.

[5] Michel P; Bogousslavsky J. Penumbra is brain: no excuse not to perfuse. *Ann. Neurol,* 2005, 58, 661-3.

[6] Hacke W; Albers G; Al-Rawi Y; Bogousslavsky J; Davalos A; Eliasziw M; Fischer M; Furlan A; Kaste M; Lees KR; Soehngen M; Warach S. The Desmoteplase in Acute Ischemic Stroke Trial (DIAS): a phase II MRI-based 9-hour window acute stroke thrombolysis trial with intravenous desmoteplase. *Stroke,* 2005, 36, 66-73.

[7] Furlan AJ; Eyding D; Albers GW; Al-Rawi Y; Lees KR; Rowley HA; Sachara C; Soehngen M; Warach S; Hacke W. Dose Escalation of Desmoteplase for Acute Ischemic Stroke (DEDAS): evidence of safety and efficacy 3 to 9 hours after stroke onset. *Stroke,* 2006, 37, 1227-31.

[8] Pulsinelli W; Jacewicz M. Animal models of brain ischemia. In: Barnett HJ; Mohr JP; Stein BM; Yatsu FM, editors. *Stroke: Pathophysiology, diagnosis, and management,* New York, NY, USA: Churchill Livingstone; 1992; 49-67.

[9] Hossmann KA. Animal models of cerebral ischemia. 1. Review of literature. *Cerebrovasc. Dis,* 1991, 1 (suppl 1), 2-15.

[10] The National Institute of Neurological Disorders and Stroke rt-PA Stroke Study Group. Tissue plasminogen activator for acute ischemic stroke. *N. Engl. J. Med,* 1995, 333, 1581-7.

[11] O'Collins VE; Macleod MR; Donnan GA; Horky LL; van der Worp BH; Howells DW. 1,026 experimental treatments in acute stroke. *Ann. Neurol,* 2006, 59, 467-77.

[12] Fisher M. Recommendations for advancing development of acute stroke therapies: Stroke Therapy Academic Industry Roundtable 3. *Stroke,* 2003, 34, 1539-46.

[13] Muir KW; Buchan A; von Kummer R; Rother J; Baron JC. Imaging of acute stroke. *Lancet Neurol,* 2006, 5, 755-68.

[14] Astrup J; Siesjo BK; Symon L. Thresholds in cerebral ischemia - the ischemic penumbra. *Stroke,* 1981, 12, 723-5.

[15] Green RA; Odergren T; Ashwood T. Animal models of stroke: do they have value for discovering neuroprotective agents? *Trends Pharmacol. Sci,* 2003, 24, 402-8.

[16] Warach S. Measurement of the ischemic penumbra with MRI: it's about time. *Stroke,* 2003, 34, 2533-4.

[17] Meng X; Fisher M; Shen Q; Sotak CH; Duong TQ. Characterizing the diffusion/perfusion mismatch in experimental focal cerebral ischemia. *Ann. Neurol,* 2004, 55, 207-12.

[18] Bardutzky J; Shen Q; Henninger N; Schwab S; Duong TQ; Fisher M. Characterizing tissue fate after transient cerebral ischemia of varying duration using quantitative diffusion and perfusion imaging. *Stroke,* 2007, 38, 1336-44.

[19] Henninger N; Sicard KM; Schmidt KF; Bardutzky J; Fisher M. Comparison of ischemic lesion evolution in embolic versus mechanical middle cerebral artery occlusion in Sprague Dawley rats using diffusion and perfusion imaging. *Stroke,* 2006, 37, 1283-7.

[20] Bardutzky J; Shen Q; Henninger N; Bouley J; Duong TQ; Fisher M. Differences in ischemic lesion evolution in different rat strains using diffusion and perfusion imaging. *Stroke,* 2005, 36, 2000-5.

[21] Maulaz A; Piechowski-Jozwiak B; Michel P; Bogousslavsky J. Selecting patients for early stroke treatment with penumbra images. *Cerebrovasc. Dis,* 2005, 20 Suppl 2, 19-24.

[22] Baron JC. Mapping the ischaemic penumbra with PET: implications for acute stroke treatment. *Cerebrovasc. Dis,* 1999, 9, 193-201.

[23] Schellinger PD; Fiebach JB; Jansen O; Ringleb PA; Mohr A; Steiner T; Heiland S; Schwab S; Pohlers O; Ryssel H; Orakcioglu B; Sartor K; Hacke W. Stroke magnetic resonance imaging within 6 hours after onset of hyperacute cerebral ischemia. *Ann. Neurol,* 2001, 49, 460-9.

[24] Latchaw RE; Yonas H; Hunter GJ; Yuh WT; Ueda T; Sorensen AG; Sunshine JL; Biller J; Wechsler L; Higashida R; Hademenos G. Guidelines and recommendations for perfusion imaging in cerebral ischemia: A scientific statement for healthcare professionals by the writing group on perfusion imaging, from the Council on Cardiovascular Radiology of the American Heart Association. *Stroke,* 2003, 34, 1084-104.

[25] Bandera E; Botteri M; Minelli C; Sutton A; Abrams KR; Latronico N. Cerebral blood flow threshold of ischemic penumbra and infarct core in acute ischemic stroke: a systematic review. *Stroke,* 2006, 37, 1334-9.

[26] Heiss WD. Best measure of ischemic penumbra: positron emission tomography. *Stroke,* 2003, 34, 2534-5.

[27] Chalela JA; Kidwell CS; Nentwich LM; Luby M; Butman JA; Demchuk AM; Hill MD; Patronas N; Latour L; Warach S. Magnetic resonance imaging and computed tomography in emergency assessment of patients with suspected acute stroke: a prospective comparison. *Lancet,* 2007, 369, 293-8.

[28] Mansfield P; Coxon R; Glover P. Echo-planar imaging of the brain at 3.0 T: first normal volunteer results. *J. Comput. Assist. Tomogr,* 1994, 18, 339-43.

[29] Shellock FG. Magnetic resonance safety update 2002: implants and devices. *J. Magn. Reson. Imaging,* 2002, 16, 485-96.

[30] Le Bihan D; Breton E; Lallemand D; Grenier P; Cabanis E; Laval-Jeantet M. MR imaging of intravoxel incoherent motions: application to diffusion and perfusion in neurologic disorders. *Radiology,* 1986, 161, 401-7.

[31] Phan TG; Wright PM; Markus R; Howells DW; Davis SM; Donnan GA. Salvaging the ischaemic penumbra: more than just reperfusion? *Clin. Exp. Pharmacol. Physiol,* 2002, 29, 1-10.

[32] Schaefer PW. Applications of DWI in clinical neurology. *J. Neurol. Sci,* 2001, 186 Suppl 1, S25-35.

[33] Gass A; Niendorf T; Hirsch JG. Acute and chronic changes of the apparent diffusion coefficient in neurological disorders--biophysical mechanisms and possible underlying histopathology. *J. Neurol. Sci,* 2001, 186 Suppl 1, S15-23.

[34] Helenius J; Soinne L; Perkio J; Salonen O; Kangasmaki A; Kaste M; Carano RA; Aronen HJ; Tatlisumak T. Diffusion-weighted MR imaging in normal human brains in various age groups. *AJNR Am. J. Neuroradiol,* 2002, 23, 194-9.

[35] Pierpaoli C; Jezzard P; Basser PJ; Barnett A; Di Chiro G. Diffusion tensor MR imaging of the human brain. *Radiology,* 1996, 201, 637-48.

[36] Weisskoff RM; Chesler D; Boxerman JL; Rosen BR. Pitfalls in MR measurement of tissue blood flow with intravascular tracers: which mean transit time? *Magn. Reson. Med,* 1993, 29, 553-8.

[37] Heiland S. Disturbed brain fuction. In: von Kummer R; Back T, editors. *Magnetic resonace imaging in ischemic stroke*, Berlin: Springer-Verlag; 2006; 103-116.

[38] Ostergaard L; Weisskoff RM; Chesler DA; Gyldensted C; Rosen BR. High resolution measurement of cerebral blood flow using intravascular tracer bolus passages. Part I: Mathematical approach and statistical analysis. *Magn. Reson. Med,* 1996, 36, 715-25.

[39] Ostergaard L; Sorensen AG; Chesler DA; Weisskoff RM; Koroshetz WJ; Wu O; Gyldensted C; Rosen BR. Combined diffusion-weighted and

perfusion-weighted flow heterogeneity magnetic resonance imaging in acute stroke. *Stroke,* 2000, 31, 1097-103.

[40] Perkio J; Soinne L; Ostergaard L; Helenius J; Kangasmaki A; Martinkauppi S; Salonen O; Savolainen S; Kaste M; Tatlisumak T; Aronen HJ. Abnormal intravoxel cerebral blood flow heterogeneity in human ischemic stroke determined by dynamic susceptibility contrast magnetic resonance imaging. *Stroke,* 2005, 36, 44-9.

[41] Hossmann KA; Hoehn-Berlage M. Diffusion and perfusion MR imaging of cerebral ischemia. *Cerebrovasc. Brain Metab. Rev,* 1995, 7, 187-217.

[42] Calamante F; Thomas DL; Pell GS; Wiersma J; Turner R. Measuring cerebral blood flow using magnetic resonance imaging techniques. *J. Cereb. Blood Flow Metab,* 1999, 19, 701-35.

[43] Heiland S; Kreibich W; Reith W; Benner T; Dorfler A; Forsting M; Sartor K. Comparison of echo-planar sequences for perfusion-weighted MRI: which is best? *Neuroradiology,* 1998, 40, 216-21.

[44] Boxerman JL; Hamberg LM; Rosen BR; Weisskoff RM. MR contrast due to intravascular magnetic susceptibility perturbations. *Magn. Reson. Med,* 1995, 34, 555-66.

[45] Heiland S; Reith W; Forsting M; Sartor K. How do concentration and dosage of the contrast agent affect the signal change in perfusion-weighted magnetic resonance imaging? A computer simulation. *Magn. Reson. Imaging,* 2001, 19, 813-20.

[46] Bruening R; Berchtenbreiter C; Holzknecht N; Essig M; Wu RH; Simmons A; Heuck A; Maschek A; Meusel M; Williams SC; Cox T; Knopp MV; Reiser M. Effects of three different doses of a bolus injection of gadodiamide: assessment of regional cerebral blood volume maps in a blinded reader study. *AJNR Am. J. Neuroradiol,* 2000, 21, 1603-10.

[47] Dijkhuizen RM; Nicolay K. Magnetic resonance imaging in experimental models of brain disorders. *J. Cereb. Blood Flow Metab,* 2003, 23, 1383-402.

[48] Furlan M; Marchal G; Viader F; Derlon JM; Baron JC. Spontaneous neurological recovery after stroke and the fate of the ischemic penumbra. *Ann. Neurol,* 1996, 40, 216-26.

[49] Schlaug G; Benfield A; Baird AE; Siewert B; Lovblad KO; Parker RA; Edelman RR; Warach S. The ischemic penumbra: operationally defined by diffusion and perfusion MRI. *Neurology,* 1999, 53, 1528-37.

[50] Masdeu JC; Irimia P; Asenbaum S; Bogousslavsky J; Brainin M; Chabriat H; Herholz K; Markus HS; Martinez-Vila E; Niederkorn K; Schellinger PD; Seitz RJ. EFNS guideline on neuroimaging in acute stroke. Report of an EFNS task force. *Eur. J. Neurol,* 2006, 13, 1271-83.

[51] Duong TQ; Fisher M. Applications of diffusion/perfusion magnetic resonance imaging in experimental and clinical aspects of stroke. *Curr. Atheroscler. Rep,* 2004, 6, 267-73.

[52] Guadagno JV; Donnan GA; Markus R; Gillard JH; Baron JC. Imaging the ischaemic penumbra. *Curr. Opin. Neurol,* 2004, 17, 61-7.

[53] Sobesky J; Zaro Weber O; Lehnhardt FG; Hesselmann V; Neveling M; Jacobs A; Heiss WD. Does the mismatch match the penumbra? Magnetic resonance imaging and positron emission tomography in early ischemic stroke. *Stroke,* 2005, 36, 980-5.

[54] Parsons MW; Barber PA; Chalk J; Darby DG; Rose S; Desmond PM; Gerraty RP; Tress BM; Wright PM; Donnan GA; Davis SM. Diffusion- and perfusion-weighted MRI response to thrombolysis in stroke. *Ann. Neurol,* 2002, 51, 28-37.

[55] Guadagno JV; Warburton EA; Aigbirhio FI; Smielewski P; Fryer TD; Harding S; Price CJ; Gillard JH; Carpenter TA; Baron JC. Does the acute diffusion-weighted imaging lesion represent penumbra as well as core? A combined quantitative PET/MRI voxel-based study. *J. Cereb. Blood Flow Metab,* 2004, 24, 1249-54.

[56] Guadagno JV; Jones PS; Fryer TD; Barret O; Aigbirhio FI; Carpenter TA; Price CJ; Gillard JH; Warburton EA; Baron JC. Local relationships between restricted water diffusion and oxygen consumption in the ischemic human brain. *Stroke,* 2006, 37, 1741-8.

[57] Guadagno JV; Warburton EA; Jones PS; Fryer TD; Day DJ; Gillard JH; Carpenter TA; Aigbirhio FI; Price CJ; Baron JC. The diffusion-weighted lesion in acute stroke: heterogeneous patterns of flow/metabolism uncoupling as assessed by quantitative positron emission tomography. *Cerebrovasc. Dis,* 2005, 19, 239-46.

[58] Baron JC; Warach S. Imaging. *Stroke,* 2005, 36, 196-9.

[59] Ay H; Koroshetz WJ; Vangel M; Benner T; Melinosky C; Zhu M; Menezes N; Lopez CJ; Sorensen AG. Conversion of ischemic brain tissue into infarction increases with age. *Stroke,* 2005, 36, 2632-6.

[60] Neumann-Haefelin T. Disturbed proton diffusion. In: von Kummer R; Back T, editors. *Magnetic resonance imaging in ischemic stroke*, Berlin: Springer-Verlag; 2006; 117-131.

[61] Tatlisumak T. Is CT or MRI the method of choice for imaging patients with acute stroke? Why should men divide if fate has united? *Stroke,* 2002, 33, 2144-5.

[62] Baird AE; Warach S. Magnetic resonance imaging of acute stroke. *J. Cereb. Blood Flow Metab,* 1998, 18, 583-609.

[63] Fiebach J; Jansen O; Schellinger P; Knauth M; Hartmann M; Heiland S; Ryssel H; Pohlers O; Hacke W; Sartor K. Comparison of CT with diffusion-weighted MRI in patients with hyperacute stroke. *Neuroradiology,* 2001, 43, 628-32.

[64] Fiebach JB; Schellinger PD; Jansen O; Meyer M; Wilde P; Bender J; Schramm P; Juttler E; Oehler J; Hartmann M; Hahnel S; Knauth M; Hacke W; Sartor K. CT and diffusion-weighted MR imaging in randomized order: diffusion-weighted imaging results in higher accuracy and lower interrater variability in the diagnosis of hyperacute ischemic stroke. *Stroke,* 2002, 33, 2206-10.

[65] Li F; Han S; Tatlisumak T; Carano RA; Irie K; Sotak CH; Fisher M. A new method to improve in-bore middle cerebral artery occlusion in rats: demonstration with diffusion- and perfusion-weighted imaging. *Stroke,* 1998, 29, 1715-9.

[66] Hjort N; Christensen S; Solling C; Ashkanian M; Wu O; Rohl L; Gyldensted C; Andersen G; Ostergaard L. Ischemic injury detected by diffusion imaging 11 minutes after stroke. *Ann. Neurol,* 2005, 58, 462-5.

[67] Lovblad KO; Laubach HJ; Baird AE; Curtin F; Schlaug G; Edelman RR; Warach S. Clinical experience with diffusion-weighted MR in patients with acute stroke. *AJNR Am. J. Neuroradiol,* 1998, 19, 1061-6.

[68] Mack WJ; Komotar RJ; Mocco J; Coon AL; Hoh DJ; King RG; Ducruet AF; Ransom ER; Oppermann M; DeLaPaz R; Connolly ES, Jr. Serial magnetic resonance imaging in experimental primate stroke: validation of MRI for pre-clinical cerebroprotective trials. *Neurol. Res,* 2003, 25, 846-52.

[69] Kelly PJ; Hedley-Whyte ET; Primavera J; He J; Gonzalez RG. Diffusion MRI in ischemic stroke compared to pathologically verified infarction. *Neurology,* 2001, 56, 914-20.

[70] Schwamm LH; Koroshetz WJ; Sorensen AG; Wang B; Copen WA; Budzik R; Rordorf G; Buonanno FS; Schaefer PW; Gonzalez RG. Time course of lesion development in patients with acute stroke: serial diffusion- and hemodynamic-weighted magnetic resonance imaging. *Stroke,* 1998, 29, 2268-76.

[71] Miyabe M; Mori S; van Zijl PC; Kirsch JR; Eleff SM; Koehler RC; Traystman RJ. Correlation of the average water diffusion constant with cerebral blood flow and ischemic damage after transient middle cerebral artery occlusion in cats. *J. Cereb. Blood Flow Metab,* 1996, 16, 881-91.

[72] Pierpaoli C; Righini A; Linfante I; Tao-Cheng JH; Alger JR; Di Chiro G. Histopathologic correlates of abnormal water diffusion in cerebral ischemia: diffusion-weighted MR imaging and light and electron microscopic study. *Radiology,* 1993, 189, 439-48.

[73] Wang L; Yushmanov VE; Liachenko SM; Tang P; Hamilton RL; Xu Y. Late reversal of cerebral perfusion and water diffusion after transient focal ischemia in rats. *J. Cereb. Blood Flow Metab,* 2002, 22, 253-61.

[74] Li F; Silva MD; Sotak CH; Fisher M. Temporal evolution of ischemic injury evaluated with diffusion-, perfusion-, and T2-weighted MRI. *Neurology,* 2000, 54, 689-96.

[75] Kazemi M; Silva MD; Li F; Fisher M; Sotak CH. Investigation of techniques to quantify in vivo lesion volume based on comparison of water apparent diffusion coefficient (ADC) maps with histology in focal cerebral ischemia of rats. *Magn. Reson. Imaging,* 2004, 22, 653-9.

[76] Li F; Han SS; Tatlisumak T; Liu KF; Garcia JH; Sotak CH; Fisher M. Reversal of acute apparent diffusion coefficient abnormalities and delayed neuronal death following transient focal cerebral ischemia in rats. *Ann. Neurol,* 1999, 46, 333-42.

[77] Olah L; Wecker S; Hoehn M. Secondary deterioration of apparent diffusion coefficient after 1-hour transient focal cerebral ischemia in rats. *J. Cereb. Blood Flow Metab,* 2000, 20, 1474-82.

[78] Li F; Liu KF; Silva MD; Omae T; Sotak CH; Fenstermacher JD; Fisher M; Hsu CY; Lin W. Transient and permanent resolution of ischemic lesions on diffusion-weighted imaging after brief periods of focal ischemia in rats: correlation with histopathology. *Stroke,* 2000, 31, 946-54.

[79] Kidwell CS; Saver JL; Mattiello J; Starkman S; Vinuela F; Duckwiler G; Gobin YP; Jahan R; Vespa P; Kalafut M; Alger JR. Thrombolytic reversal of acute human cerebral ischemic injury shown by diffusion/perfusion magnetic resonance imaging. *Ann. Neurol,* 2000, 47, 462-9.

[80] Neumann-Haefelin T; Wittsack HJ; Wenserski F; Li TQ; Moseley ME; Siebler M; Freund HJ. Diffusion- and perfusion-weighted MRI in a patient with a prolonged reversible ischaemic neurological deficit. *Neuroradiology,* 2000, 42, 444-7.

[81] Doege CA; Kerskens CM; Romero BI; Brunecker P; Junge-Hulsing J; Muller B; Villringer A. MRI of small human stroke shows reversible diffusion changes in subcortical gray matter. *Neuroreport,* 2000, 11, 2021-4.

[82] Kidwell CS; Saver JL; Villablanca JP; Duckwiler G; Fredieu A; Gough K; Leary MC; Starkman S; Gobin YP; Jahan R; Vespa P; Liebeskind DS; Alger JR; Vinuela F. Magnetic resonance imaging detection of microbleeds before thrombolysis: an emerging application. *Stroke,* 2002, 33, 95-8.

[83] Arenillas JF; Rovira A; Molina CA; Grive E; Montaner J; Alvarez-Sabin J. Prediction of early neurological deterioration using diffusion- and perfusion-weighted imaging in hyperacute middle cerebral artery ischemic stroke. *Stroke,* 2002, 33, 2197-203.

[84] Davalos A; Blanco M; Pedraza S; Leira R; Castellanos M; Pumar JM; Silva Y; Serena J; Castillo J. The clinical-DWI mismatch: a new diagnostic approach to the brain tissue at risk of infarction. *Neurology,* 2004, 62, 2187-92.

[85] Hasegawa Y; Fisher M; Latour LL; Dardzinski BJ; Sotak CH. MRI diffusion mapping of reversible and irreversible ischemic injury in focal brain ischemia. *Neurology,* 1994, 44, 1484-90.

[86] Olah L; Wecker S; Hoehn M. Relation of apparent diffusion coefficient changes and metabolic disturbances after 1 hour of focal cerebral ischemia and at different reperfusion phases in rats. *J. Cereb. Blood Flow Metab,* 2001, 21, 430-9.

[87] Fiehler J; Foth M; Kucinski T; Knab R; von Bezold M; Weiller C; Zeumer H; Rother J. Severe ADC decreases do not predict irreversible tissue damage in humans. *Stroke,* 2002, 33, 79-86.

[88] Kidwell CS; Alger JR; Saver JL. Beyond mismatch: evolving paradigms in imaging the ischemic penumbra with multimodal magnetic resonance imaging. *Stroke,* 2003, 34, 2729-35.

[89] Warach S. Stroke neuroimaging. *Stroke,* 2003, 34, 345-7.

[90] Ay H; Oliveira-Filho J; Buonanno FS; Ezzeddine M; Schaefer PW; Rordorf G; Schwamm LH; Gonzalez RG; Koroshetz WJ. Diffusion-weighted imaging identifies a subset of lacunar infarction associated with embolic source. *Stroke,* 1999, 30, 2644-50.

[91] Wessels T; Rottger C; Jauss M; Kaps M; Traupe H; Stolz E. Identification of embolic stroke patterns by diffusion-weighted MRI in clinically defined lacunar stroke syndromes. *Stroke,* 2005, 36, 757-61.

[92] Baird AE; Lovblad KO; Schlaug G; Edelman RR; Warach S. Multiple acute stroke syndrome: marker of embolic disease? *Neurology,* 2000, 54, 674-8.

[93] Kang DW; Latour LL; Chalela JA; Dambrosia J; Warach S. Early ischemic lesion recurrence within a week after acute ischemic stroke. *Ann Neurol,* 2003, 54, 66-74.

[94] Sa de Camargo EC; Koroshetz WJ. Neuroimaging of ischemia and infarction. *NeuroRx,* 2005, 2, 265-76.

[95] Grandin CB; Duprez TP; Smith AM; Oppenheim C; Peeters A; Robert AR; Cosnard G. Which MR-derived perfusion parameters are the best predictors of infarct growth in hyperacute stroke? Comparative study between relative and quantitative measurements. *Radiology,* 2002, 223, 361-70.

[96] Wittsack HJ; Ritzl A; Fink GR; Wenserski F; Siebler M; Seitz RJ; Modder U; Freund HJ. MR imaging in acute stroke: diffusion-weighted and perfusion imaging parameters for predicting infarct size. *Radiology,* 2002, 222, 397-403.

[97] Simonsen CZ; Rohl L; Vestergaard-Poulsen P; Gyldensted C; Andersen G; Ostergaard L. Final infarct size after acute stroke: prediction with flow heterogeneity. *Radiology,* 2002, 225, 269-75.

[98] Sohn CH; Sevick RJ; Frayne R. Contrast-enhanced MR angiography of the intracranial circulation. *Magn. Reson. Imaging Clin. N Am,* 2003, 11, 599-614.

[99] Reese T; Bochelen D; Sauter A; Beckmann N; Rudin M. Magnetic resonance angiography of the rat cerebrovascular system without the use of contrast agents. *NMR Biomed,* 1999, 12, 189-96.

[100] Beckmann N. High resolution magnetic resonance angiography non-invasively reveals mouse strain differences in the cerebrovascular anatomy in vivo. *Magn. Reson. Med,* 2000, 44, 252-8.

[101] Schellinger PD; Fiebach JB. Intracranial hemorrhage: the role of magnetic resonance imaging. *Neurocrit. Care,* 2004, 1, 31-45.

[102] Gomori JM; Grossman RI. Mechanisms responsible for the MR appearance and evolution of intracranial hemorrhage. *Radiographics,* 1988, 8, 427-40.

[103] Gomori JM; Grossman RI; Goldberg HI; Zimmerman RA; Bilaniuk LT. Intracranial hematomas: imaging by high-field MR. *Radiology,* 1985, 157, 87-93.

[104] Krishnamoorthy T; Fiorelli M. MR detection of intracranial hemorrhage. In: von Kummer R; Back T, editors. *Magnetic resonance imaging in ischemic stroke*, Berlin: Springer-Verlag; 2006; 159-169.

[105] Fiebach JB; Schellinger PD; Gass A; Kucinski T; Siebler M; Villringer A; Olkers P; Hirsch JG; Heiland S; Wilde P; Jansen O; Rother J; Hacke W; Sartor K. Stroke magnetic resonance imaging is accurate in hyperacute intracerebral hemorrhage: a multicenter study on the validity of stroke imaging. *Stroke,* 2004, 35, 502-6.

[106] Kidwell CS; Chalela JA; Saver JL; Starkman S; Hill MD; Demchuk AM; Butman JA; Patronas N; Alger JR; Latour LL; Luby ML; Baird AE; Leary MC; Tremwel M; Ovbiagele B; Fredieu A; Suzuki S; Villablanca JP; Davis S; Dunn B; Todd JW; Ezzeddine MA; Haymore J; Lynch JK; Davis L; Warach S. Comparison of MRI and CT for detection of acute intracerebral hemorrhage. *Jama,* 2004, 292, 1823-30.

[107] Hartmann M; Jansen O; Deinsberger W; Vogel J; Sartor K. MRI of acute experimental intracerebral hematoma. *Neurol. Res,* 2000, 22, 512-6.

[108] Peeling J; Yan HJ; Chen SG; Campbell M; Del Bigio MR. Protective effects of free radical inhibitors in intracerebral hemorrhage in rat. *Brain Res,* 1998, 795, 63-70.

[109] Strbian D; Tatlisumak T; Ramadan UA; Lindsberg PJ. Mast cell blocking reduces brain edema and hematoma volume and improves outcome after experimental intracerebral hemorrhage. *J. Cereb. Blood Flow Metab,* 2006, 27, 795-802.

[110] Kakuda W; Thijs VN; Lansberg MG; Bammer R; Wechsler L; Kemp S; Moseley ME; Marks MP; Albers GW. Clinical importance of microbleeds in patients receiving IV thrombolysis. *Neurology,* 2005, 65, 1175-8.

[111] Kim HS; Lee DH; Ryu CW; Lee JH; Choi CG; Kim SJ; Suh DC. Multiple cerebral microbleeds in hyperacute ischemic stroke: impact on prevalence and severity of early hemorrhagic transformation after thrombolytic treatment. *AJR Am. J. Roentgenol,* 2006, 186, 1443-9.

[112] Hermier M; Nighoghossian N; Derex L; Berthezene Y; Blanc-Lasserre K; Trouillas P; Froment JC. MRI of acute post-ischemic cerebral hemorrhage in stroke patients: diagnosis with T2*-weighted gradient-echo sequences. *Neuroradiology,* 2001, 43, 809-15.

[113] Knight RA; Barker PB; Fagan SC; Li Y; Jacobs MA; Welch KM. Prediction of impending hemorrhagic transformation in ischemic stroke using magnetic resonance imaging in rats. *Stroke,* 1998, 29, 144-51.

[114] Dijkhuizen RM; Asahi M; Wu O; Rosen BR; Lo EH. Delayed rt-PA treatment in a rat embolic stroke model: diagnosis and prognosis of ischemic injury and hemorrhagic transformation with magnetic resonance imaging. *J. Cereb. Blood Flow Metab,* 2001, 21, 964-71.

[115] Neumann-Haefelin C; Brinker G; Uhlenkuken U; Pillekamp F; Hossmann KA; Hoehn M. Prediction of hemorrhagic transformation after thrombolytic therapy of clot embolism: an MRI investigation in rat brain. *Stroke,* 2002, 33, 1392-8.

[116] Jiang Q; Zhang RL; Zhang ZG; Knight RA; Ewing JR; Ding G; Lu M; Arniego P; Zhang L; Hu J; Li Q; Chopp M. Magnetic resonance imaging

characterization of hemorrhagic transformation of embolic stroke in the rat. *J. Cereb. Blood Flow Metab,* 2002, 22, 559-68.

[117] Vo KD; Santiago F; Lin W; Hsu CY; Lee Y; Lee JM. MR imaging enhancement patterns as predictors of hemorrhagic transformation in acute ischemic stroke. *AJNR Am. J. Neuroradiol,* 2003, 24, 674-9.

[118] Kim EY; Na DG; Kim SS; Lee KH; Ryoo JW; Kim HK. Prediction of hemorrhagic transformation in acute ischemic stroke: role of diffusion-weighted imaging and early parenchymal enhancement. *AJNR Am. J. Neuroradiol,* 2005, 26, 1050-5.

[119] Latour LL; Kang DW; Ezzeddine MA; Chalela JA; Warach S. Early blood-brain barrier disruption in human focal brain ischemia. *Ann. Neurol,* 2004, 56, 468-77.

[120] Fenstermacher JD; Knight RA; Ewing JR; Nagaraja T; Nagesh V; Yee JS; Arniego PA. Estimating blood-brain barrier opening in a rat model of hemorrhagic transformation with Patlak plots of Gd-DTPA contrast-enhanced MRI. *Acta Neurochir. Suppl,* 2003, 86, 35-7.

[121] Jiang Q; Ewing JR; Ding GL; Zhang L; Zhang ZG; Li L; Whitton P; Lu M; Hu J; Li QJ; Knight RA; Chopp M. Quantitative evaluation of BBB permeability after embolic stroke in rat using MRI. *J. Cereb. Blood Flow Metab,* 2005, 25, 583-92.

[122] Knight RA; Nagesh V; Nagaraja TN; Ewing JR; Whitton PA; Bershad E; Fagan SC; Fenstermacher JD. Acute blood-brain barrier opening in experimentally induced focal cerebral ischemia is preferentially identified by quantitative magnetization transfer imaging. *Magn. Reson. Med,* 2005, 54, 822-32.

[123] Macleod MR; O'Collins T; Howells DW; Donnan GA. Pooling of animal experimental data reveals influence of study design and publication bias. *Stroke,* 2004, 35, 1203-8.

[124] Lees KR; Zivin JA; Ashwood T; Davalos A; Davis SM; Diener HC; Grotta J; Lyden P; Shuaib A; Hardemark HG; Wasiewski WW. NXY-059 for acute ischemic stroke. *N. Engl. J. Med,* 2006, 354, 588-600.

[125] Ehrenreich H; Hasselblatt M; Dembowski C; Cepek L; Lewczuk P; Stiefel M; Rustenbeck HH; Breiter N; Jacob S; Knerlich F; Bohn M; Poser W; Ruther E; Kochen M; Gefeller O; Gleiter C; Wessel TC; De Ryck M; Itri L; Prange H; Cerami A; Brines M; Siren AL. Erythropoietin therapy for acute stroke is both safe and beneficial. *Mol. Med,* 2002, 8, 495-505.

[126] Furlan A; Higashida R; Wechsler L; Gent M; Rowley H; Kase C; Pessin M; Ahuja A; Callahan F; Clark WM; Silver F; Rivera F. Intra-arterial prourokinase for acute ischemic stroke. The PROACT II study: a randomized controlled trial. Prolyse in Acute Cerebral Thromboembolism. *JAMA,* 1999, 282, 2003-11.

[127] Sherman DG; Atkinson RP; Chippendale T; Levin KA; Ng K; Futrell N; Hsu CY; Levy DE. Intravenous ancrod for treatment of acute ischemic

stroke: the STAT study: a randomized controlled trial. Stroke Treatment with Ancrod Trial. *JAMA,* 2000, 283, 2395-403.

[128] Grotta J. Neuroprotection is unlikely to be effective in humans using current trial designs. *Stroke,* 2002, 33, 306-7.

[129] Dirnagl U. Bench to bedside: the quest for quality in experimental stroke research. *J. Cereb. Blood Flow Metab,* 2006, 26, 1465-78.

[130] Fisher M; Tatlisumak T. Use of animal models has not contributed to development of acute stroke therapies: con. *Stroke,* 2005, 36, 2324-5.

[131] Savitz SI; Fisher M. Future of neuroprotection for acute stroke: In the aftermath of the SAINT trials. *Ann. Neurol,* 2007, [Epub ahead of print].

[132] Fisher M; Henninger N. Translational research in stroke: taking advances in the pathophysiology and treatment of stroke from the experimental setting to clinical trials. *Curr. Neurol. Neurosci. Rep,* 2007, 7, 35-41.

[133] Savitz SI. A critical appraisal of the NXY-059 neuroprotection studies for acute stroke: A need for more rigorous testing of neuroprotective agents in animal models of stroke. *Exp. Neurol,* 2007, [Epub ahead of print].

[134] Fisher M; Ratan R. New perspectives on developing acute stroke therapy. *Ann. Neurol,* 2003, 53, 10-20.

[135] Warach S; Kaufman D; Chiu D; Devlin T; Luby M; Rashid A; Clayton L; Kaste M; Lees KR; Sacco R; Fisher M. Effect of the Glycine Antagonist Gavestinel on cerebral infarcts in acute stroke patients, a randomized placebo-controlled trial: The GAIN MRI Substudy. *Cerebrovasc. Dis,* 2006, 21, 106-11.

[136] Rudin M; Beckmann N; Porszasz R; Reese T; Bochelen D; Sauter A. In vivo magnetic resonance methods in pharmaceutical research: current status and perspectives. *NMR Biomed,* 1999, 12, 69-97.

[137] Lo EH; Matsumoto K; Pierce AR; Garrido L; Luttinger D. Pharmacologic reversal of acute changes in diffusion-weighted magnetic resonance imaging in focal cerebral ischemia. *J. Cereb. Blood Flow Metab,* 1994, 14, 597-603.

[138] Minematsu K; Fisher M; Li L; Davis MA; Knapp AG; Cotter RE; McBurney RN; Sotak CH. Effects of a novel NMDA antagonist on experimental stroke rapidly and quantitatively assessed by diffusion-weighted MRI. *Neurology,* 1993, 43, 397-403.

[139] Yenari MA; Palmer JT; Sun GH; de Crespigny A; Mosely ME; Steinberg GK. Time-course and treatment response with SNX-111, an N-type calcium channel blocker, in a rodent model of focal cerebral ischemia using diffusion-weighted MRI. *Brain Res,* 1996, 739, 36-45.

[140] Kucharczyk J; Mintorovitch J; Moseley ME; Asgari HS; Sevick RJ; Derugin N; Norman D. Ischemic brain damage: reduction by sodium-calcium ion channel modulator RS-87476. *Radiology,* 1991, 179, 221-7.

[141] Muller TB; Haraldseth O; Jones RA; Sebastiani G; Lindboe CF; Unsgard G; Oksendal AN. Perfusion and diffusion-weighted MR imaging for in vivo evaluation of treatment with U74389G in a rat stroke model. *Stroke,* 1995, 26, 1453-8.

[142] Tatlisumak T; Carano RA; Takano K; Meiler MR; Li F; Sotak CH; Arndts D; Pschorn U; Fisher M. Broad-spectrum cation channel inhibition by LOE 908 MS reduces infarct volume in vivo and postmortem in focal cerebral ischemia in the rat. *J. Neurol. Sci,* 2000, 178, 107-13.

[143] Tatlisumak T; Takano K; Carano RA; Fisher M. Effect of basic fibroblast growth factor on experimental focal ischemia studied by diffusion-weighted and perfusion imaging. *Stroke,* 1996, 27, 2292-7.

[144] Tatlisumak T; Carano RA; Takano K; Opgenorth TJ; Sotak CH; Fisher M. A novel endothelin antagonist, A-127722, attenuates ischemic lesion size in rats with temporary middle cerebral artery occlusion: a diffusion and perfusion MRI study. *Stroke,* 1998, 29, 850-7.

[145] Ebisu T; Katsuta K; Fujikawa A; Aoki I; Umeda M; Naruse S; Tanaka C. Early and delayed neuroprotective effects of FK506 on experimental focal ischemia quantitatively assessed by diffusion-weighted MRI. *Magn. Reson. Imaging,* 2001, 19, 153-60.

[146] Shi J; Bui JD; Yang SH; He Z; Lucas TH; Buckley DL; Blackband SJ; King MA; Day AL; Simpkins JW. Estrogens decrease reperfusion-associated cortical ischemic damage: an MRI analysis in a transient focal ischemia model. *Stroke,* 2001, 32, 987-92.

[147] Warach S; Pettigrew LC; Dashe JF; Pullicino P; Lefkowitz DM; Sabounjian L; Harnett K; Schwiderski U; Gammans R. Effect of citicoline on ischemic lesions as measured by diffusion-weighted magnetic resonance imaging. Citicoline 010 Investigators. *Ann. Neurol,* 2000, 48, 713-22.

[148] Sacco RL; DeRosa JT; Haley EC, Jr.; Levin B; Ordronneau P; Phillips SJ; Rundek T; Snipes RG; Thompson JL. Glycine antagonist in neuroprotection for patients with acute stroke: GAIN Americas: a randomized controlled trial. *Jama,* 2001, 285, 1719-28.

[149] Lees KR; Asplund K; Carolei A; Davis SM; Diener HC; Kaste M; Orgogozo JM; Whitehead J. Glycine antagonist (gavestinel) in neuroprotection (GAIN International) in patients with acute stroke: a randomised controlled trial. GAIN International Investigators. *Lancet,* 2000, 355, 1949-54.

[150] Del Zoppo GJ. Thrombolysis: from the experimental findings to the clinical practice. *Cerebrovasc. Dis,* 2004, 17 Suppl 1, 144-52.

[151] Busch E; Kruger K; Allegrini PR; Kerskens CM; Gyngell ML; Hoehn-Berlage M; Hossmann KA. Reperfusion after thrombolytic therapy of embolic stroke in the rat: magnetic resonance and biochemical imaging. *J. Cereb. Blood Flow Metab,* 1998, 18, 407-18.

[152] Takano K; Carano RA; Tatlisumak T; Meiler M; Sotak CH; Kleinert HD; Fisher M. Efficacy of intra-arterial and intravenous prourokinase in an embolic stroke model evaluated by diffusion-perfusion magnetic resonance imaging. *Neurology,* 1998, 50, 870-5.

[153] Jiang Q; Zhang RL; Zhang ZG; Ewing JR; Divine GW; Chopp M. Diffusion-, T2-, and perfusion-weighted nuclear magnetic resonance imaging of middle cerebral artery embolic stroke and recombinant tissue

plasminogen activator intervention in the rat. *J. Cereb. Blood Flow Metab,* 1998, 18, 758-67.

[154] Hjort N; Butcher K; Davis SM; Kidwell CS; Koroshetz WJ; Rother J; Schellinger PD; Warach S; Ostergaard L. Magnetic resonance imaging criteria for thrombolysis in acute cerebral infarct. *Stroke,* 2005, 36, 388-97.

[155] Warach S; Wardlaw J. Advances in imaging 2005. *Stroke,* 2006, 37, 297-8.

[156] Butcher KS; Parsons M; MacGregor L; Barber PA; Chalk J; Bladin C; Levi C; Kimber T; Schultz D; Fink J; Tress B; Donnan G; Davis S. Refining the perfusion-diffusion mismatch hypothesis. *Stroke,* 2005, 36, 1153-9.

[157] Albers GW; Thijs VN; Wechsler L; Kemp S; Schlaug G; Skalabrin E; Bammer R; Kakuda W; Lansberg MG; Shuaib A; Coplin W; Hamilton S; Moseley M; Marks MP. Magnetic resonance imaging profiles predict clinical response to early reperfusion: the diffusion and perfusion imaging evaluation for understanding stroke evolution (DEFUSE) study. *Ann. Neurol,* 2006, 60, 508-17.

In: Drug Design Research Perspectives
Editor: Stanley P. Kaplan, pp. 213-249

ISBN: 978-1-60021-823-1

Chapter VIII

Understanding the Interaction and the Structure–Activity Correlation of HIV-1 RT Inhibitors of Efavirenz Derivatives and WT and K103N HIV-1 RT Using Molecular Docking, 3D-QSAR Approaches and Quantum Chemical Calculations

***Pornpan Pungpo*[1*], *Auradee Punkvang*[1], *Patchreenart Saparpakorn*[2] and *Supa Hannongbua*[2]**

[1] Faculty of Science, Ubonratchathani University, Ubonratchathani, Thailand, 34190

[2] Faculty of Science, Kasetsart University, Bangkok, Thailand, 10900

Abstract

Molecular docking and 3D-QSAR analyses were performed to understand the interaction between a series of efavirenz derivatives with WT and K103N HIV-1 RT. To model the potential binding modes of efavirenz derivatives in the binding pocket of WT and K103N HIV-1 RT, molecular docking approaches by using GOLD and Autodock 3.05 programs were performed. The results show that the docking results obtained from both methods reveal a good ability to reproduce the X-ray bound conformation with rmsd less than 1.0 Å for both WT and mutant enzymes. The docking calculations of all efavirenz derivatives in the data set were, consecutively, performed to elucidate their orientations in the binding pockets. The results derived from docking analysis give additional information and further probes the inhibitor-enzyme interactions. The correlation of the results obtained from docking models and the inhibitory activities validate each other and lead to better understanding of the structural requirements for the activity. Therefore, these results are informative to improve the development of more efficient HIV-1 RT inhibitors, especially, active against mutant enzyme. Based on the molecular alignment of conformations obtained from molecular docking procedures, the high predictive 3D-QSAR models were produced by using CoMFA and CoMSIA approaches. The CoMFA models reveal the importance of steric and electrostatic interactions through contour maps. The resulting CoMSIA models enhance the understanding of steric, electrostatic, hydrophobic, electron donor and acceptor requirements for ligands binding to the K103N HIV-1 RT. Moreover, quantum chemical calculations were carried out to analyze the interaction energies of the selected inhibitors with the individual amino acids in the binding pocket. The obtained results show the important interactions of inhibitors with the enzyme at the residue level. Consequently, the results obtained from structure-based, ligand-based design approaches and quantum chemical calculations can be integrated to identify the structural requirements of HIV-1 RT inhibitors in the class of efavirenz compounds. The principle derived from the present study provides a gainful guideline to design and predict new and more potent compounds active against WT and K103N HIV-1 RT.

Introduction

Human immunodeficiency virus type-1 (HIV-1) is the primary etiological agent of the Acquired Immune Deficiency Syndrome (AIDS). HIV-1 reverse transcriptase (RT) plays an essential role for virus replication from single-stranded RNA viral genome into a double-stranded proviral DNA. As RT is virus specific and acting before viral integration into the host chromosome, it is an attractive target for several antiviral therapeutic agents used for current anti-HIV therapy. Over the past decade, numerous HIV-1 RT inhibitors have been developed and classified into two major classes [1,2]. The nucleoside RT inhibitors (NRTI) such as AZT, ddC and ddI have been widely used to treat AIDS patients. NRTI are attractive drug candidates in that their binding site is unique to HIV-1 RT. Thus these derivatives are high cellular toxicity and several side effects result from the disruption of normal DNA polymerase activity. The other class of HIV-1 RT inhibitors is the non-nucleosides RT inhibitors (NNRTI). NNRTI are highly specific for HIV-1 RT by binding an allosteric site in a noncompetitive manner with respect to the substrate and displacing the polymerase active site catalytic residues [3]. The NNRTI have been classified into first and second generation NNRTI. The first generation NNRTI such as HEPT, TIBO and dipyridodiazepinone or nevirapine [4] are much less toxic than the NRTI. However, a serious problem with the NNRTI is the emergence of drug-resistant variants of HIV-1, both in cell culture and in patients, leading to the decreased therapeutic efficacy of NNRTI [5,6]. The primary cause of this viral resistance to NNRTI is the mutation of the enzyme, which changes the shape of the pocket, where most of the reported NNRTI are bound [7]. A Common features among these are mutations of Y181C, Y188C and K103N which confer some degree of resistance to all of the NNRTI [8,9]. In clinical trials and therapeutic use of NNRTI, the most commonly observed mutation within RT is K103N [10]. Therefore, the second generation NNRTI such as efavirenz, carboxanilides, and quinoxalines is identified and discovered for high binding affinity in both the absence and presence of specific mutations. The second generation NNRTI have a more favorable resistance profile, showing small losses of activity against many common drug-resistance mutation.

Efavirenz, (-)-6-chloro-4-cyclopropylethynyl-4-trifluoromethyl-1,4-dihydro-2*H*-3,1-benzoxazin-2-one, is one of 20 anti-HIV drugs approved for clinical use [11]. Although HIV-

1 infection is generally well controlled by efavirenz-containing regimens, one of the difficulties encountered clinically during antiviral therapy has been the emergence of resistance mutant viral forms. 90% of patients who have rebounded after efavirenz-inclusive HAART possess the K103N mutation [12]. Like other NNRTI, the K103N mutation is the most frequent mutation observed within RT resulting from the therapeutic interventions involving NNRTI [13-15]. Since the current available drug could not cover enough to completely eradicate HIV within the AIDS patients, to provide the optimum chemical benefit, the inhibitor-enzyme interactions and the structure–activity correlation of efavirenz derivatives and the WT and K103N HIV-1 RT are necessary to identify.

There is considerable interest in SAR investigations of efavirenz derivatives [16-24]. Moreover, the X-ray crystallographic structures of efavirenz in complex with both wild-type and K103N HIV-1 RT were determined leading to an understanding of some structural factors that confer resilience to drug-resistance mutation [25-26]. Molecular modeling and computer-aided molecular design approaches have been applied to develop informative models for rational drug design. A review about the structural information and drug–enzyme interaction of the NNRTI based on computational chemistry approaches has been recently reported [27]. Moreover, the activity prediction of efavirenz analogues with the K103N RT was examined by MC/ELR calculations identifying controlling factors for binding with K103N mutant [28]. The effect of the K103N mutation of HIV-1 RT on the activity of efavirenz analogues was studied via MC/FEP calculations. The relative fold resistance energies indicate that efavirenz binds to K103N RT in a manner similar to the wild-type enzyme. The improved performance of the quinazolinones against the mutant enzyme is attributed to formation of a more optimal hydrogen-bonding network with bridging water molecules between the ligands and E138 [29]. A series of targeted MD simulations were studied on the effect that the common K103N mutation in HIV-1 reverse transcriptase has on the binding of NNRTI, including efavirenz [30]. The obtained results suggest that resilience to the insidious effects of the irksome K103N mutation can be attained by designing ligands that are able to disrupt the N103 – Y188 interaction. Molecular mechanics PBSA ligand binding energy and interaction of efavirenz derivatives with HIV-1 RT were studied based on molecular docking and MD simulations [31]. The results indicate that the tight association of the ligand to the HIV-1 RT binding pocket was based on hydrogen bonding between efavirenz's N1 and the oxygen of backbone of K101. Moreover, electrostatic interaction was mainly contributed by two amino acid residues in the binding site; K101 and H235. The hypothetical 3D pharmacophore model for NNRTI was developed by using a combined ligand- and structure-based molecular modeling approach [32]. The obtained information leads to the design and synthesis a new potential class of NNRTI.

As the X-ray crystallographic data of HIV-1 RT complexed with NNRTI are available, it is possible to apply structure-based drug design techniques using molecular docking calculations to predict the potential binding mode of ligand in a receptor binding site and also yield important information for other ligands binding to the same target [27,33-41]. The docking analysis provides additional structural information and further probes the interactions of inhibitors and the HIV-1 RT binding site. Three dimensional quantitative structure-activity relationship (3D-QSAR) method by using comparative molecular field analysis (CoMFA) [42] and comparative molecular similarity indices analysis (CoMSIA) [43] are highly active areas of research in drug design. A number of QSAR studies were reported to identify important structural features responsible for the inhibition by NNRTI [27,37,38, 44-49]. As the molecular alignment is a crucial step to a successful 3D-QSAR model by CoMFA and CoMSIA, the alignment scheme generated from the docking study shows a promising molecular alignment producing reliable QSAR models for the NNRTI [38, 50-52].

Recently, quantum chemical calculations were successfully used to calculate the interaction energies and the binding energies of the NNRTI, such as TIBO and nevirapine derivatives, in the HIV-1 RT binding pocket. These results reveal the effects of surrounding amino acid residues to the NNRTI [53, 54]. Based on the B3LYP/6-31G(d,p) and ONIOM2 methods, particular interaction between efavirenz and the HIV-1 RT binding site was investigated. The stability of this complex system indicates that hydrogen bond interactions play an important role in the bound efavirenz/HIV-1 RT complex [55]. The calculated interaction energies using MP2/6-31G(d,p) level of theory could confirm the new designed nevirapine derivatives having the hydrogen bond interaction either N103 or C181 amino acid residues [36]. The results are in agreement with the previous study reported that density functional method can not handle for the H-π interaction calculation, whereas MP2 method of calculations has taken into account the H-π interaction [54] .

In the present study, structure-based drug design approaches using molecular docking calculations were performed to explore the potential binding mode of efavirenze derivatives in the WT and K103N binding pockets and to elucidate the inhibitor-enzyme interaction of the inhibitors in the corresponding binding sites. Then, the binding conformations obtained from docking calculations were subjected to 3D-QSAR studies by using CoMFA and CoMSIA approaches to determine the different structural requirements for the WT and K103N inhibition activities. Additionally, quantum chemical calculations were carried out to calculate the interaction energies of selected inhibitors with the individual amino acid residues in the binding pockets.

Methods of Calculation

Biological Data

The chemical structures of various efavirenz derivatives and their inhibitory activities against both wild type (WT RT) and mutant type (K103N) are reported [19, 20, 22]. All presented biological data are from the same laboratory and were estimated under the same experimental conditions. The potency has been defined as log (1/C), where C is the ability of efavirenz derivatives to inhibit wild-type and K103N RF strain of HIV-1 (IC_{90}) [56]. In the present analysis, 56 efavirenz derivatives were separated into two groups; 49 compounds served as the training set and 7 compounds serve as the test set, sampling from structurally diverse molecules possessing various range of logarithm unit of activities, reported in table 1.

Structures of 56 efavirenz derivatives and experimental biological activities against both WT RT and K103N RT

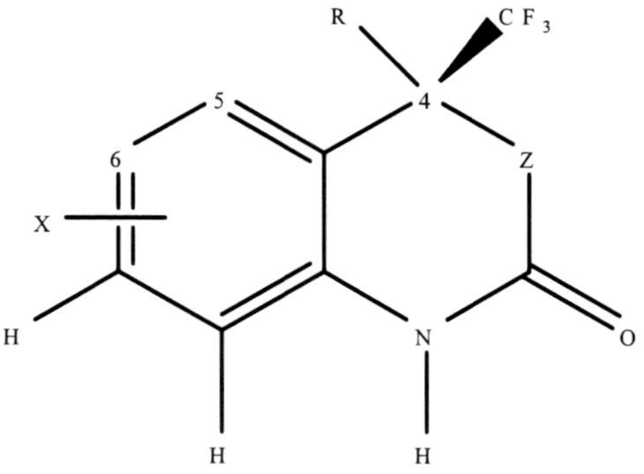

Compound no.	X	R	Z	Experimental log(1/C)	
				WT	K103N
01	6-Cl	CC-cyclopropyl	O	8.77	7.19
02	6-Cl	CC-2-pyridyl	O	8.27	5.96
03[a]	6-Cl	CC-3-pyridyl	O	8.40	6.94
04	6-Cl	CC-2-furanyl	O	8.42	6.82
05	6-Cl	CC-3-furanyl	O	8.42	6.49
06	5,6-diF	CC-3-pyridyl	O	8.65	7.23
07	6-F	CC-3-furanyl	O	8.60	6.43
08	6-F	CC-3-pyridyl	O	8.59	6.48
09	5,6-diF	CC-3-furanyl	O	8.49	6.55
10	5,6-diF	CC-2-thienyl	O	8.65	6.81
11	5,6-diF	CC-3-thienyl	O	8.63	6.86
12	6-Cl	$OCH_2CH_2CH_2CH_3$	O	7.99	6.00
13	6-Cl	$OCH_2CH_2CH(CH_3)_2$	O	8.00	6.54
14	6-Cl	$OCH_2CHCH(CH_3)$cis	O	8.36	6.63
15	6-Cl	$OCH_2CHCH(CH_3)$tran	O	8.25	6.39
16	6-Cl	$OCH_2CHC(CH_3)_3$	O	8.5 7	7.08
17	6-Cl	OCH_2CCCH_3	O	8.50	6.51
18	6-Cl	OCH_2CHCCl_2	O	8.02	6.62
19[a]	6-F	$OCH_2CHC(CH_3)_2$	O	8.53	6.97
20	6-F	$OCH_2CHCH(CH_3)$tran	O	8.05	5.94
21	5,6-diF	$OCH_2CHC(CH_3)_2$	O	8.81	7.19
22[a]	5,6-diF	OCH_2CHCH_2	O	8.19	5.79
23	5,6-diF	OCH_2CHCCl_2	O	8.20	6.74
24	5,6-diF	CC-ethyl	NH	8.82	7.85
25	5-F	CC-cyclopropyl	NH	8.85	7.05
26	5-Cl,6-F	CC-isopropyl	NH	8.52	7.82
27	5-Cl	CC-cyclopropyl	NH	8.60	7.2
28	5,6-diF	CC-cyclopropyl	NH	8.68	7.89

Table 1. (Continued)

Compound no.	X	R	Z	Experimental log(1/C)	
				WT	K103N
29	5,6-diF	CC-isopropyl	NH	8.68	7.85
30	6-F	CC-cyclopropyl	NH	8.70	7.32
31	5,6-diF	CC-2-pyridyl	NH	8.70	6.96
32	6-F	CC-ethyl	NH	8.60	7.15
33	5-Cl,6-F	CC-cyclopropyl	NH	8.57	7.74
34	6-MeO	CC-cyclopropyl	NH	8.54	7.4
35	6-F	CC-2-pyridyl	NH	8.30	6.32
36	5-F,6-Cl	CC-cyclopropyl	NH	8.32	7.74
37[a]	5-Cl,6-F	CC-2-pyridyl	NH	8.64	7.14
38	6-Cl	CC-cyclopropyl	NH	8.57	7.66
39[a]	6-MeO	CC-isopropyl	NH	8.42	7.25
40	6-MeO	CC-phenyl	NH	8.49	6.55
41	5,6-diF	CC-phenyl	NH	8.21	6.72
42[a]	6-F	CC-phenyl	NH	8.18	6.49
43	6-Cl	CC-2-pyridyl	NH	8.47	6.8
44	6-Cl	CC-ethyl	NH	8.48	7.59
45	6-Cl	CC-phenyl	NH	8.15	6.6
46	6-F	CC-isopropyl	NH	8.59	7.57
47[a]	5,6-diCl	CC-cyclopropyl	NH	8.10	7.74
48	6-Cl	CC-isopropyl	NH	8.52	7.66
49	6-MeO	CC-2-pyridyl	NH	8.09	6.47
50	5-MeO,6-Cl	CC-cyclopropyl	NH	8.46	8.12
51	5-MeO,6-Cl	CC-phenyl	NH	8.10	6.95
52	5-MeO,6-Cl	CC-3-pyridyl	NH	8.15	6.86
53	5-OH,6-Cl	CC-cyclopropyl	NH	8.44	7.55
54	6-Cl	CHCO-cyclopropyl	NH	8.44	7.12
55	6-Cl	CHCO-phenyl	NH	8.09	7.34
56	6-Cl	CHCO-3-pyridyl	NH	8.34	7.12

[a] seven compounds comprising of compounds 03,19,22,37,39,42 and 47, used as the test set

Molecular Modeling

As the availability of enzyme-inhibitor complexes (pdb codes 1fk9 and 1fko for the efavirenz/WT HIV-1 RT complex and the efavirenz/K103N HIV-1 RT complex, respectively), the starting geometries of efavirenz structures are obtained from the crystallographic structures [25]. All efavirenz derivatives were constructed and modified by ALCHEMY 2000 program [57]. Then, full geometry optimization, based on *ab-initio* molecular orbital method at the HF/3-21G level of theory, of all structures was carried out. The calculations were performed by the GAUSSIAN 03 program [58]. Partial atomic charges required for the calculations of the electrostatic interactions were then derived.

Docking Study

To investigate the potential binding modes of efavirenz derivatives, all derivatives were docked to WT (pdb code 1fk9) and K103N (pdb code 1fko) HIV-1 RT by using docking programs. Two docking programs, GOLD [59] and Autodock 3.05 [60], were applied to determine the ability of these programs for reproducing the efavirenz's X-ray orientation in its binding pocket for the efavirenz/RT complexes of WT and K103N mutation. GOLD (Genetic Optimization for Ligand Docking) is a docking method using a genetic algorithm (GA) for docking flexible ligands into protein binding sites. In this study, GA parameters were set to 7-8 times speed-up setting. The number of chromosome in each population and the number of operation were set to 100 and 10,000, respectively. Docked conformation which is the highest GoldScore was selected. Autodock 3.05 keeps the macromolecule rigid, while allows torsional flexibility for the ligand. Docking ligand to macromolecule was carried out using Lamarckian Genetic Algorithm (LGA), with an initial population of 50 randomly placed individuals. Fifty independent docking runs were carried out for each ligand. Docked conformation represented the lowest final docked energy was selected. The root-mean-square deviation (rmsd) of the top-scored pose from the X-ray pose was used to determine the validation of the docking methods.

Molecular Alignment and Comfa and Comsia Set up

The alignment rule, i.e., the positioning of the molecular structure within a fixed lattice, is the key to successfully set up CoMFA and CoMSIA models. For these reasons, in the present study, the molecular alignments derived from the molecular docking procedures were used to construct CoMFA and CoMSIA models.

To obtain CoMFA and CoMSIA descriptors fields, a three dimensional cubic lattice, with 2 Å grid spacing, was generated automatically around these molecules and it was ensured that the grid extended the molecular dimensions by 4.0 Å in all directions, performed by Sybyl 7.0 [61]. CoMFA descriptors were computed using a sp^3 carbon atom with +1 charge (default probe atom in Sybyl), served as the probe atom to generate steric (Lennard–Jones 6-12 potential) field energies and electrostatic (Coulombic potential) fields. The probe atom was placed at each lattice point and their steric and electrostatic interactions with each atom in the molecule were all calculated with CoMFA standard scaling and then compiled in a CoMFA QSAR table. The energy cutoff values of 30 kcal/mol were selected for both electrostatic and steric fields.

CoMSIA similarity index descriptors were derived with the same lattice box as used for the CoMFA calculations, with a grid spacing of 2 Å using a common C probe atom of 1 Å radius, as well as a charge. CoMSIA similarity indices ($A_{F,K}$) for a molecule j with atoms i at a grid point are calculated according to the following equation;

$$A_{F,k}^{q}(j) = \sum_{i=1}^{n} w_{probe,k} w_{ik} e^{-\alpha r_{iq}^{2}}$$

Five physicochemical properties *k* (steric, electrostatic, hydrophobic, hydrogen bond donor and hydrogen bond acceptor) were evaluated using the probe atom. A Gaussian-type distance was used between the grid point *q* and atom *i* of the molecule. The default value of 0.3 was used as the attenuation factor ($\propto$). In CoMSIA, the steric indices are related to the third power of the atomic radii, the electrostatic descriptors are derived from atomic partial charges, the hydrophobic fields are derived from atom-based parameters [62]. and the hydrogen donor and acceptor indices are obtained by the rule-based method derived from the experimental values [63].

The CoMFA and CoMSIA descriptors were used as independent variables and log (1/C) values were used as dependent variables to derive the 3D QSAR model. Partial least square technique (PLS) was employed to derive a linear relationship and cross-validation was performed using the leave-one-out method with a 2.0 kcal/mol column filter to minimize the influence of the noisy columns. The analyses were carried out with a maximum of six components, and subsequently, using the number of component (noc) at which the difference in the r^2_{cv} value to the next one was less than 0.05 [64]. Consequently, a non cross-validated analysis was performed using the optimal number of components previously identified and was then employed to analyze the results.

Predictive Ability

Q^2 or r^2_{cv} values were used to evaluate the overall predictive ability of the model. r^2_{cv} is calculated according to the following equation;

$$r^2_{cv} = (SSY - PRESS)/SSY$$

where SSY represents the variance of the biological activities of molecule around the mean value and PRESS is the prediction error sum of squares derived from the leave-one-out method. The uncertainty of the prediction is defined as;

$$S_{PRESS} = [PRESS/(n-k-1)]^{1/2}$$

where k is the number of variables in the model and n is the number of compounds used in the study.

Interaction Energy Calculations

For clearly elucidate the interaction of efavirenz derivatives in the binding pocket, the individual interaction between some derivatives and the surrounding amino acid in the WT and K103N binding pocket are investigated by using quantum chemical calculations. In this study, X-ray structure of efavirenz, docked conformation of efavirenz (compound 01), compounds which have the highest and less activity against WT RT (compounds 25 and 12), and compounds which have the highest and the less activity against K103N RT (compounds

50 and 22) are selected to calculate their interaction energies formed with the surrounding amino acids in the WT and K103N binding pocket. The geometries of the 14 amino acids surrounding within 6 Å from efavirenz are taken from the protein structures, which are starting geometries for docking. Each amino acid is terminated with hydrogen atoms. Compound with each amino acid are then performed single point calculation at B3LYP/6-31G(d) and MP2/6-31G(d) levels of theory by using Gaussian03 program. The interaction energy (INT) of the compound with each amino acid is defined as follows:

$$INT = E_{(compound + each\ amino\ acid)} - [\ E_{(compound)} + E_{(each\ amino\ acid)}\ ]$$

where $E_{(compound + each\ amino\ acid)}$ is the energy of the complex structure between compound and each amino acid. $E_{(compound)}$ and $E_{(each\ amino\ acid)}$ are the energies of compound and each amino acid, respectively.

Results and Discussion

Docking Analysis

Validation of the Docking Method

Two docking methods, GOLD and Autodock, were applied to dock the efavirenz compound back into the WT and K103N HIV-1 RT binding pockets. The rmsd values of the docked posed from the X-ray pose of efavirenz obtained from GOLD method are 0.94 Å and 1.0 Å for WT and K103N RT binding pocket, respectively. Based on Autodock results, the rmsd values are 0.35 Å and 0.53 Å for WT and K103N RT binding pocket, respectively, reported in table 2. The results show that both methods reveal a good ability to reproduce the X-ray bound conformation with rmsd less than 1.0 Å for both WT and mutant enzymes. The binding mode of the docked efavirenz shows a very similar interaction with the X-ray crystallographic structure, depicted in figure 1. Therefore, GOLD and Autodock programs with the used parameters could be extended to search and evaluate the binding conformations of efavirenz derivatives in the corresponding binding pockets.

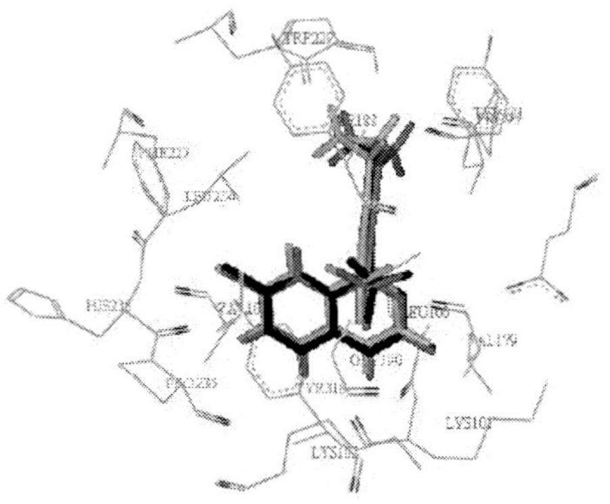

(a)

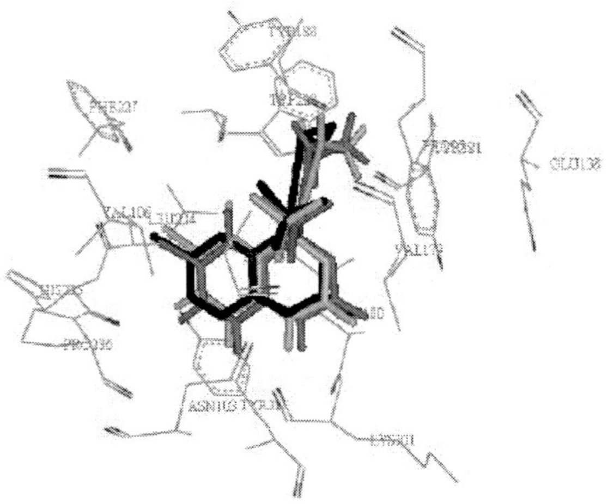

(b)

Figure 1. The conformations of docked efavirenz by using GOLD (grey) and Autodock (dark grey) and compared with the orientation of X-ray pose (black); (a) in the WT binding pocket (pdb code 1fk9) and (b) in the K103N binding pocket (pdb code 1fko).

Table 2. The rmsd (Å) of the docked pose from the X-ray pose of efavirenz by using GOLD and Autodock methods

Docking method	Rmsd (Å)	
	WT (pdb code 1fk9)	K103N (pdb code 1fko)
GOLD	0.94	1.00
Autodock	0.35	0.53

Docking Analysis of Efavirenz Derivatives in the Binding Pocket

To evaluate the binding orientation of efavirenz derivatives in the WT and K103N HIV-1 RT binding pocket, 56 efavirenz compounds, shown in table 1, were docked into the binding pocket using GOLD and Autodock methods. The docked orientations of all efavirenz derivatives are laid to be similarly butterfly like shape as the efavirenz compound in the corresponding binding pocket, as shown in figure 2. It shows that the common ring structures of all derivatives were well superimposed each other. To elucidate the potential binding mode of compounds in the binding pocket relating to their inhibitory activities, the docked results of highest and less active compounds against both WT and K103 N HIV-1 RT were considered. Comparing with the X-ray pose orientation of efavirenz, the docked conformations of compounds 25 and 12, the highest and less active compounds against WT HIV-1 RT, respectively, are depicted in figure 3. The docked conformations of compound 25 produced from both docking methods show similar orientation in the WT binding pocket. The conformations could form the hydrophobic interaction between the cyclopropyl group and the hydrophobic pocket of Y181, Y188 and W229, whereas the bulky alkoxy group of compound 12 may lead to conflict with these amino acids. The hydrogen bond between 6-H position of the benzoxazin-2-one ring with carbonyl group of backbone H235 could be possibly formed, instead of the Van der waals interaction of 6-Cl in efavirenz. Moreover, there is possibility for the Van der waals interaction between 5-F of the ring and V106. This may explain why compounds 25 and 12 shows higher potency and less potency against WT HIV-1 RT comparing with efavirenz compound, respectively. Figure 4 shows the docked conformation of compound 25 in the K103N HIV-1 RT. The conformation reveals that the intermolecular hydrogen bond of hydrogen atom with H235 was eliminated. Moreover, the hydrophobic interaction between the cyclopropyl group with the Y181, Y188 and W229 was lost. This information obtained from docking conformations is in agreement with the experimental results that compound 25, showing the highest potency for WT HIV-1 RT inhibition, has diminished activity against K103N HIV-1 RT.

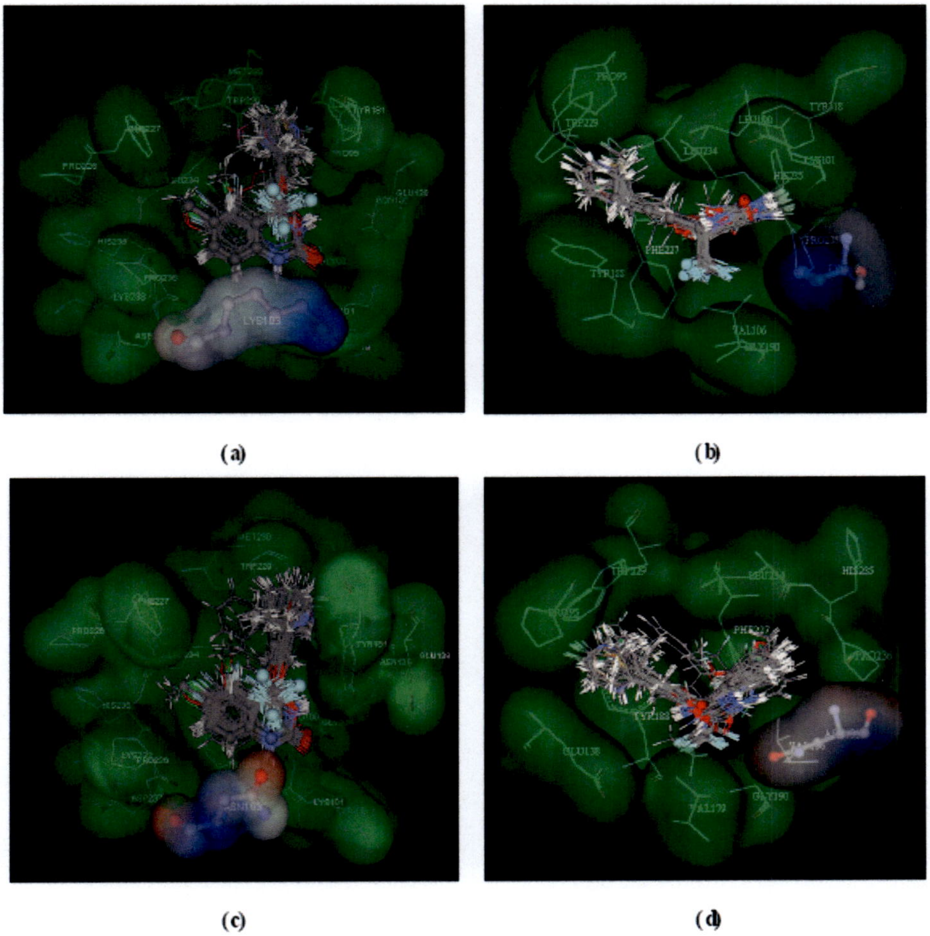

Figure 2. The orientation of docked efavirenz derivatives by using Autodock in the WT binding pocket (a. top view and b. side view) and in the K103N binding pocket (c. top view and d. side view).

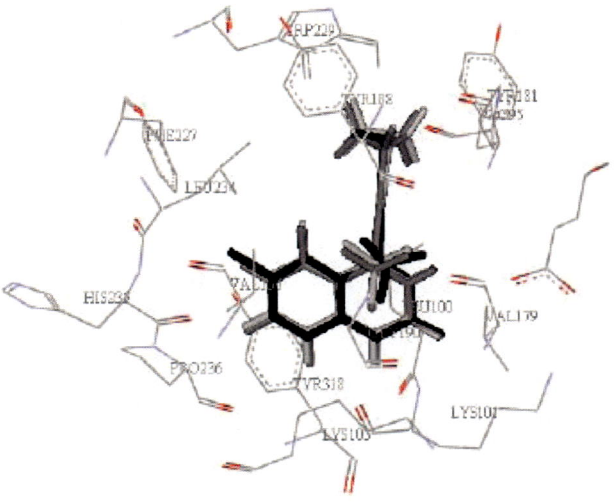

(a)

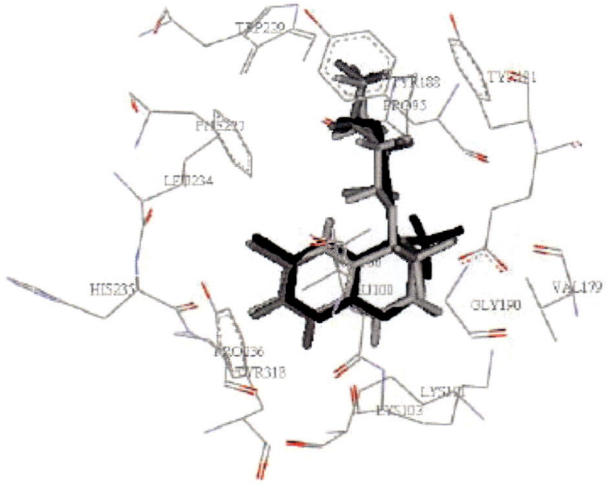

(b)

Figure 3. The conformations of docked efavirenz derivatives by using GOLD (grey) and Autodock (dark grey) and compared with the orientation of X-ray pose (black) in the WT HIV-1 RT; (a) Compound 25 and (b) Compound 12.

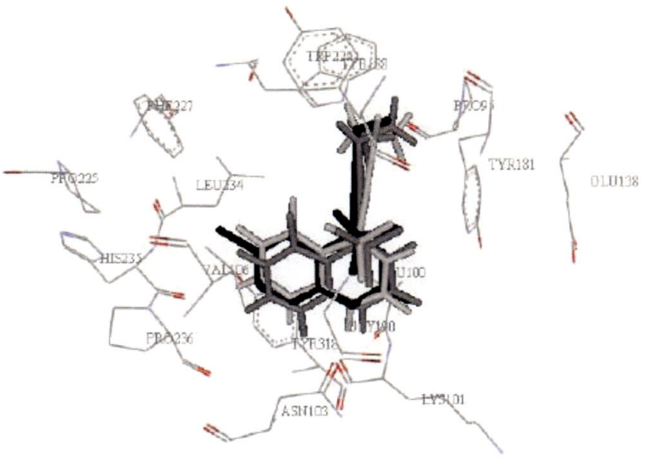

Figure 4. The conformations of docked efavirenz derivative (compound 25) by using GOLD (grey) and Autodock (dark grey) compared with the orientation of X-ray pose (black) in the K103N HIV-1 RT.

The docked conformations of compounds 50 and 22, the highest and less active compounds against K103N HIV-1 RT, were further analyzed. The docked conformations obtained from GOLD and Autodock methods comparing with the X-ray pose orientation of efavirenz are shown in figure 5. Numerous favorable interactions found for compound 50 are described as follows. The Van der waals interaction could be formed between methoxy group substituted at C5 position of the benzoxazin-2-one ring and the sidechain of L234, F227 and Y188. There are additional Van der waals interaction between 6-Cl and the sidechain of L234, V106. The hydrophobic interaction between the cyclopropyl group with the aromatic sidechain of Y188 and W229 could be formed. It is interesting to note that the hydrogen atom attached to the nitrogen atom on the Z substituent could increase hydrogen-pi (H-π) interaction with the aromatic sidechain of Y181, whereas compound 22 and efavirenz could not be found these interactions. Moreover, compound 22 with smaller substituents made the loss of the hydrophobic interactions with the aromatic ring of W229. These results may explain why compound 50 shows the highest activity and compound 22 has less inhibitory activity against K103N HIV-1 RT. It is interesting to note that, for compound 50, the docked conformation in the WT binding reveals the possibility of the hydrogen bond formation. The Van der waals interactions could be also found. Although the H-π interaction was eliminated in the pocket, the cyclopropyl group could match more to the hydrophobic pocket of Y188, W229 and Y181 in the WT binding pocket as show in figure 6. This may be the reason that compound 50 shows excellent inhibition against both WT and K103N HIV-1 RT.

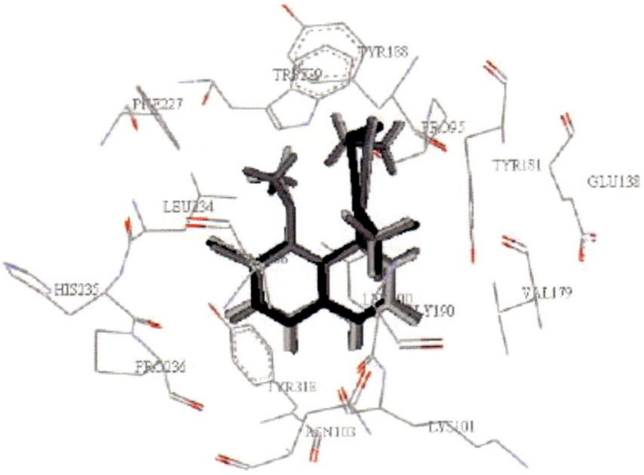

(a)

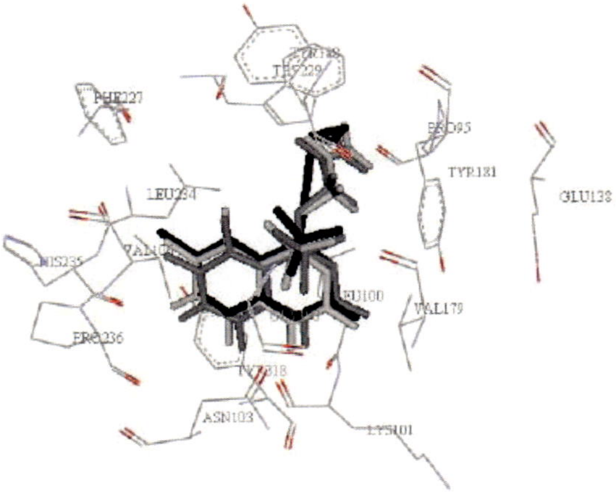

(b)

Figure 5. The conformations of docked efavirenz derivatives by using GOLD (grey) and Autodock (dark grey) compared with the orientation of X-ray pose (black) in the K103N HIV-1 RT; (a) Compound 50 (b) Compound 22.

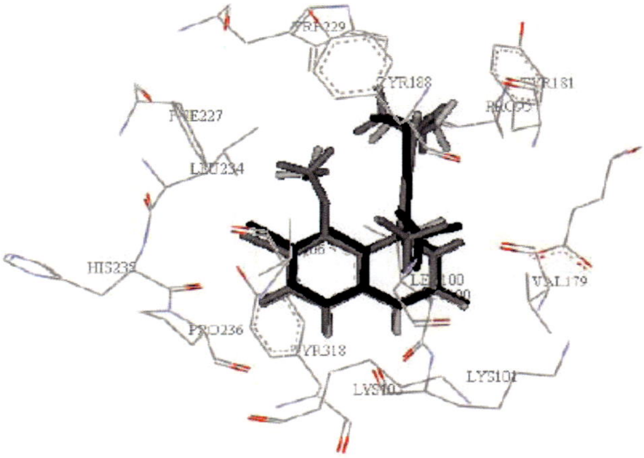

Figure 6. The conformations of docked efavirenz derivative (compound 50) by using GOLD (grey) and Autodock (dark grey) compared with the orientation of X-ray pose (black) in the WT HIV-1 RT.

Molecular docking approaches successfully prove as powerful tools to determine the potential binding mode and to elucidate inhibitor-enzyme interactions of efavirenz compound and its derivatives in the HIV-1 RT binding pocket. The important interactions of efavirenez derivative active against the WT and K103N HIV-1 RT can be drawn as follows. With respect to the WT HIV-1 inhibition, the hydrogen bond between the polar atom (H and oxygen carbonyl) on benzoxazin-2-one ring with the main sidechains of H235, K101 and P236 are favorable to enhance the inhibitory activity. In particular, the additional hydrogen bond at 6-substituent position seems to be important to active against HIV-1 RT. Another important interaction of WT HIV-1 RT inhibitors is the hydrophobic interaction between the R substituent attached to C4 on benzoxazin-2-one ring with the side chains of Y181, T188 and W229. It should be noted that regarding the K103N HIV-1 RT, the hydrogen-π interaction of the NH group on Z position with the aromatic side chain of Tyr181, the Van der waals interactions on 5,6-substituent position with surrounding amino acid residues (F227, V106, L234 and Y188) and the hydrophobic interactions of the R group on benzoxazin-2-one ring with the aromatic side chain of W229 and Y188 serve as significant interactions for the particular inhibition.

The docked conformations of efavirenz derivatives derived from GOLD and Autodock methods reveal very similar orientations in the WT and K103N RT binding pockets. However, based on lower rmsd of the docked pose from the X-ray pose of efavirenz, the docked conformations derived from Autodock method has been selected for further study on 3D-QSAR analysis and interaction energy estimation using quantum chemical calculations.

Qsar Analysis

The conformation with the lowest final docked energy was selected for the structural alignment in 3D-QSAR analyses. CoMFA and CoMSIA studies were applied to determine relationships between structural properties and HIV-1 inhibitions, based on the docked conformations.

CoMFA Models

Two different types of biological activities were considered, WT and K103N HIV-1 RT inhibitory activities. The models obtained from the analyses include steric and electrostatic field contributions. The statistical parameters of CoMFA models of compounds are summarized in table 3. Regarding WT HIV-1 RT inhibition, the best predictive ability QSAR model with r^2_{cv} of 0.662, s-press = 0.145 and noc = 3, is obtained. The steric and electrostatic contributions of this model are 63.8 % and 36.2 %, respectively. Other statistical results of the best CoMFA model are the conventional r^2 (r^2 = 0.936) and the standard error of estimation (0.066). F is 83.441 and the probability (P) of obtaining this value of F if r^2 is actually zero (probability of $r^2 = 0$) is lower than 0.001. The plot between predicted and experimental WT HIV-1 RT inhibitory affinities of the non-cross-validated analysis of the model is presented in figure 7a.

For further analysis, the CoMFA results with respect to K103N HIV-1 RT inhibition were investigated. The high predictive model with r^2_{cv} of 0.755, s-press = 0.302 and noc = 6, is produced, shown in table 3. The model has 51.2 % contribution from the steric field and 48.8 % contribution from electrostatic field, showing approximately equal contributions for explanation of the K103N inhibitory activities. The statistical parameters obtained are that the conventional r^2 is 0.944, the standard error of estimation is 0.144, F is equal to 107.318 and the probability (P) of obtaining this value of F if r^2 is actually zero (probability of $r^2 = 0$) is lower than 0.001. The plot between predicted and experimental K103N HIV-1 RT inhibitory affinities of the non-cross-validated analysis of the model is presented in figure 7b.

Table 3. Summary of CoMFA models for the WT and K103 N HIV-1 RT inhibition

Statistical results	WT inhibition	K103 N inhibition
r^2_{cv}	0.662	0.755
noc	3	6
s-press	0.145	0.302
r^2	0.936	0.944
S	0.066	0.144
F	83.441	107.318
Field Contribution		
Steric	63.8	51.2
Electrostatic	36.2	48.8

Prediction for Compounds in the Test Set

As the obtained CoMFA models, listed in table 3, show reasonable predictive power for WT and K103N RT inhibition, both models were used to predict the inhibitory activities of 7 efavirenz derivatives in the test set. The comparison of predicted and observed biological activities of these compounds is listed in table 5 and plotted in figures 7a and 7b, respectively. Based on the residual value, the inspection of the data reveals the usefulness of the models for the prediction of the activities of the efavirenz compounds which are not included in the training set.

CoMSIA Analysis

The results of CoMSIA studies were presented in table 4. CoMSIA analysis was performed using five different property fields (steric, electrostatic, hydrophobic, hydrogen bond donor and hydrogen bond acceptor fields). In the present study, the propose of using five different descriptors is not to increase the significance and the predictive ability of the 3D QSAR models but to partition the various properties into the spatial location where they play a decisive role in determining biological activity. Comparing with the obtained CoMFA model, the best CoMSIA model corresponding to the WT HIV-1 RT inhibition reveals improved predictive ability than CoMFA model with a high predictive model (r^2_{cv} of 0.708, s-press = 0.142 and noc = 6). The other statistical results are that the conventional r^2 value is 0.894, the standard error of estimation is 0.085 and F is equal to 50.844. The corresponding field contributions of these five descriptor variables are 12.2, 19.5, 24.1, 16.8 and 27.3, respectively. The experimental and calculated affinities derived from the CoMSIA model for the WT inhibitory affinity is plotted in figure 8a.

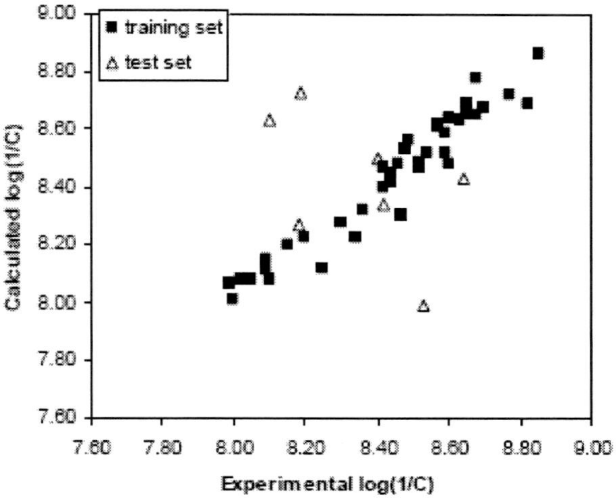

(a)

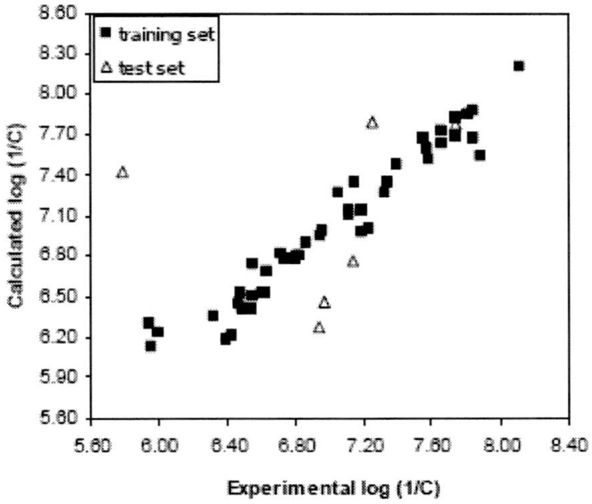

(b)

Figure 7. The plots between predicted and experimental inhibitory affinities of the non-cross-validated analysis of the CoMFA model; (a) WT inhibitory affinities and (b) K103N inhibitory affinities.

According to the K103N HIV-1 RT inhibition, the best CoMSIA model with the combination of all fields yielded high predictive ability with r^2_{cv} of 0.773, s-press = 0.286, noc = 3, the conventional r^2 = 0.938, the standard error of estimation = 0.155 and F = 93.344. The steric, electrostatic, hydrophobic, hydrogen bond donor and hydrogen bond acceptor field contributions of the model are 13.4, 22.3, 20.5, 26.5 and 17.3, respectively. The experimental and calculated affinities derived from the CoMSIA model for the K103N inhibitory affinity is plotted in figure 8b.

Table 4. Summary of CoMSIA models for the WT and K103N HIV-1 RT inhibition

Statistical results	WT inhibition	K103 N inhibition
r^2_{cv}	0.708	0.773
noc	6	3
s-press	0.142	0.286
r^2	0.894	0.938
S	0.085	0.155
F	50.844	93.344
Field Contribution		
Steric	12.2	13.4
Electrostatic	19.5	22.3
Hydrophobic	24.1	20.5
Hydrogen donor	16.8	26.5
Hydrogen acceptor	27.3	17.3

Table 5. Predicted log (1/C) WT and K103N HIV-1 RT inhibitory affinities of the tested efavirenz compounds derived from CoMFA models

Compound No.	WT HIV-1 RT inhibitory affinity			K103N HIV-1 RT inhibitory affinity		
	Expt.log (1/C)	Calc.log (1/C)[a]	Residual	Expt.log (1/C)	Calc. log (1/C)[b]	Residual
03	8.40	8.50	-0.1	6.94	6.28	0.66
19	8.53	7.99	0.54	6.97	6.46	0.51
22	8.19	8.73	-0.54	5.79	7.42	-1.63
37	8.64	8.43	0.21	7.14	6.76	0.38
39	8.42	8.34	0.08	7.25	7.79	-0.54
42	8.18	8.27	-0.09	6.49	6.51	-0.02
47	8.10	8.63	-0.53	7.74	7.78	-0.04

[a] calculated by CoMFA model for the WT inhibition

[b] calculated by CoMFA model for the K103N inhibition.

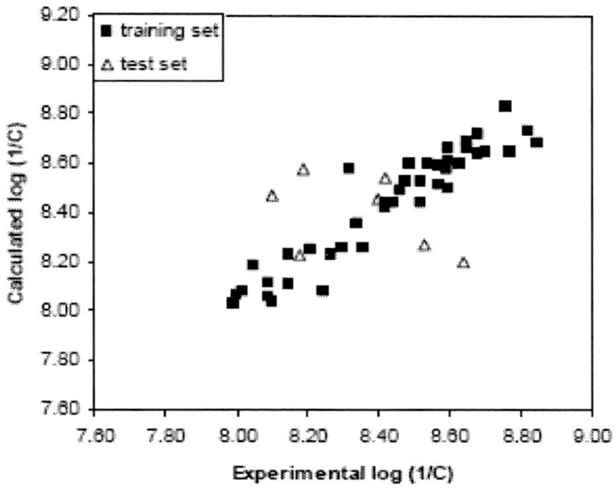

(a)

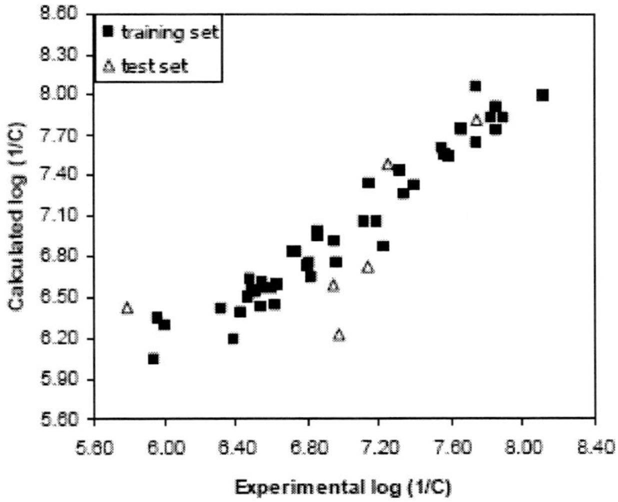

(b)

Figure 8. The plots between predicted and experimental inhibitory affinities of the non-cross-validated analysis of the CoMSIA model; (a) WT inhibitory affinities and (b) K103N inhibitory affinities.

Prediction for Compounds in the Test Set

In order to evaluate the predictive ability of the resulting CoMSIA models, the same 7 efavirenz derivative tested set as used in CoMFA study was employed to predict the inhibitory activities. The comparison of predicted and observed biological activities of these compounds for WT and K103N inhibition, is reported in table 6 and plotted in figures 8a and 8b, respectively. The good correlation shows the validation of the obtained CoMSIA models for efavirenz derivatives excluded from the training set.

Table 6. Predicted log (1/C) WT and K103N HIV-1 RT inhibitory affinities of the tested efavirenz compounds derived from CoMSIA models

Compound No.	WT HIV-1 RT inhibitory affinity			K103N HIV-1 RT inhibitory affinity		
	Expt.log (1/C)	Calc.log (1/C)[a]	Residual	Expt.log (1/C)	Calc. log (1/C)[b]	Residual
03	8.40	8.46	-0.06	6.94	6.58	0.36
19	8.53	8.27	0.26	6.97	6.23	0.74
22	8.19	8.58	-0.39	5.79	6.43	-0.64
37	8.64	8.20	0.44	7.14	6.72	0.42
39	8.42	8.54	-0.12	7.25	7.48	-0.23
42	8.18	8.23	-0.05	6.49	6.54	-0.05
47	8.10	8.47	-0.37	7.74	7.81	-0.07

[a] calculated by CoMSIA model for the WT inhibition
[b] calculated by CoMSIA model for the K103N inhibition.

Graphical Interpretation of Fields

CoMFA Contributions

The 3D–QSAR contour maps obtained from CoMFA results for all efavirenz derivatives illustrate clearly the steric and electrostatic contributions for ligand binding. To better understand the field interactions between the enzyme and inhibitors, the amino acid residues surrounding efavirenz compound in the binding pocket were merged into the contour mapsu. Sterically unfavorable regions are depicted in yellow, whereas favorable regions are in green. Electrostatic positive favorable regions correspond to the blue areas and the negative favorable regions correspond to the red areas.

The steric and electrostatic field contributions to WT HIV-1 RT inhibition is shown in figure 9a. Compound 25, the most potent compound active against WT HIV-1 RT of the series, is used as a reference structure. There is a prominent blue contour located between the C6 and C7 positions on the benzoxazin-2-one ring. It reveals that incorporation of electropositive group at this position would be preferable for the activity. This suggestion is supported by experimental data that, as clearly examples, compound 25 having hydrogen atom attached to the C6 position produces higher potency compared to compounds with halogen groups attached to the similar position (compounds 28, 30, 33, 34, 36, 38, 47, 50 and 53). Particularly, compound 25 shows the highest potency for the WT inhibition in the data set. There are large green contours corresponding to the location of the R substituent attached to the C4 position on the benzoxazin-2-one ring. However, the tolerated steric requirements of this region are highlighted by yellow contours presented around that region, especially a yellow one overlapping with the Tyr188 residue. It is indicated that an additional bulky group at this site would be favorable for the affinity, but the size and the dimension of the substituents should not be too large. This means that steric occupancy with too bulky groups of the sidechain leads to a steric conflict resulting in diminished favorable interactions with the aromatic ring of Y188. These suggestions agree well with the trend observed experimentally, reported in table 1, that the cyclopropyl acetylene group is the optimum sidechain for this position. Compound 1 with the cyclopropyl acetylene group attached at the C4 substituent shows higher potency. Compounds 2-5 and 12-18, occupying the C4 site with too large substituents, result in significantly reduced potency against WT RT. A large blue electrostatic contour area close to the cyclopropyl group of the R position suggests that electropositive groups are predicted to increase activity in those areas. This suggestion agrees with the experimental data that compounds containing small heterocyclic rings having more electron rich properties, such as pyridyl, furanyl or thienyl groups (compounds 2-9), have diminished affinities against WT RT, compared to compound 1.

The contributions of steric and electrostatic fields to K103N RT inhibition is shown in figure 9b. The most potent compound active against K103N HIV-1 RT of the series, compound 50, is put in the contour as a reference structure. Interestingly, the large blue contour located close to the C6 position on the phenyl ring is completely absent, compared to the WT inhibition. There are two red contours presented in the vicinity of the substituent attached to C6 position instead. A sterically favored green area is located in the region of the C5 substituent attached to the phenyl ring. It can be summarized that, in contrast to the WT inhibition, compounds having a bulky group attached to the C5 position and the preference of electronegative groups at the C6 position are more favorable for the K103N inhibition. This is in agreement with the experimental observation that compounds having sterically 5,6-halogen substituents (compounds 28, 33 and 36) confer higher potency against the K103N mutant compared to 5-halogen analogues (compounds 25 and 27). Compounds 50,51 and 53, occupying the C5 position with bulkier groups such as methoxy or hydroxyl group and the C6 position with chlorine atom, yield higher potency compared to compounds 38 and 45, respectively. In particular, compound 50 shows the highest activity against K103N HIV-1 RT in the data set. This could be the reason why compound 25, the highest active compound against WT HIV-1 RT, shows a significantly reduced potency against K103N RT. A blue contour close to Z substituent of the benzoxazin-2-one ring suggests that a group with low electron density would play a favorable role in activity. This suggestion is consistent with the experimental data that the replacement of an oxygen atom with the NH group at this position always yields much more potent derivatives active against K103N RT, as exemplified by compounds 1, 2 compared to compounds 38 and 43, respectively. At the region close to the R position, a large green contour is still presented for the K103N RT inhibition and the yellow one overlapping with Y188 is also located close to approximately similar region. The structural requirements for this region could be explained in similar manner as those for WT RT inhibition. The structural requirements obtained from the CoMFA models for WT and K103N HIV-1 RT inhibition show high correspondence to molecular docking results previously described and QSAR studies reported [37,38].

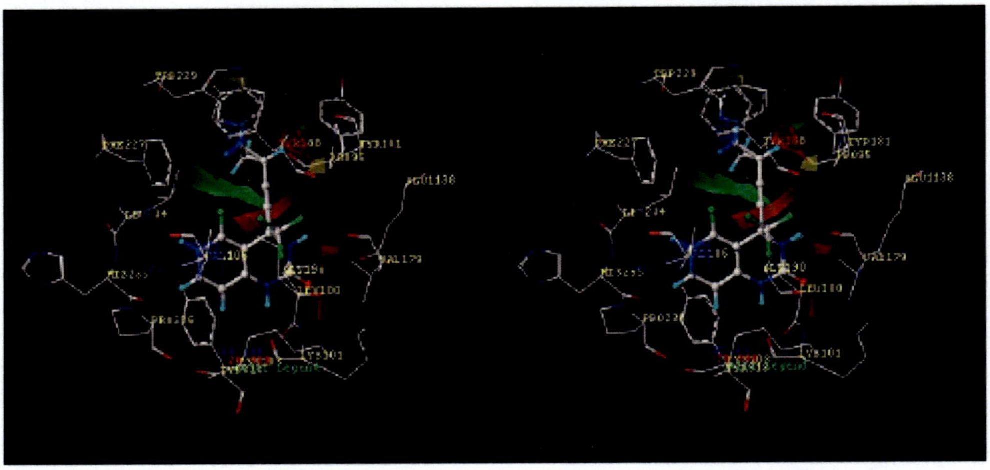

(a)

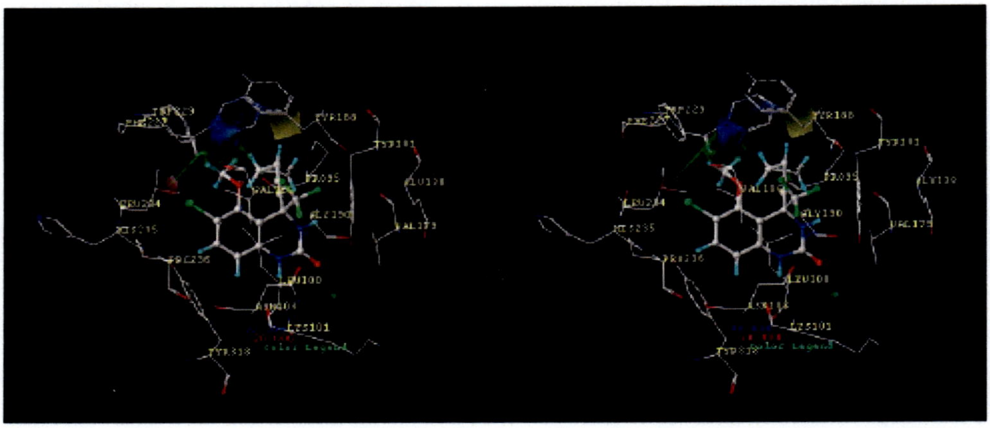

(b)

Figure 9. Stereoview of CoMFA steric and electrostatic STDEV*COEFF contour plots based on WT and K103N HIV-1 RT inhibitions from the analysis of CoMFA models with non-cross-validation, as shown in (a) and (b), respectively. Green contours refer to sterically favored regions; yellow contours indicate disfavored area. Compound 25 and 50 are displayed inside the fields as ball and stick presentation in (a) and (b), respectively.

CoMSIA Contributions

Additional to the steric and electrostatic fields used in CoMFA study, hydrophobic, hydrogen bond donor and hydrogen bond acceptor fields contributing to binding affinities can be derived from the CoMSIA models. To better understand the field interactions between the enzyme and inhibitors, the amino acid residues surrounding the efavirenz compound in the binding pocket were merged into the contour maps. In the present study, since the CoMSIA steric and electrostatic contours (not shown) are similarly placed as those of the CoMFA models, only hydrophobic, hydrogen bond donor and hydrogen bond acceptor field contributions were discussed further. The magenta and white regions highlight areas where hydrophilic and hydrophobic properties are preferable, respectively. The cyan contours indicate region where hydrogen bond donor groups increase activity, whereas purple contours indicate region where hydrogen bond donor groups decrease activity. Hydrogen bond acceptors are favored in areas indicated by orange contours, whereas white contours are areas where hydrogen bond acceptors are disfavored.

The contribution of hydrophobic fields to the WT inhibition is shown in figure 10a. A predominant feature of the hydrophobic contour plot is the presence of a large pink contour corresponding to the locations of the groups attached to C5 and C6 positions. It reflects that introducing hydrophobic substituents at the positions would enhance the biological activity. As demonstrated, halogen groups attached to C5 or C6 positions on the benzoxazin-2-one ring (compounds 25-33, 36 and 38) yield increasing binding affinities. On the other hand, methoxy or hydroxyl groups attached to C5 or C6 positions (compounds 34, 39-40, 49-53) would not be preferable for binding affinities. Moreover, the favorable hydrophobic region is located close to the vicinity of the R substituent, attached to the C4 position. As examples, more hydrophilic derivatives with alkoxy groups substituted at C4 (compounds 12-18) are less active for WT inhibition compared with compound 1. In contrast to WT inhibition, the contour map of hydrophobic properties highlights different areas for the K103N inhibition, as shown in figure 10b. A white contour is presented near the substituent attached to C5 position. With the combination of the steric contour map obtained from CoMFA models, it suggests that the favorable hydrophilic together with steric substituents at the position could enhance the K103N RT inhibitory activity. This indicates that the bulky group with low hydrophobicity would be favorable to the biological activity, as demonstrated by compound 50, the highest active compound against K103N HIV-1 RT. A pink contour in the vicinity of the R4 substituent shows the importance of the hydrophobic component with an optimum steric bulk of the sidechain at the R4 position. This suggestion agrees well with the experimental report that the optimum sidechain for this position is the acetylene cyclopropyl group.

Figures 11a and 11b show the contributions of the hydrogen donor fields for the WT and K103N inhibition, respectively. For WT inhibition, two cyan contours located at the NH group of the benzoxazin-2-one ring and presented in the vicinity of Z substituent represents that placement of hydrogen-bond donors at the positions is beneficial for the receptor binding. This could be attributed to the hydrogen bond interaction between the main chain carbonyl oxygen of K101 and the amide NH of efavirenz compound [25]. For the K103N inhibition shown in figure 11b, the cyan contour located at the NH group of the benzoxazin-2-one ring is completely absent, while a cyan contour close to the Z substituent is presented. The minor cyan contour corresponds to the hydrogen donor substituent preferable at this position. This could be the reason that compounds having NH group as the Z substituent are more preferable for the K103N inhibitory activity than that of O atom. The contribution is consistent with the docking study and the electrostatic contribution derived from the CoMFA model.

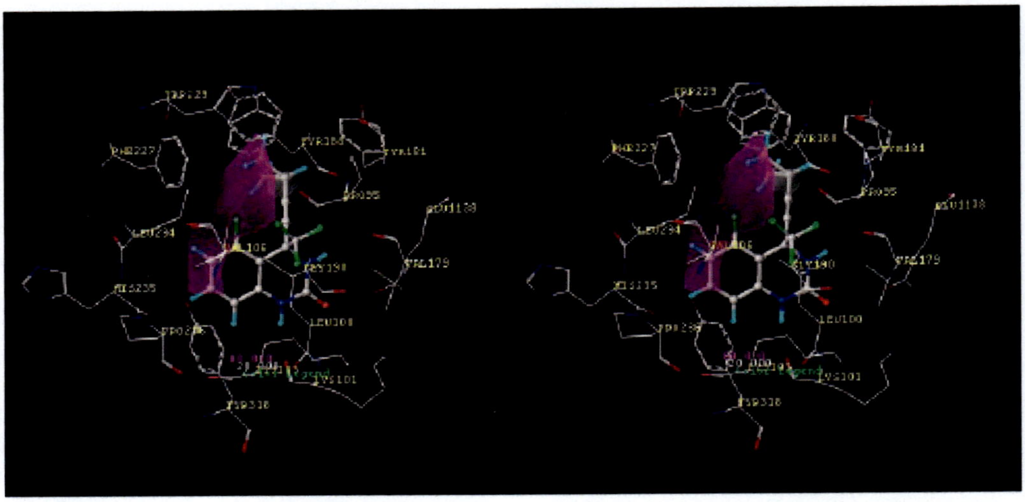

(a)

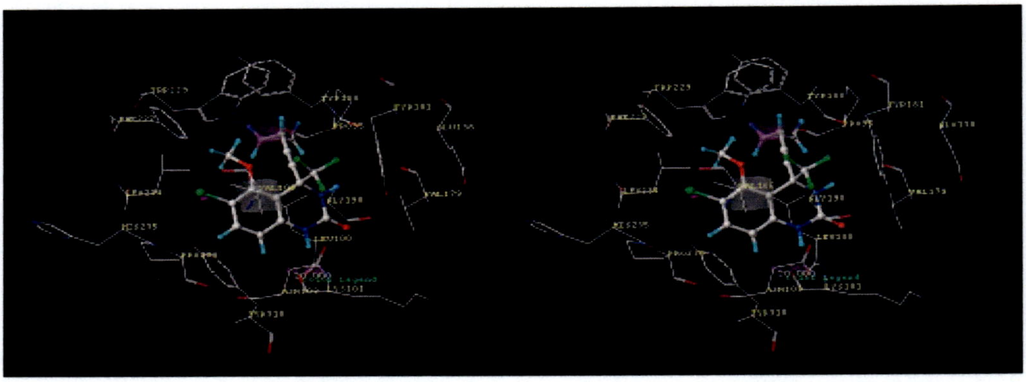

(b)

Figure 10. Stereoview of CoMSIA hydrophobic STDEV*COEFF contour plots based on WT and K103N HIV-1 RT inhibitions from the analysis of CoMSIA models with non-cross-validation, as shown in (a) and (b), respectively. Magenta contours refer to positive hydrophobic favoring areas; white contours indicate hydrophilic favoring areas. Compound 25 and 50 are displayed inside the fields as ball and stick presentation in (a) and (b), respectively.

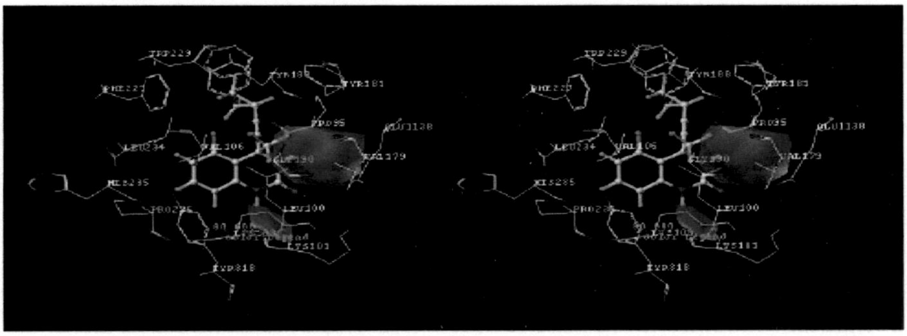

(a)

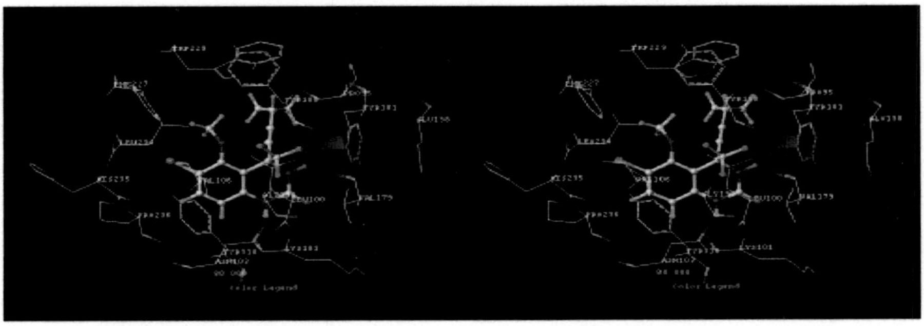

(b)

Figure 11. Stereoview of CoMSIA hydrogen donor STDEV*COEFF contour plots based on WT and K103N HIV-1 RT inhibitions from the analysis of CoMSIA models with non-cross-validation, as shown in (a) and (b), respectively. Cyan contours refer to hydrogen donor fields favoring areas; purple contours indicate hydrogen donor fields disfavoring areas. Compound 25 and 50 are displayed inside the fields as ball and stick presentation in (a) and (b), respectively.

The hydrogen bond acceptor field contribution for the WT and K103N inhibition are shown in figures 12a and 12b, respectively. According to WT inhibition, a prominent orange contour is presented in the region of R substituent attached to the C4 site suggesting the preference of hydrogen bond acceptors at the position. It is in agreement with the observation that compounds containing acetylenic sidechain as the R substituent possessing hydrogen acceptor property, generally enhance the inhibitory activities. Compounds having alkoxy groups rather than acetylenic sidechain at the R substituent show moderate to less binding affinities. An orange contour close to the carbonyl oxygen of the benzoxazin-2-one ring reveals that acceptor functions of the ligand directing to the location would be preferable for the binding affinity. This suggestion is in consistent with the obtained docking results described previously. Similar to the contribution obtained from the WT inhibition, a large orange contour corresponding to the location of the R group substituted to the C4 position shown in figure 12b, suggests that hydrogen bond acceptors at the position could be beneficial for K103N binding affinity. Based on the obtained CoMSIA results, it could be confirmed that not only steric and electrostatic interactions contribute to the WT and K103N inhibitory activities, hydrophobic, hydrogen donor and acceptor fields are important to explain the variance of the data.

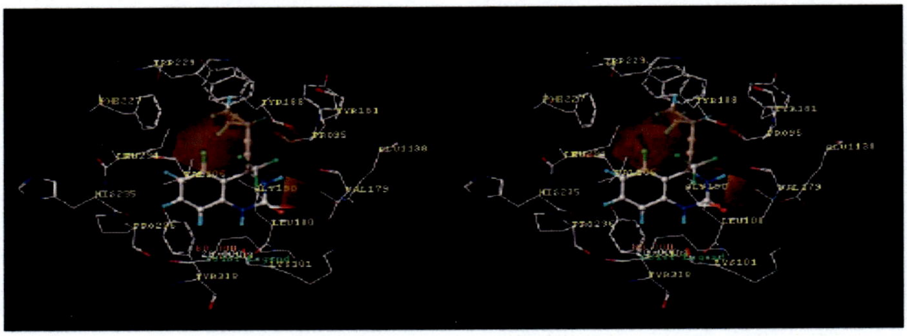

(a)

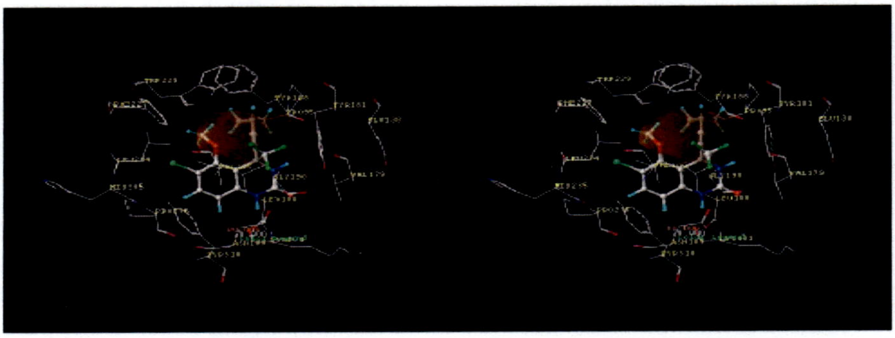

(b)

Figure 12. Stereoview of CoMSIA hydrogen acceptor STDEV*COEFF contour plots based on WT and K103N HIV-1 RT inhibitions from the analysis of CoMSIA models with non-cross-validation, as shown in (a) and (b), respectively. Orange contours refer to hydrogen acceptor fields favoring areas; white contours indicate hydrogen acceptor fields disfavoring areas. Compound 25 and 50 are displayed inside the fields as ball and stick presentation in (a) and (b), respectively.

As the availability of the crystal structure of efavirenz complexed with HIV-1 RT [25,26], there are numerous contacts involving a series of hydrophobic contacts; the aromatic sidechains of Y181, Y188, W229 and F227 in the binding pocket with the bound inhibitor. The hydrogen bond between the benzoxain-2-one NH and the mainchain carbonyl oxygen of K101 is formed. There is also a Van der Waals contact between the γ-C of the sidechain of K103 and the nitrogen of benzoxain-2-one ring of efavirenz. An analysis of the structure of efavirenz complexed with the K103N RT reveals that the substitution of a charged and linear lysine for the uncharged and branched asparagine at the position 103 resulting in a drastic change in the chemical environment in the proximity of the mutation and a significant conformational rearrangement within the drug binding pocket compared with the wild type complex. Mainly, there are two consequences for the binding of NNRTI: changed hydrophobic and electrostatic properties of the binding pocket. It is evident that favorable hydrophobic interactions of the amino acid in the binding pocket with the bound inhibitors and the hydrogen bond interaction are eliminated. Accordingly, the inhibitory affinities of some inhibitors are drastically reduced. Based on the obtained results, it could be suggested the methoxy group attached to the C5 position could be possible to generate favorable interactions compensating the hydrophobic interactions lost. The presence of NH group as the Z substituent in the benzoxazin-2-one ring possibly contributes to form the H-π interaction with the phenyl ring sidechain of Y181. The interactions should compensate the loss of significant interactions between the bound inhibitor and surrounding amino acid residues in the binding pocket. These results show high consistency to the obtained docking and CoMFA and CoMSIA models in the present study, highlighting the structural difference required for WT and K103N inhibitions of efavirenz derivatives.

Interaction Energy Calculations

Interaction Energy of X-Ray Structure and Docked Efavirenz in the WT and K103N Binding Pockets

The interaction energies between efavirenz from the X-ray structure and each amino acid surrounding in the WT and K103N RT binding pockets are shown in table 7. In the WT binding pocket, the interaction energies of efavirenz are -16.1 to 1.0 kcal/mol at the B3LYP/6-31G(d) level of theory. In case of interaction energies at the MP2/6-31G(d) levels of theory, all attractive interaction energies are obtained (-18.6 to -0.9 kcal/mol). Interaction energies from both levels of calculation shows in the same trend and the major attractive interaction is the interaction with K101 (-16.1 and -18.6 kcal/mol at the B3LYP/6-31G(d) and MP2/6-31G(d) levels of theory, respectively. From the complex structure, amino and carbonyl groups of benzoxazin-2-one ring of efavirenz compound could form two H-bonding with carbonyl group (1.98 Å) and amino group (3.17 Å) of backbone K101, respectively. The interaction energies at the MP2/6-31G(d) levels of theory also reveal the important interaction of efavirenz to the WT binding pocket with L100, K103, Y181, Y188, H325, and P236.

Table 7. Interaction energies between efavirenz from X-ray structure and each amino acid surrounding in the WT and K103N binding pockets

Amino acid residue	Interaction energy (kcal/mol)			
	WT binding pocket		K103N binding pocket	
	B3LYP/6-31G(d)	MP2/6-31G(d)	B3LYP/ 6-31G(d)	MP2/6-31G(d)
P95	-0.3	-0.9	-0.3	-0.9
L100	-0.6	-7.6	0.2	-7.7
K101	-16.1	-18.6	-10.1	-12.1
K103/N103	-1.8	-5.3	-1.4	-3.2
V106	1.0	-2.4	-0.1	-3.2
Y181	-0.6	-3.8	1.2	-4.3
Y188	-1.0	-5.7	-1.6	-6.6
G190	-1.0	-1.7	-1.0	-1.9
F227	0.1	-1.0	0.4	-1.0
W229	-0.7	-2.5	-1.3	-3.3
L234	0.2	-1.8	0.0	-2.5

Table 7 (Continued)

P236	-0.9	-3.1	-1.1	-2.7
Y318	-0.3	-2.7	-0.2	-3.3

For the K103N binding pocket, the interaction energies of efavirenz are -10.1 to 1.2 kcal/mol and -12.1 to -0.9 kcal/mol at the B3LYP/6-31G(d) and MP2/6-31G(d) levels of theory, respectively. Although K101 is the major attractive interaction of efavirenz in K103N binding pocket, the results clearly show the loss of attractive interaction with K101 about 5.0-6.5 kcal/mol, as compared with the WT binding pocket. Their interaction energies in the K103N binding pocket with K101 are -10.0 and -12.1 kcal/mol at the B3LYP/6-31G(d) and MP2/6-31G(d) levels of theory, respectively. Because of the effect of amide group at N103, the efavirenz position has slightly changed from the position in WT binding pocket and caused the longer H-bonding with carbonyl group of backbone K101. Amino and carbonyl groups of benzoxazin-2-one ring of efavirenz formed two H-bonding with carbonyl group (2.75 Å) and amino group (2.93 Å) of backbone K101, respectively. As compared to the interaction with K103 in the WT binding pocket, the interaction energies to N103 are decreased about 0.4 and 2.1 kcal/mol at the B3LYP/6-31G(d) and MP2/6-31G(d) levels of theory, respectively. These may cause from the loss of H-bonding between sidechain K103 between the WT binding pocket and efavirenz. For the interaction with other amino acids in the K103N binding pocket, the interaction energies are about ±1 kcal/mol in both levels of theory as compared with those in the WT binding pocket.

In order to investigate the interaction energies of the docked conformation of other efavirenz derivatives, the interaction energies of docked efavirenz (compound.1) in WT and K103N binding pockets are firstly studied in comparison to those of X-ray structure. The interaction energies of compound 1 with the WT and K103N binding pockets are shown in table 8 and table 9, respectively. For the WT binding pocket, the interaction energies of compound 1 are -17.1 to 0.9 kcal/mol and -19.9 to -0.9 kcal/mol at the B3LYP/6-31G(d) and MP2/6-31G(d) levels of theory, respectively. Except K103, the interaction energies of compound 1 with other amino acids from both levels of theory are similar to those of X-ray structure with ±1.5 kcal/mol. The interaction energies between compound 1 and K103 are more attractive than X-ray structure and K103 which the different of interaction energies are 3.6 and 2.9 kcal/mol at the B3LYP/6-31G(d) and MP2/6-31G(d) levels of theory, respectively. This more attractive interaction is caused from the stronger attractive interaction formed between fluoro atom of trifluoro group in efavirenz and γ-C of sidechain K103. The distances are 3.81 Å and 3.51 Å in X-ray structure and compound 1, respectively. Because this substituted group (trifluoro group) does not vary in this study, more attractive interaction with K103 may occur to other efavirenz derivatives with the same trend. According to this explanation, compound 1 can be used to compare the interaction with other efavirenz derivatives. In case of compound 1 in the K103N binding pocket, the interaction energies are -7.3 to 4.6 kcal/mol and -10.3 to -1.3 kcal/mol at the B3LYP/6-31G(d) and MP2/6-31G(d) levels of theory, respectively. The interaction energy differences between X-ray and compound 1 are ±1.7 kcal/mol, except L100 and K101. Because of a little movement of compound no. 1 to form more attractive interaction with Y181, attractive interaction with L100 and K101 are lost about 1.8 and 2.7 kcal/mol at MP2/6-31G(d), respectively. More attractive interaction with Y181 could occur with other efavirenz derivatives in the same manner. Therefore, to determine the interaction energies of other docked efavirenz derivatives, compound 1, which is the docked conformation of efavirenz, is used to compare in the further step.

Table 8. Interaction energies between docked conformations (docked efavirenz (compound 01), compounds 25 and 12) and each amino acid surrounding in the WT binding pocket

Amino acid residue	Interaction energy (kcal/mol)					
	compound 01		compound 25		compound 12	
	B3LYP/ 6-31G(d)	MP2/ 6-31G(d)	B3LYP/ 6-31G(d)	MP2/ 6-31G(d)	B3LYP/ 6-31G(d)	MP2/ 6-31G(d)
P95	-0.1	-0.9	-0.1	-0.9	-0.3	-0.8
L100	0.9	-6.3	0.4	-7.0	3.6	-3.2
K101	-17.1	-19.9	-17.2	-20.0	-15.3	-18.0
K103	-5.4	-8.2	-7.3	-10.2	-4.6	-7.7
V106	0.7	-2.5	-1.7	-4.1	1.2	-2.2
Y181	-0.1	-4.1	-0.1	-4.0	2.1	-1.9
Y188	-0.7	-5.2	-1.2	-6.1	1.0	-4.6
G190	-1.4	-2.1	-1.5	-2.3	-1.3	-2.3
F227	-0.1	-1.1	0.0	-0.4	-0.2	-1.2
W229	-1.4	-2.5	-0.4	-2.3	0.3	-2.5
L234	0.2	-1.8	-0.1	-1.4	0.1	-1.8
H235	-1.5	-2.9	-2.2	-3.0	-0.4	-2.0
P236	-1.4	-3.6	-0.8	-2.7	-1.2	-3.7
Y318	-0.3	-3.7	0.3	-3.4	0.1	-3.7

Table 9. Interaction energies between docked conformations (docked efavirenz (compound 01), compounds 50 and 22) and each amino acid surrounding in the K103N binding pocket

Amino acid residue	Interaction energy (kcal/mol)					
	compound 01		compound 50		compound 22	
	B3LYP/ 6-31G(d)	MP2/ 6-31G(d)	B3LYP/ 6-31G(d)	MP2/ 6-31G(d)	B3LYP/ 6-31G(d)	MP2/ 6-31G(d)
P95	-0.3	-1.3	-0.1	-1.1	-0.1	-0.7
L100	4.6	-5.0	0.8	-7.8	2.8	-5.1
K101	-7.3	-10.3	-14.8	-17.3	-6.9	-9.8
N103	0.0	-1.5	0.3	-1.2	-0.1	-2.0
V106	0.7	-2.0	0.3	-3.1	0.1	-3.0
Y181	0.4	-5.9	-2.1	-8.9	1.5	-5.0
Y188	-1.6	-5.1	-1.9	-7.5	-0.5	-4.1
G190	-1.3	-2.2	-1.5	-2.5	-1.0	-2.0
F227	-0.3	-1.3	0.7	-1.4	-0.4	-0.9

Table 9 (Continued)

W229	-1.0	-3.8	-1.1	-4.0	-0.7	-1.6
L234	-0.2	-2.0	0.0	-2.9	0.3	-0.4
H235	-1.5	-2.3	-1.9	-2.9	-1.7	-2.1
P236	-1.4	-2.6	-1.0	-2.5	-1.8	-3.1
Y318	0.3	-3.3	0.2	-3.4	0.0	-3.0

Interaction Energy of Compounds 25 and 12 in WT Binding Pocket

The individual interaction is also observed with the compounds 25 and 12, the highest and less WT activity, respectively. Their interaction energies compared with compound 1 are shown in table 8. The interaction energies of compound 25 in WT binding pocket are -17.2 to 0.4 kcal/mol and -20.0 to -0.4 kcal/mol at the B3LYP/6-31G(d) and MP2/6-31G(d) levels of theory, respectively. As compared with compound 1, the results confirm the docking results that slightly more attractive interaction to compound 25 could be formed to L100, K101, Y188, G190, and H235 with ±1.0 kcal/mol from both levels of theory. K103 and V106 are formed more attractive interactions within 2.0 kcal/mol. More attractive interaction with V106 is formed with the 5-F of the ring and this cause a slight shift of the fluoromethyl group to come closer to K103. For the less WT activity compound, interaction energies of compound 12 are -15.3 to 3.6 kcal/mol and -18.0 to -0.8 kcal/mol the B3LYP/6-31G(d) and MP2/6-31G(d) levels of theory, respectively. The decreased attractive interaction, as compared with compound 1, is found between compound 12 and most amino acids. Because of the bulky group attached at R position, steric interaction occurs with L100, K101, and Y181. The individual interaction results support and agree well with the experimental results that compounds 25 and 12 show higher and lower potency than compound 1, respectively.

Interaction Energy of Compound 50 and 22 in K103N Binding Pocket

In the case of the K103N RT binding pocket, compounds 50 and 22, the highest and less potency against K103N RT, respectively, are selected to study. Their interaction energies are shown in table 9. The interaction action energies of compound 50 in K103N binding pocket are -14.8 to 0.8 kcal/mol and -17.3 to -1.1 kcal/mol at the B3LYP/6-31G(d) and MP2/6-31G(d) levels of theory, respectively. From both levels of theory, more attractive interaction has significantly formed to L100, K101, Y181, and Y188. The attached substituted groups in compound 50 cause more fit of its conformation in K103N binding pocket than compound 1. Stronger attractive interaction is also formed with L100 and K101. The nitrogen atom on the Z substituent of compound 50 make the attractive interaction increased. The H-π interaction to the aromatic sidechain of Y181 could be formed. The methoxy group attached at C5-position of phenyl ring has fitted with the hydrophobic pocket consisting of V106, Y188, and L234. Most of other amino acids in the binding pocket also slightly form more attractive interaction with compound 50. For the less active compound against the K103N RT, the interaction energies of compound 22 are -6.9 to 2.8 kcal/mol and -9.8 to -0.7 kcal/mol at the B3LYP/6-31G(d) and MP2/6-31G(d) levels of theory, respectively. From their docked conformation, it can be seen that the substituting group at R-position has caused the steric effect with the pocket around R-position. From the interaction energies, this substituting group at R-position of compound 22 has caused the less attractive interaction with P95,

Y188, W229, and L234. The individual interaction results are in agreement with the experimental results that compounds 50 and 22 shows highest and less potency against the K103N RT, respectively.

Based on the calculated results, more attractive interaction are found by using the MP2/6-31G(d) level of theory, as compare with B3LYP/6-31G(d) level of theory. The obtained interaction energies agree well as compared with the results previously reported [36,54]. Therefore, to ensure that H-π interaction is included in the calculation, MP2 method should be appropriated applied in the study. The interaction energies could be helpful for understanding the individual interaction between ligand and the binding pocket and suggesting for the guideline of the new more potent inhibitor design.

Conclusion

The molecular docking calculations and 3D-QSAR analyses were successfully combined to investigate the interaction and the relationship between structural requirements of efavirenz derivatives for WT and K103N HIV-1 RT. The potential binding orientation of the inhibitors in the binding pockets could be identified, by using docking studies. The docking results provide additional insight into essential inhibitor-enzyme interactions for different types of wild type and mutant type of HIV-1 RT. Based on the docking conformations, the reliable and predictive CoMFA and CoMSIA models of efavirenz derivatives for the WT and K103N RT inhibition were derived. The QSAR models are successfully used to discriminate between the structural requirements for WT and K103N inhibitory activities. Moreover, the interaction energy trend calculated from quantum chemical calculations of the inhibitors and individual amino acid residues in the binding pockets is informative to highlight particular ligand-receptor interaction in molecular level. The results derived from all approaches validate each other and agree well with the ligand-receptor complex interaction derived from the X-ray crystallographic data. Evidently, in the present study, molecular modeling with the combination of structure-based and ligand-based drug design approaches integrated with quantum chemical calculations has been proven as attractive and efficient tools for better understanding of the key structural element for enhancing the interaction between efavirenz compounds and the WT and K103N RT. Consequently, the obtained results enable to provide beneficial guidelines to design novel compounds with higher anti-HIV–1 RT activities against WT and K103N RT.

Acknowledgements

The authors thank Prof. Peter Wolschann for helpful discussion and support. The reasearch grants provided by the Thailand Research Fund (TRG4880001 and BRG4780007) are gratefully acknowledged. We thank LCAC, the postgraduate on education and research in petroleum and petrochemical technology, the high performance computing centers of Kasetsart University, the National Electronics and Computer Technology (HPCC/NECTEC) and the computer center of the University of Vienna for providing computational resources.

References

[1] Arnold, E; Das, K; Yadav, PNS; Hsiou, Y; Boyer, PL; Hughes, SH. Targeting HIV Reverse Transcriptase for Anti-AIDS Drug Design. *Drug. Des. Disc.* 1996, 13, 29-47.

[2] De Clercq, E. The Role of Nonnucleoside Reverse Transcriptase Inhibitors (NNRTI) in the Therapy of HIV-1 Infection. *Anti.Viral Res.* 1998, 38, 153-179.

[3] Evans, DB; Brawn, K; Deibel, MR; Tarpley, WG; Sharma SK. A Recombinant Ribonuclease H Domain of HIV-1 Reverse Transcriptase that is Enzymatically Active. *J. Biol. Chem.* 1991, 266, 20583-20585.

[4] Young, SD. Non-nucleoside Inhibitors of HIV-1 reverse Transcriptase. *Perspect. Drug Desc. Des.* 1993, 1, 181-192.

[5] Richman, D; Shih, CK; Lowy, I; Rose, J; Prodanovich, P; Goff, S; Griffin, J. Human Immunodeficiency Virus Type 1 Mutants Resistant to Nonnucleoside Inhibitors of Reverse Transcriptase Arise in Tissue Culture. *Proc. Natl. Acad. Sci. U.S.A.* 1991, 88, 11241–11245.

[6] Balzarini, J; Karlsson, A; Sardana, VV; Emini, EA; Camarasa, MJ; De Clercq, E. Human Immunodeficiency Virus 1 (HIV-1)-Specific Reverse Transcriptase (RT) Inhibitors may Suppress the Replication of Specific Drug-Resistant (E138K) RT HIV-1 Mutants or Select for Highly Resistant (Y181C–.C181I) RT HIV-1 Mutants. *Proc. Natl. Acad. Sci. U.S.A.* 1994, 91, 6599–6603.

[7] De Clercq, E. Non-nucleoside Reverse Transcriptase Inhibitors (NNRTI) for the Treatment of Human Immunodeficiency Virus Type 1 (HIV-1) Infections: Strategies to Overcome Drug Resistance Development. *Med. Res. Rev.* 1996, 16, 125–157.

[8] Ren, J; Esnouf, R; Garman, E; Somers, D; Ross, C; Kirby, I.; Keeling, J; Darby, G; Jones, Y; Stuart, D; Stammers, D. High Resolution Structures of HIV-1 RT from Four RT-Inhibitor Complexes. *Nat. Struct. Biol.* 1995, 2, 293–302

[9] Ding, J; Das, K; Moereels, H; Koymans, L; Andries, K; Janssen, PA; Hughes, SH; Arnold, E. Structure of HIV-1 RT/TIBO R 86183 Complex Reveals Similarity in the Binding of Diverse Nonnucleoside Inhibitors. *Nat. Struct. Biol.* 1995, 2, 407–415.

[10] Schinazi, RF; Larder, BA; Mellors, J. Mutations in Retroviral Genes Associated with Drug Resistance. *International Antiviral News*. 1997, *5*, 129–135.

[11] Young, SD; et al., Emini, EA. L-743, 726 (DMP-266): A Novel, Highly Potent nonnucleoside Inhibitor of the Human Immunodeficiency Virus Type 1 Reverse Transcriptase. *Antimicrob. Agents Chemother.* 1995, 39, 2602–2605.

[12] Bacheler, L; Wieslow, O; Snyder, S; Hanna, G; D'Aquila, R. The SUSTIVA Resistance Study Team. *Abstract, presented at the 12th World AIDS Conference, Geneva, Switzerland.* 1998.

[13] Bacherler, L; Jeffrey, S; Hanna, G; D'Aquila, R; Wallace, L; Logue, K; Cordova, B; Hertogs, K; Larder, B; Buckery, R; Baker, D; Gallagher, K; Scarnati, H; Tritch, R; Rizzo, C. Genotypic Correlate of Phenotypic Resistance

to Efavirenz in Virus Isolates from Patients Failing Nonnucleoside Reverse Transcriptase Inhibitor Therapy. *J. Virol.* 2001, 75, 4999-5008.

[14] Miller, V; Sturmer, M; Staszewski, S; Groschel, B; Hertogs, K; de Bethune, MP; Pauwels, R; Harrigan, PR; Bloor, S; Kemp, SD; Larder, BA. The M184V Mutation in HIV-1 Reverse Transcriptase (RT) Conferring Lamivudine Resistance does not Result in Broad Cross-resistance to Nucleoside Analogue RT Inhibitors. *AIDS.* 1998, 12, 705-712.

[15] Torti, C; Pozniak, A; Nelson, M; Hertogs, K; Gazzard, BG. Distribution of K103N and/or Y181C HIV-1 Mutations by Exposure to Zidovudine and Non-nucleoside Reverse Transcriptase Inhibitors. *Antimicrob. Chemother.* 2001, 48, 113-116.

[16] Patel, M; Ko, SS; McHugh, RJ; Markwalder, JA; Srivastana, AS; Cordova, BC; Klabe, RM; Erickson-Viitanen, S; Trainor, GL; Seitz, SP. Synthesis and Evaluation of Analogs Efavirenz (SUSTIVA^{TM}) as HIV-1 Reverse Transcriptase Inhibitors. *Bioorg. Med. Chem. Lett.* 1999, 9, 2805-2810.

[17] Patel, M; McHugh, RJ; Jr.; Cordova, BC; Klabe, RM; Erickson-Viitanen, S; Trainor, GL; Ko, SS. Synthesis and Evaluation of Benzoxazinones as HIV-1 Reverse Transcriptase Inhibitors. Analogs of Efavirenz (SUSTIVA^{TM}). *Bioorg. Med. Chem. Lett.* 1999, 9, 3221-3224.

[18] Patel, M; McHugh, RJ; Jr; Cordova, BC; Klabe, RM; Erickson-Viitanen, S; Trainor, GL; Rodgers, JD. Synthesis and Evaluation of Quinoxalinones as HIV-1 Reverse Transcriptase Inhibitors. *Bioorg. Med. Chem. Lett.* 2000, 10, 1729-1731.

[19] Corbett, JW; Ko, SS; Rodgers, JD; Gearhart, LA; Magnus, NA; Bacheler, LT; Diamond, S; Jeffrey, S; Klabe, RM; Cordova, BC; Garber, S; Logue, K; Trainor, GL; Anderson, PS; Erickson-Viitanen, SK. Inhibition of Clinically Relevant Mutant Variants of HIV-1 by Quinazolinone Non-Nucleoside Reverse Transcriptase Inhibitors. *J. Med. Chem.* 2000, 43, 2019-2030.

[20] Corbett, JW; Pan, S; Markwalder, JA; Cordova, BC; Klabe, RM; Garber, S; Rodgers, JD; Erickson-Viitanen, SK. 3,3a-Dihydropyrano[4,3,2,-*de*]quinazolin-2(1*H*)-ones are Potent Non-nucleoside Reverse Transcriptase Inhibitors. *Bioorg. Med. Chem. Lett.* 2001, 11, 211-214.

[21] Markwalder, JA; Christ Abdul Mutlib, DD; Cordova, BC; Klabe, RM; Seitz, SP. Synthesis and Biological Activities of Potential Metabolites of the Non-nucleoside Reverse Transcriptase Inhibitor Efavirenz. *Bioorg. Med. Chem. Lett.* 2001, 11, 619-622.

[22] Cocuzza, AJ; Chidester, DR; Cordova, BC; Jeffrey, S; Parsons, RL; Bacheler, LT; Erickson-Viitanen, S; Trainor, GL; Ko, SS. Synthesis and Evaluation of Efavirenz (Sustiva^{TM}) Analogues as HIV-1 Reverse Transcriptase Inhibitors: Replacement of the Cyclopropylacetylene Side Chain. *Bioorg. Med. Chem. Lett.* 2001, 11, 1177-1179.

[23] Cocuzza, AJ; Chidester, DR; Cordova, BC; Klabe, RM; Jeffrey, S; Diamond, S; Weigelt, CA; Ko, SS; Bacheler, LT; Erickson-Viitanen, S; Rodgers, JD. 4,1-

Benzoxazepinone Analogues of Efavirenz (Sustiva^{TM}) as HIV-1 Reverse Transcriptase Inhibitors. *Bioorg. Med. Chem. Lett.* 2001, 11, 1389-1392.

[24] Patel, M; McHugh, RJ; Jr.; Cordova, BC; Klabe, RM; Bacheler, LT; Erickson-Viitanen, S; Rodgers, JD. Synthesis and Evaluation of Novel Quinoxalinones as HIV-1 Reverse Transcriptase Inhibitors. *Bioorg. Med. Chem. Lett.* 2001, 11, 1943-1945.

[25] Ren, J; Milton, J; Weaver, KL; Short, SA; Stuart, DI; Stammers, DK. Structural Basis for the Resilience of Efavirenz (DMP-266) to Drug Resistance Mutations in HIV-1 Reverse transcriptase. *Structure.* 2000, 8, 1089-1094.

[26] Lindberg, J; Sigurosson, S; Lowgren, S; Andersson, HO; Sahlberg, C; Noreen, R; Fridborg, K; Zhang, H; Unge, T. Structural Basis for Inhibitory Efficacy of Efavirenz (DMP-266), MSC194 and PNU142721 Towards the HIV-1 RT K103N Mutant. *Eur. J. Biochem.* 2002, 269, 1670-1677.

[27] Hannongbua, S. Structural Information and Drug–Enzyme Interaction of the Non-Nucleoside Reverse Transcriptase Inhibitors Based on Computational Chemistry Approaches. *Top. Heterocycl. Chem*. 2006, 4, 55-84.

[28] Udier-Blagovic, M; Watkins, EK; Tirado-Rives, J; Jorgensen, WL. Activity Predictions for Efavirenz Analogues with the K103N Mutant of HIV Reverse Transcriptase. *Bioorg. Med. Chem. Lett.* 2003, 13, 3337-3340.

[29] Udier-Blagovic, M; Tirado-Rives, J; Jorgensen, WL. Structural and Energetic Analyses of the Effects of the K103N Mutation of HIV-1 Reverse Transcriptase on Efavirenz Analogues. *J. Med. Chem.* 2004, 47, 2389-2392.

[30] Rodriguez-Barrios, F; Balzarini, J; Gago, F. The Molecular Basis of Resilience to the Effect of the Lys103Asn Mutation in Non-nucleoside HIV-1 Reverse Transcriptase Inhibitors Studied by Targeted Molecular Dynamics Simulations. *J. AM. Chem. Soc.* 2005, 127, 7570-7578.

[31] Weinzinger, P; Hannongbua, S; Wolschann, P. Related Articles, Links Molecular Mechanics PBSA Ligand Binding Energy and Interaction of Efavirenz Derivatives with HIV-1 Reverse Transcriptase. *J. Enzyme. Inhib. Med.* Chem. 2005, 20, 129-34.

[32] Barreca, ML; Rao, A; Luca, LD; Zappala, M; Monforte, AM; Maga, G; Pannecouque, C; Balzarini, J; Clercq, ED; Chimirri, A; Monforte, P. Computational Strategies in Discovering Novel-nucleoside Inhibitors of HIV-1 RT. *J. Med. Chem.* 2005, 48, 3433-3437.

[33] D'Cruz, OJ; Uckun, FM. Novel tight binding PETT, HEPT and DABO-based non-nucleoside inhibitors of HIV-1 reverse transcriptase. *J. Enzyme Inh. and Med. Chem*. 2006, 21, 329-350.

[34] Xie, Q; Tang, Y; Li, W; Wang, HX; Qiu ZB. Investigation of the Binding Mode of (-)-Meptazinol and Bis-Meptazinol Derivatives on Acetylcholinesterase Using a Molecular Docking Method. *J. Mol. Model.* 2006, 12, 390-397.

[35] Rawal, RK; Kumar, A; Siddiqi, MI; Katti, SB. Molecular Docking Studies on 4-Thiazolidinones as HIV-1 RT Inhibitors. *J. Mol. Model.* 2007, 13, 155–161.

[36] Saparpakorn, P; Hannongbua, S; Rognan, D. Design of Nevirapine Derivatives Insensitive to the K103N and Y181C HIV-1 Reverse Transcriptase Mutants. *SAR QSAR Environ Res*. 2006, 17, 183-94.

[37] Pungpo, P; Saparpakorn, P; Wolschann, P; Hannongbua, S. Computer-Aided Molecular Design of Highly Potent HIV-1 RT Inhibitors: 3D QSAR and Molecular Docking Studies of Efavirenz Derivatives. *SAR and QSAR in Environmental Research*. 2006, 17, 353-370.

[38] Zhang, ZM; Zheng, LD; Shen, J; Luo, X; Zhu, W; Hualiang, J. Towards Discovering Dual Functional Inhibitors Against Both Wild Type and K103N Mutant HIV-1 Reverse Transcriptases: Molecular Docking and QSAR Studies on 4,1-Benzoxazepinone Analogues. *J. Comp. Aided Mol. Des.* 2006, 20, 281–293.

[39] D'Cruz, OJ; Uckun, FM. Novel Broad-Spectrum Thiourea Non-Nucleoside Inhibitors for the Prevention of Mucosal HIV Transmission. *Current HIV Research*. 2006, 4, 329-345.

[40] Ragno, R; Artico, M; De Martino, G; La Regina, G; Coluccia, A; Di Pasquali, A; Silvestri, R. Docking and 3-D QSAR Studies on Indolyl Aryl Sulfones. Binding Mode Exploration at the HIV-1 Reverse Transcriptase Non-Nucleoside Binding Site and Design of Highly Active N-(2-Hydroxyethyl)carboxamide and N-(2-Hydroxyethyl) carbohydrazide Derivatives. *J. Med. Chem.* 2005, 48, 213-223.

[41] Medina-Franco, JL; Rodriguez-Morales, S; Juarez-Gordiano, C; Hernandez-Campos, Al; Jimenez-Barbero, J; Castillo, R. Flexible Docking of Pyridinone Derivatives into the Non-Nucleoside Inhibitor Binding Site of HIV-1 Reverse Transcriptase. *Bio. Med. Chem.* 2004, 12, 6085-6095.

[42] Cramer, RD; Patterson, DE; Bunce, JD. Comparative Molecular Field Analysis (CoMFA). 1. Effect of Shape on Binding of steroids to Carrier Proteins. *J. Am. Chem. Soc.* 1988, 110, 5959-5967.

[43] Klabe, G; Abraham, U; Mietzner, T. Molecular Similarity Indices in a Comparative Analysis (CoMSIA) of Drug Molecules to Correlate and Predict their Biological Activity. *J. Med. Chem.* 1994, 37, 4130-4146.

[44] Kubinyi, H. 3D QSAR in Drug Design: Theory, Methods and Applications. *ESCOM Science Publishers: Leiden*. 1993.

[45] Hannongbua, S; Nivesanond, K; Lawtrakul, L; Pungpo, P; Wolschann, P. 3D-Quantitative Structure-activity Relationships of HEPT Derivatives as HIV-1 Reverse Transcriptase Inhibitors, Based on Ab Initio Calculations. *J. Chem. Inf. Compul. Sci.* 2001, 41, 848-855.

[46] Hannongbua, S; Pungpo, P; Wolschann, P. Quantitative Structure-activity Relationships and Comparative Molecular Field Analysis of the HIV-1 Reverse Transcriptase Inhibitor of TIBO Derivatives. *J. Comp. Aided Mol. Des.* 1999, 13, 563-577.

[47] Pungpo, P; Hannongbua, S. Three-dimensional Quantitative Structure-activity Relationships Study on HIV-1 Reverse Transcriptase Inhibitors in the Class of

Dipyridodiazepinone Derivatives, Using Comparative Molecular Field Analysis. *J. Mol. Grap. Mod.* 2000, 18, 581-590.

[48] Zhang, H; Li, H; Liu, C. CoMFA, CoMSIA and Molecular Hologram QSAR Studies of Novel Neuronal nAChRs Ligands-Open Ring Analogues of 3-Pyridyl Ether. *J. Chem. Inf. Model.* 2005, 45, 440-448.

[49] Barreca, ML; Carotti, A; Carrieri, A; Chimirri, A; Monforte, AM; Calace, MP; Rao, A. Comparative Molecular Field Analysis (CoMFA) and Docking Studies of Non-Nucleoside HIV-1 RT Inhibitors (NNIs). *Bio. Med. Chem.* 1999, 7, 2283-2292.

[50] Chen, HF; Yao, XJ; Li, Q; Yuan, SG; Panaye, A; Doucet, JP; Fan, BT. Comparative Study of Nonnucleoside Inhibitors with HIV-1 Reverse Transcriptase Based on 3D-QSAR and Docking. *J. Comp. Aided Mol. Des.* 2003, 14, 455-474.

[51] Medina-Franco, JL; Rodriguez-Morales, S; Juarez-Gordiano, C; Hernandez-Campos, A; Castillo, R. Docking-based CoMFA and CoMSIA Studies of Non-Nucleoside Reverse Transcriptase Inhibitors of the Pyridinone Derivative Type. *J. Comp. Aided. Mol. Des*. 2004, 18, 345-360.

[52] Zhou, Z; Madura, JD. CoMFA 3D-QSAR Analysis of HIV-1 RT Nonnucleoside Inhibitors, TIBO Derivatives Based on Docking Conformation and Alignment. *J. Chem. Inf. Comput. Sci*. 2004, 44, 2167-2178.

[53] Sea-oon, S; Kuno, M; Hannongbua, S. Binding Energy Analysis for Wild-Type and Y181C Mutant HIV-1 RT/8-Cl TIBO Complex Structures: Quantum Chemical Calculations Based on the ONIOM Method. *Proteins*. 2005, 61, 859-869.

[54] Kuno, M; Hannongbua, S; Morokuma, K. Theoretical Investigation on Nevirapine and HIV-1 Reverse Transcriptase Binding Site Interaction, Based on ONIOM Method. *Chem. Phys Lett.* 2003, 380, 456-463.

[55] Nunrium, P; Kuno, M; Saen-oon, S; Hannongbua, S. Particular Interaction between Efavirenz and the HIV-1 Reverse Transcriptase Binding Site as Explained by the ONIOM2 Method. *Chem. Phys. Lett.* 2005, 405, 198-202.

[56] Bacheler, LT; Paul, M; Jadhav, PK; Otto, M; Miller, J. An Assay for HIV RNA in Infected Cell Lysates and Its Use for the Rapid Evaluation of Antiviral Efficacy. *Antiviral. Chem. Chemother.* 1994, 5, 111-121.

[57] Alchemy 2000, Tripos Associates Inc. St., Louis. MO. 1998.

[58] Frisch, MJ; Trucks, GW; Schlegel, HB; Gill, PMW; Jonhson, BG; Robb, MA; Cheeseman, JR; Keith, T; Petersson, GA; Montgomery, JA; Raghavachari, K; Al-Laham, MA; Zakrzewski, VG; Ortiz, JV; Foresman, JB; Peng, CY; Alaya, PY; Chen, W; Wong, WM; Andres, JL; Replogle, ES; Gomperts, R; Martin, RL; Fox, DJ; Binkley, JS; Defrees, DJ; Baker, J; Stewart, JP; Head-Gordon, M; Gonzales, C; Pople, JA. GAUSSIAN 94, Revision B.3, Gaussian, Inc. Pittsburgh PA. 1995.

[59] Jones, G; Wilett, P; Glen, RC; Leach, AR; Taylor, R. Development and validation of a genetic algorithm for flexible docking. *J. Mol. Biol.* 1997, 267, 727-748.

[60] Morris, GM; Goodsell DS; Olson, AJ. The Scripps Research Institute, *AutoDock* 3.05, 2000.
[61] SYBYL 7.0, Tripos Associates Inc., 1699 South Hanley Road, Suite 303, St. Louis, Missouri 63144, USA.
[62] Viswanadhan, VN; Ghose, AK; Revenkar, GR; Robins, R. Atomic and Physichochemical Parameters for Three-dimensional Structure-directed Quantitative Structure-activity Relationships. 4. Additional Parameters for Hydrophobic and Dispersive Interactions and Their Application for an Automated Superposition of Certain Naturally Occurring Antibiotics. *J. Chem. Inf. Comput. Sci.* 1989, 29, 163-172.
[63] Klabe, G. The Use of Composite Crystal-field Environments in Molecular Recognition and the De Novo Design of Protein Ligands. *J. Mol. Biol.* 1994, 237, 212-235.
[64] SYBYL Molecular Modeling Software, Version 6.3, SYBYL Ligand Base Design, Tripos Associates, Inc., St.Louis, Missouri 63144, USA, 1996, 299.

In: Drug Design Research Perspectives
Editor: Stanley P. Kaplan, pp. 251-273
ISBN: 978-1-60021-823-1

Chapter IX

Virtual Screening and the Discovery of Bioactive Molecules

Káthia M. Honório[1] and Adriano D. Andricopulo[2*]
[1] Escola de Artes, Ciências e Humanidades, Universidade de São Paulo, Av. Arlindo Bettio 1000, 03828-000, São Paulo - SP, Brazil.
[2] Laboratório de Química Medicinal e Computacional, Centro de Biotecnologia Molecular Estrutural Instituto de Física de São Carlos, Universidade de São Paulo, Av. Trabalhador São-Carlense 400, 13560-970, São Carlos - SP, Brazil

Abstract

Drug discovery is a complex process that involves a combination of classical methods, modern technologies and innovative strategies addressed to the design of new chemical entities as drug candidates against a variety of biological targets. Drug discovery has moved toward more rational strategies based on our increasing understanding of the fundamental principles of protein-ligand interactions. The combination of available knowledge of several 3D protein structures with hundreds of thousands of small-molecule drug candidates have attracted the attention of scientists from all over the world for the application of structure- and ligand-based drug design approaches. The investigation of new potential biological targets is often limited by the absence of initial compounds that bind to a specific site in the protein, possessing reasonable chemical and biological profiles for further lead optimization. Biological screening requires testing large numbers of compounds in standard models to assess potency, selectivity and affinity. Virtual screening (VS) is a computational high-performance approach to analyze large databases of chemical compounds in order to prioritize novel potential hits for biological screening. Molecular docking is used to predict the binding affinities of each compound in the active site of a biological target. Similarity searching is another important component of VS methods that involves finding molecules in a database having the strongest similarity to a user-defined target structure.

* Corresponding author: Tel: +55 16 3373-8095. Fax: +55 16 3373-9881. E-mail: aandrico@if.sc.usp.br

Typically, the target will be known to have some biological activity of interest and the objective of the search is to prioritize a dataset to optimize the pharmacodynamic and pharmacokinetic properties. The contribution in this book chapter will briefly outline our view on the present scope and future advances of VS based on some instructive examples from recent research describing this emerging strategy that is providing new valuable and effective routes for the traditional high-throughput screening (HTS) in drug discovery.

Introduction

Drug discovery is an incredibly challenging and highly rewarding task, which demands integrated efforts in several relevant aspects, involving innovation, knowledge, information, technologies, expertise, R and D investments and management skills. The shift from traditional to genomics- and proteomics-based drug research has fundamentally transformed key R and D strategies in the pharmaceutical industry. The available subset of druggable and clinical relevant targets coded by the human genome is having a profound impact on drug discovery programs in terms of the new attractive opportunities for therapeutic intervention [1-4]. The modern drug discovery paradigm has changed dramatically over the past decade and continues to evolve in response to new discoveries, technologies, and increasing demand for more drug candidates. Current and future challenges will be directed to both validating new drug targets and finding promising drug candidates for clinical development. One approach to this problem is to rapidly identify active low-molecular-weight ligands that interact and selectively modify the physiological function of macromolecular targets (e.g., enzymes, receptors) [5,6].

The identification of promising hits and the generation of high-quality lead compounds in the early stages of the drug discovery process are becoming increasingly crucial. Recent advances in medicinal chemistry at the interface of chemistry and biology have created an important foundation in the search for new drug candidates possessing a combination of optimized pharmacodynamic and pharmacokinetic properties. Despite the impressive technological and scientific advances, drug discovery has become more expensive and time consuming over the same period of time [7,8].

Traditionally, in the early stages of the drug discovery process, small-molecule chemical libraries (e.g., natural and synthetic compounds) are biollogicaly screened for the identification of new active compounds. These libraries directly vary in size and complexity, and have been used for a wide range of drug targets in a variety of biochemical assays. This process could invariably be compared to finding a needle in a haystack, where the needle is a hit or lead and the haystack represents the complex multidimensional chemical space. In general, identification of the very small selective fraction of compounds with sufficient promise as hits for further optimization is a major problem faced by the pharmaceutical laboratories involved in this time consuming and risk process [4-6].

The widespread use of combinatorial chemistry and high-throughput screening (HTS) for the discovery of lead compounds has created a large demand for small organic molecules that act on specific drug targets. These technologies focus on the simultaneous generation of large numbers of molecules and integrated biological screening of large number of samples. In contemporary drug discovery, there is a clear paradigm shift from the random screening of collections of compounds to a more rational process, which would directly affect the success rate of lead identification. The definition and assessment of chemical and biological space have revitalized the paradigm of the screening process in drug discovery and have emphasized the importance of exploring the intrinsic complementary nature of classical and modern methods in medicinal chemistry. Some factors that directly and decisively influence this process are: careful reproducibility of the chemistry; understanding of the sensitivity of the biological assays; selection of compound libraries for biological screening; and, the correct data analysis of the screening results. A critical step that can be added in this complex context is the rational design of experiments during the definition of a proper medicinal chemistry strategy. A broad variety of R and D strategies and computational methods are valuable from this point of view to assist the identification and selection of hits and leads, and the optimization of potency, affinity and selectivity.

Virtual screening (VS) is a large-scale computational technique that has been integrated to modern drug discovery programs as a very useful alternative to the massive library generation of chemically diverse molecules as well as for the complex screening paradigm. Considering the high costs involved in experimental HTS, VS has been increasingly applied to reduce the number of compounds going into *in vitro* biological evaluation [9-15]. VS technologies are widely available to both academic institutions and pharmaceutical companies around the world. The time and costs associated with VS are much lower than those required in HTS approaches. In the context of the conceptual model and scientific approach in medicinal chemistry, VS and HTS are rather distinct in nature. HTS leads to discovery of real hits, however, no information about the mechanism of action or even about the target site of binding are provided. This important knowledge is obtained when structural biology, molecular modeling or kinetics are used to study the possible binding modes and mechanisms involved. On the other hand, VS approaches provide the identification of virtual hits rather than real bioactive molecules. These virtual hits should be acquired or synthesized and then submitted to biological screenings to ensure real bioactive molecules. The difference here is that these molecules are expected to act in a designed target site and probably, in a preferential mechanism of action. The possible interplay between HTS and VS techniques is a key element in modern drug research and should play a more important role in the process of selecting promising hits and leads for a variety of biological targets.

1. Virtual Screening

Over the past few decades, HTS was the main technique employed in drug discovery for the identification of new hits and lead compounds employing compound libraries generally consisting of hundreds of thousands of compounds against a selected biological target (typically a receptor or enzyme) [16]. Nowadays, one problem involved in HTS is related to combinatorial chemistry, as it provides an enormous number of new molecules and the cost factor often prohibits biological screening to such a vast extend. In recent years, the use of computational tools to identify a reduced number of molecules with increased potential for bioactivity to be evaluated experimentally has emerged both as a complementary and new alternative to HTS. In this context, VS technologies have largely enhanced the impact of computational methods applied to chemistry (chemoinformatics) and the goal of applying such methods is to reduce large compound databases *in silico* and to select a limited number of candidate molecules for drug design [17-19].

VS methods have rapidly become an essential component of the drug discovery process. High-performance hardware and specialized software, combined with advanced knowledge of protein structure and small molecule binding modes, have made this technology a useful complement, and in some cases, a reasonable alternative to HTS. VS can be defined as the process of reducing a library containing an unmanageable number of compounds (real or virtual) to a limited number of potentially promising compounds for the target of interest by means of computational methods that exploit existing knowledge [20,21]. In VS methodology, large libraries of molecules are docked into receptor structures and ranked by the relative binding affinity. Although the energy calculations are crude, the compounds in the library are readily available, making experimental testing easy and false-positives tolerable [22]. VS approaches have two different stages: (i) docking stage, where ligands from a chemical library are posed within the binding target site; and, (ii) scoring stage, where those complexes are scored by some measure that attempts to predict the likelihood of the ligand binding to the target [23]. Figure 1 displays the main steps involved in a VS process.

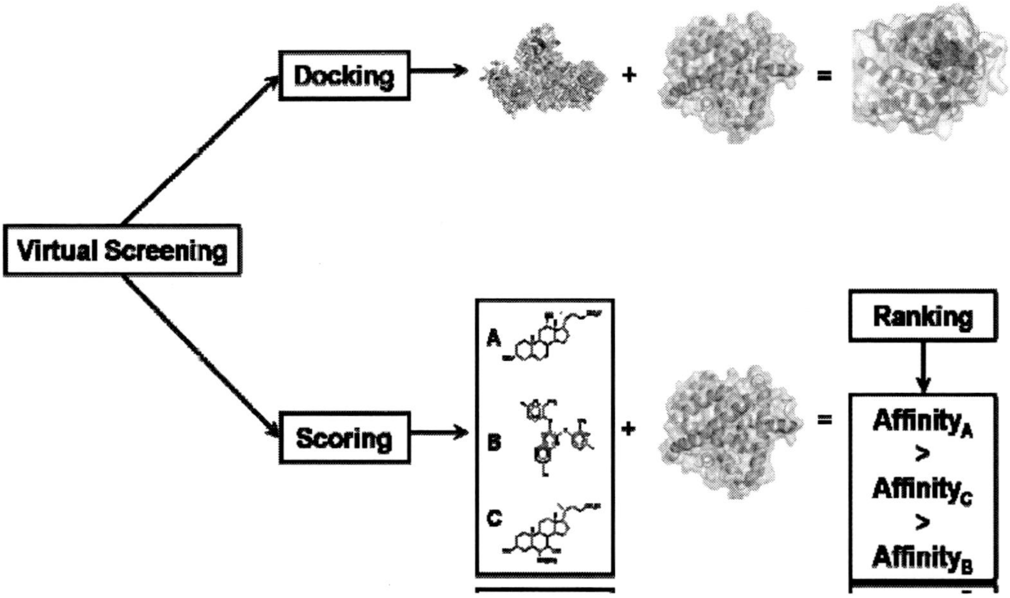

Figure 1. Main steps involved in a VS process.

Traditionally, VS can be performed according to two scenarios: ligand-based and structure-based. In ligand-based approaches, the main aim is to identify compounds with molecular similarities (descriptor-based, pharmacophore, topological, etc.) to known active molecules, using 2D and 3D database searches [24,25]. On the other hand, in structure-based approaches, 3D structures of the target proteins are required for the molecular docking of potential ligands into a known binding site [26,27]. These approaches will be discussed in more details in the next topics.

1.1. Structure-Based Virtual Screening

In structure-based virtual screening, knowledge about the 3D structure of the target and its binding site are essential to perform a protein-ligand high-throughput docking [28]. In this approach, the 3D structure of the target protein has been previously determined, usually by X-ray crystallography, NMR, or from homology modeling [29-31]. For the generation of a structure-based 3D pharmacophore model, the conformation of the ligand in the protein's binding site is additionally required. This information is provided by numerous co-crystal structures that are stored in the Protein Data Bank (PDB) [32].

The major focus of structure-based methods is the search for new ligands whose orientation and conformation achieves the highest degree of complementarity with respect to all details of the receptor's steric constraints and interaction geometries. Some difficulties are associated to structure-based methods: (i) treatment of both ligand and receptor flexibility; (ii) large number of protein-inhibitor complexes available for a single drug target, necessitating tools to compare and contrast their similarities and differences; and, (iii) the quality of protein-ligand complex structures [33]. Besides the difficulties involved in structure-based virtual screening, many successful stories have been reported in the literature. For example, scientists from Novartis Pharma AG employed the structure-based approach in their studies on protein kinase CK2 (Figure 2a) as a drug target for the development of new antitumor agents, searching for promising enzyme inhibitors [34]. So, they used high-throughput docking of their corporate compound collection in the ATP binding site of a homology model of human CK2, using an appropriate protocol. Finally, a 7-substitued indoloquinazoline compound (Figure 2b) was identified as a novel inhibitor of protein kinase CK2 by virtual screening of 400.000 compounds. Importantly, this compound is the most potent inhibitor of CK2 discovered, with an IC_{50} value of 80 nM. Its high potency, associated with considerable enzyme selectivity, provides a valuable tool for the study of the biological function of CK2 [34].

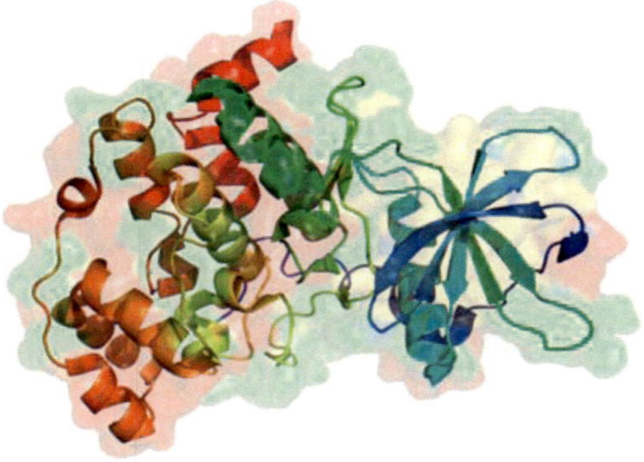

a.

Figure 1: Continued on next page

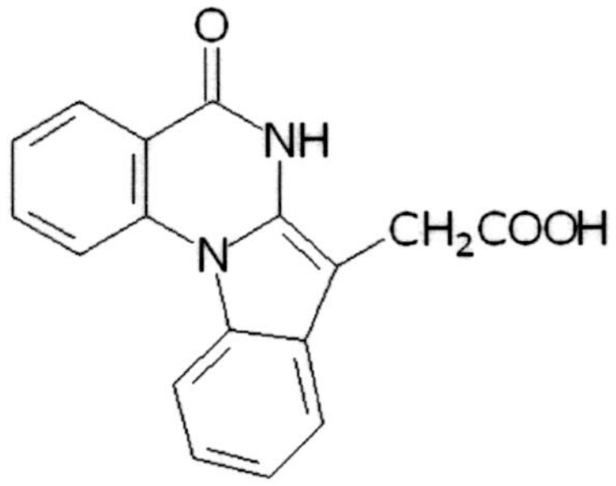

b.

Figure 2. (a) protein kinase CK2 (PDB code: 1DAW); (b) novel kinase CK2 inhibitor discovered by virtual screening.

1.2. Ligand-Based Virtual Screening

When no 3D protein structures are available for important drug targets, an alternative way to explore active molecules is the use of ligand-based virtual screening methods, which can be divided into three basic classes: similarity searching, pharmacophore pattern matching and quantitative structure–activity relationship (QSAR) modeling [35-37]. In general, these methods employ information about the molecular structure of a single compound or series of active molecules (against a biological target) as input for the identification of new bioactive molecules from databases containing thousands of compounds. A recent case of success related to ligand-based methods involved the biological targets known as ERG2, emopamil binding protein (EBP), and sigma-1 receptor (σ1), which are enzymes of sterol metabolism and an enzyme-related protein, respectively [38]. These targets share high affinity for various structurally diverse compounds. In this study, the authors employed pharmacophore models obtained with the Catalyst program, using a series of chemically diverse compounds exhibiting K_i values ranging from high-affinity (10 pM) to low-affinity (100 μM) for the three proteins. In virtual screening experiments, they retrieved drugs that were previously reported to bind to one or more proteins and also tested 11 new hits experimentally, of which three (among them raloxifene), exhibited affinities for σ1 or EBP of $<$ 60 nM. When used to search a database of 3,525 biochemicals of intermediary metabolism, a slightly modified ERG2 pharmacophore model successfully retrieved 10 substrate candidates among the top 28 hits. Their results indicate that inhibitor-based pharmacophore models for σ1, ERG2, and EBP can be used to screen drug and metabolite databases for chemically diverse compounds and putative endogenous ligands [38].

1.2.1. Similarity Searching

Similarity searching techniques are becoming increasingly important for the ligand-based virtual screening of large compound collections. These techniques are based on the observation that similar chemical structures often display similar biological activity toward unique targets (specific enzyme or receptor). Similarity searching methods are easy to handle and usually fast, and they do not require other information besides the chemical structure of the molecules. Thus, these methods in medicinal chemistry are valuable to define the chemical space of new hits and to follow up on actives from HTS, as well as for library design from virtual combinatorial libraries [39,40]. Fingerprints (i.e., vectors where each element encodes some aspects of molecular structures) are commonly used as the input for similarity calculations since, once defined and generated, they can be quickly compared by using a similarity index. Notably, the efficiency of similarity searching depends not only on the fingerprints employed, but also on the chosen similarity index [41]. In Figure 3 it can be seen the general representation of a general procedure employed in similarity searching.

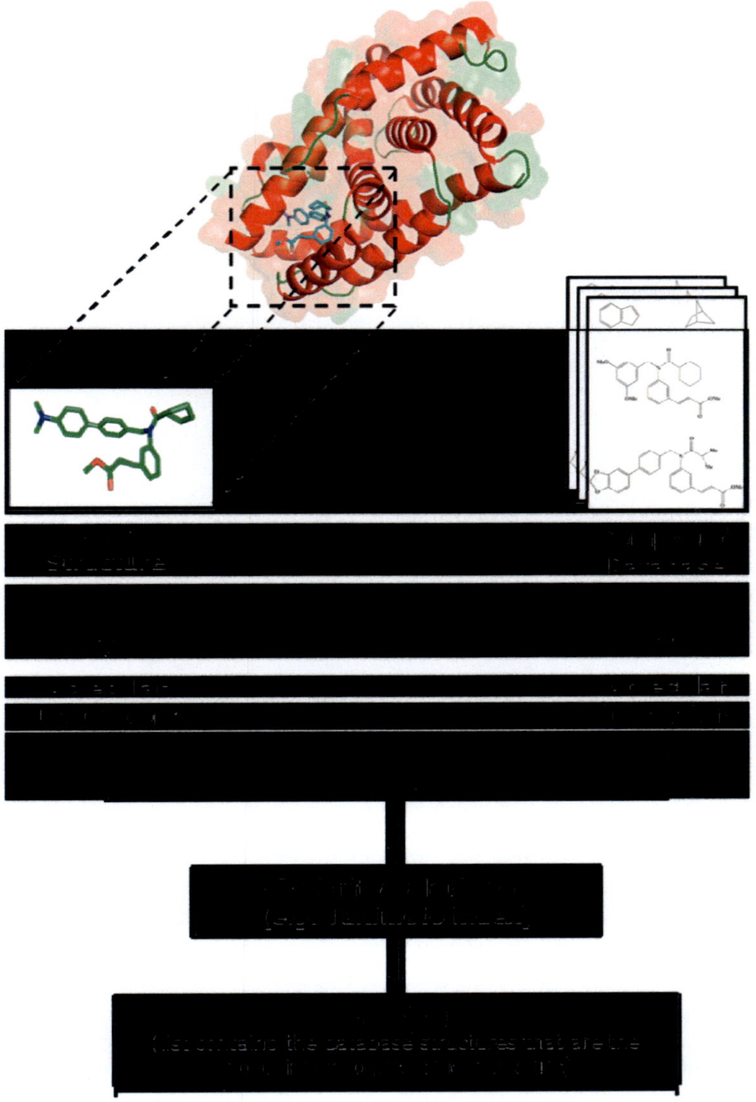

Figure 3. General procedure in a similarity searching.

1.2.2. Pharmacophore Pattern Matching

The prior use of pharmacophore model in the biological screening of candidate molecules is considered an efficient procedure, since it quickly eliminates molecules that do not possess the required structural features of the model, leading to a dramatic enrichment when compared to a purely random screening experiment [42]. In the pharmacophore concept, all ligands, regardless of chemotype, present similar steric and electrostatic features that are recognized at the target binding site and are responsible for the biological activity. A pharmacophore model represents the spatial arrangement of chemical groups or so-called "features" in a molecule that possess known activity. These "features" represent chemical functionalities related to specific electrostatic interactions, H-bonding and hydrophobic interactions [43,44].

A success example of the pharmacophore-based methods is the discovery of potent and selective T-type calcium channel blockers [45]. In this study, the authors carried out a virtual screening of commercially available compounds by using a 3D pharmacophore previously developed for T-type calcium channel blockers. Biological evaluation of 25 virtual hits resulted in the discovery of a highly potent compound, VH04 (Figure 4), with IC_{50} value of 10 nM, eight times more potent than midefradil (Posicor®), a known selective T-type calcium channel blocker.

Search for similar compounds yielded several hits with micro-molar IC_{50} values and high T-type calcium channel selectivity [45].

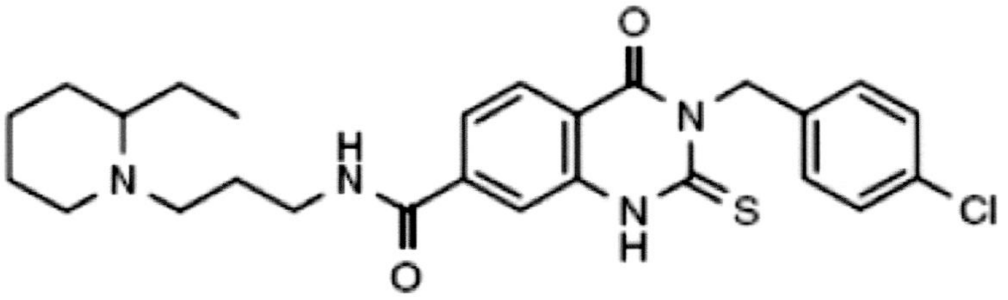

Figure 4. T-type calcium channel blocker discovered by ligand-based 3D pharmacophore-based virtual screening.

2. Integration between VS and HTS

Modern drug discovery involves the integration of a wide variety of technologies and expertise in multidisciplinary research teams. The influence of the synergy effects between HTS and VS on the selection of hits and lead compounds is a good example of the integration of advanced technologies in drug discovery programs. It is worth noting that the two approaches are quite distinct in nature and procedure. From a practical standpoint, HTS (an experimental procedure) is currently a costly and labor-intensive process, whereas VS (a computational method) offers economy, speed and flexibility to drive drug discovery projects. However, given their similar ultimate goal, we believe that these approaches are much more complementary to each other than often thought [46,47]. An important difference among these techniques is related to the knowledge about the mechanism of action of ligands upon biological receptors. In HTS, no information is provided about the possible target binding site or mechanism of action of the selected ligands. In this situation, structural and kinetic studies would be required to elucidate them. On the other hand, VS methods are capable of selecting molecules based on a specific 3D target binding site, therefore, useful information about the binding mode and mechanism of action should also be explored [48]. The use of VS techniques previous to or in parallel with HTS methods has three fundamental objectives: (i) to extend the scope of the screening to external databases; (ii) to identify a large number of structurally diverse hits; and, (iii) to reduce the assay-to-lead attrition rate observed from HTS [49]. Although HTS remains the dominant technique, VS has been increasingly used over the last years few years to facilitate the discovery of promising hits in medicinal chemistry research programs [50,51]. It is important to emphasize that, in contrast to HTS, VS methods are more easily accessible to academic laboratories around the world, without the need for expensive materials, equipments, and complex infrastructure that an empirical screen demands.

3. VS: Opportunities and Limitations

VS methods provides multiple opportunities for the discovery of novel hits in drug discovery projects. The speed at which a VS process can be completed makes it effective and quite attractive in a number of settings, including the screening of large databases of compounds for virtually any target protein of known 3D structure. Additionally, it is highly cost-effective since it circumvents the need for robotics, reagent acquisition or production, compound storage facilities, and a numbers of scientists managing the system. The relatively low resource requirements for carrying out VS studies provide accessibility to the technology to small companies and academia in addition to the traditional large pharmaceutical companies [52]. There have been many success stories describing the identification of novel ligands via VS approaches, as it is illustrated in Table 1.

Several fundamental challenges are involved in a VS process, such as sampling of various conformations related to flexible molecules and the calculation of absolute binding energies in an aqueous environment [60]. The literature reflects that VS protocols capture low-resolution information about ligand binding, but are limited in their ability to accurately predict measured binding affinities. In addition to problems associated with scoring of compound conformations, other complications exist that make it challenging to accurately predict binding conformations and compound activity. These include, among others, limited resolution of crystallographic targets, inherent flexibility, induced fit or other conformational changes that occur on binding, and the participation of water molecules in protein–ligand interactions [61,62].

Table 1. Successful applications of VS in drug discovery

Biological Target	Method(s)	Initial Database	Hits	Biological Activity	Reference
Retinoic Acid Receptors	ICM	150.000	2	200 nM	[53]
TβRI kinase	Catalyst/	200.000	1	0.005 μM	[54]
Eubacterial tRNA-guanine transglycosylase (TGT)	Selector/ Unit/ FlexX	800.000	9	0.25-249 μM	[55]
Metallopeptidase Angiotensin Converting Enzyme 2 (ACE2)	Catalyst/ eHiTS/ Q-fit	3.800.00 0	6	62-179 μM	[56]
Parasitic cysteine proteases	GOLD	241.000	4	1-63 μM	[57]
Estrogen receptor β	Catalyst/ GOLD	25.000	3	0.68-5 μM	[58]
Neurokinin-1 (NK1) receptor	Selector/ Unity/ FlexX-Pharm	827.000	1	Submicromolar range	[59]

4. Steps in a VS Process

From the considerations presented in the previous topics, it is now clear that the VS process is scientifically complex, involving several steps and parameters that can influence its performance. Basically, some commonly recognized issues in VS processes are: (i) preparation of the target protein structure; (ii) organization of databases of compounds; (iii) molecular docking; (iv) scoring; and, (v) evaluation of the results.

4.1. Target Protein Structure

The process of VS of compound libraries by docking and scoring can be carried out only when a 3D structural model is available for the target protein (e.g., protein structures, small molecule-protein co-complex structures). High-resolution crystal structures are generally preferred, and currently there is a large number of protein structures available in the Protein Data Bank (http://www.rcsb.org/pdb/) [63]. The ligand-protein co-complexes can provide useful information when employed as models or training sets for the assessment and validation of the docking methods and results. In the situation that 3D structures are not available, homology models can be derived and used as a valuable alternative [64]. In contrast, when many different structures are available, a detailed analysis of the docking performance of crystal structures of holo (ligand-bound), apo (without ligand-bound) and modeled proteins is required for comparison in order to evaluate the quality of the docking calculations, which is affected by the particular representation of the receptor used in the virtual screening. Although the holo structure usually provides the best VS results, the homology model yields significant enrichment over random [65].

4.2. Compound Selection (Database)

4.2.1. Database

Another crucial step in a VS procedure is the organization of the compound database. The compounds of the databases can be collected from commercial and non-commercial databases, catalogs, websites, in-house collections of natural products and synthetic compounds, and so forth. The database molecules can be further exploited *in silico* for drug-like properties or for a specific (pharmacokinetic (or physiochemical) profile on a project-by-project basis, if necessary [66,67]. Currently, there are many sources of databases of commercially available compounds for virtual screening, as illustrated in Table 2.

Table 2. Databases employed in virtual screening

Database	Number of Compounds	Link
ACD-SC	4.500.000	http://www.mdli.com
iResearch^{TM}	32.400.000	http://www.chemnavigator.com
Sigma-Aldrich	90.000	http://www.sigma-aldrich.com
NCI	500.000	http://dtp.nci.nih.gov
ComGenex	250.000	http://www.comgenex.com
SPECS	500.000	http://www.specs.net
Asinex Platinum	100.000	http://www.asinex.com
Asinex Gold	200.000	http://www.asinex.com
ChemBridge	435.000	http://chembridge.com/chembridge
Maybridge	56.000	http://www.maybridge.com
ChemDiv	1.000.000	http://www.chemdiv.com
ZINC	4.600.000	http://blaster.docking.org/zinc
Ligand.Info	1.000.000	http://ligand.info

The research community, particularly in academia, has largely been limited to expensive commercial databases. ZINC [68], a free database of commercially available compounds, has been attracting a great deal of attention of researchers from all around the world due to free access to the state-of-the-art and high-quality resources for virtual screening. The 2007 ZINC release (ZINC7) contains over 4.6 million compounds in ready-to-dock 3D format. The commercially available compounds in ZINC are broadly sub-categorized into several libraries, such as drug-like, lead-like, and fragment-like molecules, useful for different docking application scenarios. In addition, users may also create their own focused libraries via a web interface [68].

4.2.2. Filters for VS

A useful strategy in VS is to reduce the size of the chemical libraries by applying molecular property filters. Filtering methods eliminate compounds based on predetermined molecular and physicochemical properties and are widely used in VS for the enrichment of focused libraries. In addition, certain functional groups that contribute to toxicity can be identified and then eliminated using chemoinformatics tools. However, it is important to realize that imposing too many filters at early stages of the process could result in reducing the structural diversity of the chemical library at a critical level.

The research at initial stages of the drug discovery process focuses on the selection of compounds with good lead properties rather than compounds with good drug properties. Compounds with appropriate drug profile do not necessarily make the best leads for further optimization [69-71]. Although not mutually exclusive, leadlike and drug like properties are quite different (not mentioning that leads should be structurally less complex than drugs). A comparative study of the physicochemical properties of drugs and leads has shown that lead compounds are generally smaller and more polar than drugs, and as potency and selective are optimized through the process, lead compounds become larger and more hydrophobic. Over the past decade, the pharmaceutical industry has been active in defining leadlike and druglike properties. Lipinski and his Pfizer collaborators identified a series of properties commonly found in orally active drugs [72,73]. These molecular properties (molecular weight < 500, number of hydrogen bond donors < 5, number of hydrogen bond acceptors < 10, and Clog P < 5) are referred to as "Lipinski rule of 5", which can be used to indicate whether a molecule is likely to present appropriate oral absorption and general cell permeability. These property rules have been extensively applied in the identification and selection of compounds for lead discovery.

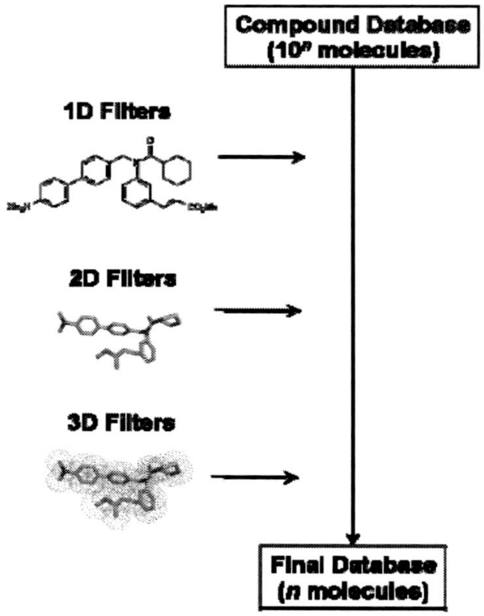

Figure 4. Database filters employed in VS.

In addition to traditional 1D and 2D filters, 3D filters can be also used to efficiently screen very large databases of compounds, and reduce the number of compounds for the docking and scoring steps [74]. Shape similarity methods employ 3D filters, but these filters are usually limited to screen databases that contain single molecular conformations [75]. In addition, pharmacophore-based screening can be carried out when pre-defined chemical and geometric molecular features are matched [76]. In general, these descriptors are highly conformation-dependent as well, and therefore of limited importance for databases containing single molecular conformations. Figure 5 illustrates the main types of filters that can be employed in database filtering.

4.3. Docking

Molecular docking algorithms attempt to generate and identify the most complementary match between a ligand and its macromolecular target. The docking process involves the prediction of ligand conformation and orientation within a target binding site [77,78]. In general, the main aims of docking studies are: accurate structural modeling and correct prediction of biological activity. So, there are essentially three steps to follow: (i) perception of the target and ligand; (ii) generation of putative complexes (sampling); and, (iii) evaluation of the "fitness" of the complex [79]. The docking process begins with the challenging application of docking algorithms that pose (dock) small molecules in the active site. It is necessary to observe that even relatively simple molecules can contain many conformational degrees of freedom. Sampling these degrees of freedom must be performed with sufficient accuracy to identify the conformation that best matches the receptor structure, and must be fast enough to permit the evaluation of thousands of compounds in a given docking run [80]. Algorithms are complemented by scoring functions that are designed to predict the biological activity through the evaluation of interactions between compounds and potential targets. Pre-selected conformers are often further evaluated using more complex scoring schemes with more detailed treatment of electrostatic and van der Waals interactions, and inclusion of at least some solvation or entropic effects [81]. It should also be noted that ligand-binding events are driven by a combination of enthalpic and entropic effects, and that either entropy or enthalpy can dominate specific interactions. This often presents a conceptual problem for contemporary scoring functions, because most of them are much more focused on capturing energetic than entropic effects [82].

Although termed "docking programs", the programs used currently are designed to carry out both tasks. The distinction between docking and scoring defines also the two major technical challenges: (i) to predict the binding mode of a molecule correctly (pose prediction); and, (ii) to predict the binding affinity of compounds (relative rank-ordering for a number of compounds) in a reliable manner [83]. Most docking algorithms assume the protein structure to be rigid, due to the high computational cost that the demand for flexibility implicates. As flexible docking clearly outperforms rigid body matches, most of the current docking programs consider the ligand as a flexible molecule [84]. Nowadays, the most widely employed algorithms can be grouped into four categories, as shown in Table 3.

It is interesting to say that docking studies reported by several research groups lead to different results depending on the settings used and this is not surprising, as docking accuracy is sensitive to the algorithm and scoring function parameters, as well as the to target structure and the compound selection protocol used during the process [85].

Table 3. Molecular docking algorithms

Category	Examples	Characteristics
Force-field-based methods	Molecular Dynamics (MD), Monte Carlo (MC) simulations.	The most accurate methods for flexible docking, allowing movements of both the target and the ligand. MC-based docking programs like QXP, ICM, and Glide are commonly used.
Evolutionary methods	Genetic Algorithms (GAs), Evolutionary Programming (EP), Tabu Search (TB).	GA, EP, and TB algorithms have all been implemented in PRO-LEADS. The most well-known docking programs based on GA are GOLD and Autodock.
Fragment-based incremental methods	DOCK, FlexX, Hammerhead, Surflex.	These methods divide the ligand into pieces that are docked in an incremental way.
Shape complementary-based methods	LigandFit, FRED.	LigandFit explores grids to compare shapes of the ligand and the target while FRED uses Gaussian functions to fit the ligand shape to the negative image of the protein shape.

4.3.1. Ligand Flexibility

The incorporation of ligand flexibility during the docking process is a matter of significant interest [86]. Methods that treat ligand flexibility can be divided into three groups: (i) systematic methods (incremental construction, conformational search, databases); (ii) random or stochastic methods (Monte Carlo, genetic algorithms, tabu search); and, (iii) simulation methods (molecular dynamics, energy minimization).

4.3.2. Protein Flexibility

Induced molecular flexibility is fundamental to understand the molecular recognition principles between ligand and biological receptor. Upon ligand binding, many systems undergo rearrangements, which range from local motions of side chains to large domain movements. In any event, receptor flexibility might have a dramatic impact in the ligand docking and VS process. It has been shown that even small changes in the receptor conformation can be important in docking results and, consequently, in the binding affinities calculations [87,88]. An alternative approach is to keep at least part of the receptor flexible in order to reproduce the so-called "induced fit", although relatively few docking programs take this into account so far. Clearly, the treatment of protein flexibility is less advanced than ligand flexibility, but various approaches have been applied to flexibly model at least part of the target, including molecular dynamics and Monte Carlo calculations, rotamer libraries and protein ensemble grids.

Nowadays, many software companies have addressed the topic of protein flexibility and improved their core docking procedures or developed additional protocols or software solutions on top of them. The way this is implemented differs from program to program. Some examples of programs that address the protein flexibility are FlexX/FlexE [89], Glide/Prime [90], GOLD [91], QXP/Flo+ [92], and ICM/IFREDA [93].

4.4. Scoring

The evaluation and ranking of predicted conformations are crucial components of the VS process. Even when binding conformations are correctly predicted, the calculations ultimately do not succeed if they do not differentiate correct poses from incorrect ones, and if real ligands cannot be identified. So, the design of reliable scoring functions is also of fundamental importance. In sum, scoring functions are used to estimate the binding affinity of novel structures or fragments in a given position inside the receptor pocket [94].

Scoring functions are needed for two purposes. During the docking process, they serve as fitness functions in the optimization of ligand orientation and conformation. Additionally, they are responsible for the estimation of binding affinity of each docked molecule [95]. Although, in principle, different functions can be used for these two purposes, in most applications the same function has been used. There are various criteria to assess the quality of a scoring function: (i) its ability to identify the correct binding mode of a ligand out of alternative docking solutions; (ii) its ability to rank related ligands with respect to their binding affinity; and, (iii) its ability to select a number of inhibitors out of a large database of inactive compounds [96].

Scoring functions must be very fast to evaluate a large number of protein–ligand complexes and, therefore, invariably neglect many terms that are part of the full thermodynamic cycle defining a binding free energy in solution. Given this limitation, scoring functions cannot be expected to give accurate affinity predictions. Nevertheless, they should recognize solutions displaying good steric and electrostatic complementarity between receptor and ligand and give lower ranks to other solutions with unlikely relative orientations of ligand and receptor groups [97]. Essentially, three classes of scoring functions are currently applied in VS process, and the main features of each one of these classes are shown in Table 4. Scoring usually proceeds in two steps: (i) possible conformations and poses of the ligand in the pocket are computationally generated (the docking step); and, (ii) these are ranked according to how well they fit into the pocket (the scoring step). In addition, results from the scoring step are often applied to compare and rank different chemical structures (called database enrichment or focusing) [98]. Scoring functions are applied in all of these stages. In many cases, the same scoring function is used in all steps, although separate functions have increasingly been employed. An alternative to using single scoring schemes is to combine several scoring functions and to assign a “consensus score” [99]. In this case, hit lists from multiple docking programs can be compared and those molecules that pass them all (or the majority) can be selected, leading to a marked improvement in enrichment [99]. This procedure has been employed, in the majority of cases, as a post-processing step after docking runs. As previously cited, consensus scoring is usually applied for virtual library enrichment, predictions of binding poses and binding affinity [99].

Table 4. Molecular docking algorithms

Class	Examples	Principles
Force-field-based scoring functions	D-Score, G-Score, GOLD, AutoDock and Dock.	Calculation of classical molecular mechanics energy functions, approximating the binding free energy of protein–ligand complexes by a sum of van der Waals and electrostatic interactions.
Empirical scoring functions	LUDI, F-Score, ChemScore, SCORE and Fresno.	Computation of the binding free energy by summing interaction terms derived from weighted structural parameters. The weights are obtained by fitting the scoring function to experimental binding constants of a training set of protein–ligand complexes.
Knowledge-based scoring functions	PMF, DrugScore and SMoG.	They represent the binding affinity as a sum of protein–ligand atom-pair interactions. These potentials are derived from the protein–ligand complexes with known structures, where probability distributions of interatomic distances are converted into distance-dependent interaction free energies of protein–ligand atom pairs using the "inverse" Boltzmann law.

In general, scoring functions describe binding affinity as a sum of independent terms and, thus, they all suffer from the same disadvantages: (i) size dependence, that is, the larger the molecule, the higher the probability of a favorable score; (ii) they usually ignore entropic effects (rigid receptor and single ligand binding modes with no ensemble averages); and, (iii) some scoring functions ignore solvation and desolvation effects [97]. Current scoring functions are actually quite similar to each other, even for those that belong to different categories. This fact is understandable, as many scoring functions formulate their equations in a similar way. Nevertheless, it is encouraging to observe that the available scoring functions have been applied to a wide range of protein–ligand systems with substantial success. However, improvements are still need to reliably predict the binding affinity of a ligand-protein complex even when an experimental 3D structure is available.

5. Conclusion

In spite of all the recent important advances, drug discovery remains a difficult and complex task. Beyond emphasizing the importance of a synergy between fundamental aspects of chemistry and biology, the several resource-intensive steps in modern drug discovery requires the integration of technological and scientific innovation, with the end result depending on a proper balance of experimental and computational design, compound selection and optimization, biological assays, data analysis and management. Although the trial-and-error nature is inherent in drug research, rational concepts and advanced computational methods should be considered as valuable tools during the stages of lead selection and optimization. The development of HTS was expected to solve the problem of compound selection for a variety of specific biological targets by exposing all of biological space to all possible chemical space. However, this fundamental point remains critical in medicinal chemistry programs and new methods are required in order to provide new alternatives for the pharmaceutical industry. VS methods have become an integral part of the drug design process for many academic laboratories and pharmaceutical companies as a timely and cost-effective tool for the finding of new hits and leads. VS strategies are influencing traditional strategies to analyze large databases of chemical compounds in order to identify possible lead candidates with good pharmacodynamic and pharmacokinetic profiles. In sum, a computational or medicinal chemist must be prepared to understand the limits of applicability of any computational method, and to distinguish between those opportunities which are appropriate to apply the VS technology to purse a new drug candidate.

References

[1] Hajduk, P. J. and Greer, J. (2007). A Decade of Fragment-Based Drug Design: Strategic Advances and Lessons Learned. *Nat. Rev. Drug Discov.*, *6*, 211-219.

[2] Hopkins, A. L. and Groom, C. R. The Druggable Genome. (2002). *Nat. Rev. Drug Discov.*, *1*, 727-730.

[3] Overington, J. P., Al-Lazikani, B. and Hopkins, A. L. (2006). How Many Drugs Are There? *Nat. Rev. Drug Discov.*, *5*, 993-996.

[4] Andricopulo, A. D. and Montanari, C. A. (2005). Structure-Activity Relationships for the Design of Small-Molecule Inhibitors. *Mini-Rev. Med. Chem.*, *5*, 585-593.

[5] Ghose, A. K., Herbertz, T., Salvino, J. M. and Mallamo, J. P. (2006). Knowledge-Based Chemoinformatic Approaches to Drug Discovery. *Drug Discov. Today*, *11*, 1107-1014.

[6] Stahura, F. L. and Bajorath M. (2005). New Methodologies for Ligand-Based Virtual Screening. *Curr. Pharm. Des.*, *11*, 1189-1202.

[7] DiMasi, J. A., Hansen, R. W. and Grabowski, H. G. (2003). The Price of Innovation: New Estimates of Drug Development Costs. *J. Health Econ.*, *22*, 151-185.

[8] Berndt, E. R., Gottschalk, A. H., Philipson, T. J. and Strobeck, M. W. (2005). Industry Funding of the FDA: Effects of PDUFA on Approval Times and Withdrawal Rates. *Nat. Rev. Drug Discov.*, *4*, 545-554.

[9] Oprea, T. I. and Matter, H. (2004). Integrating Virtual Screening in Lead Discovery. *Curr. Opin. Chem. Biol.*, *8*, 349-358.

[10] Klebe, G. (2006). Virtual Ligand Screening: Strategies, Perspectives and Limitations. *Drug. Discov. Today*, *11*, 580-594.

[11] Stahura, F. L. and Bajorath, J. (2004). Virtual Screening Methods that Complement HTS. *Comb. Chem. High Throughput Screen., 7,* 259-269.

[12] Yoon, S., Smellie, A., Hartsough, D. and Filikov, A. (2005). Surrogate Docking Structure-Based Virtual Screening at High Throughput Speed. *J. Comp.-Aided Mol. Des., 19*, 483-497.

[13] Langer, T. and Kroyat, E. M. (2003). Chemical Feature-Based Pharmacophores and Virtual Library Screening for Discovery of New Leads. *Curr. Opin. Drug Discov. Dev., 6,* 370-376.

[14] Jalaie, M. and Shanmugasundaram, V. (2006). Virtual Screening: Are We There Yet? *Mini-Rev. Med. Chem., 6,* 1159-1167.

[15] Klebe, G. (1995). Toward a More Efficient Handling of Conformational Flexibility in Computer-Assisted Modelling of Drug Molecules. *Persp. Drug Discov. Des., 3,* 85-105.

[16] Fox, S., Farr-Jones, S., Sopchak, L., Boggs, A. and Comley, J. (2004). High-Throughput Screening: Searching for Higher Productivity. *J. Biomol. Screen., 9*, 354-358.

[17] Walters, W. P., Stahl, M. T. and Murcko, M. A. (1998). Virtual Screening - An Overview. *Drug Discov. Today, 3*, 160-178.

[18] Langer, T. and Hoffmann, R. D. (2001). Virtual Screening: an Effective Tool for Lead Structure Discovery. *Curr. Pharm. Design,* 7, 509-527.

[19] Posner, B. A. (2005). High-Throughput Screening-Driven Lead Discovery: Meeting the Challenges of Finding New Therapeutics. *Curr. Drug Discov. Dev., 8,* 487-494.

[20] Bajorath, J. (2002). Virtual Screening in Drug Discovery: Methods, Expectations and Reality. *Curr. Drug Discov.,* 24-28.

[21] Shoichet, B. K., McGovern, S. L., Wei, B. and Irwin, J. J. (2002). Lead Discovery Using Molecular Docking. *Curr. Opin. Chem. Biol., 6*, 439-446.

[22] Schneider, G. and Bohm, H. J. (2002). Virtual Screening and Fast Automated Docking Methods. *Drug. Discov. Today, 7*, 64-70.

[23] Taylor, R. D., Jewsbury, P. J. and Essex, J. W. (2002). A Review of Protein-Small Molecule Docking Methods. *J. Comput. Aided Mol. Des., 16,* 151-166.

[24] Kitchen, D. B., Decornez, H., Furr, J. R. and Bajorath, J. (2004). Docking and Scoring in Virtual Screening for Drug Discovery: Methods and Applications. *Nat. Rev. Drug Discov., 3,* 935-949.

[25] Shoichet, B. K. (2004). Virtual Screening of Chemical Libraries. *Nature, 432,* 862-865.

[26] Branson, K. M. and Smith, B. J. (2004). A Role of Virtual Screening in Computer Aided Structure-Based Drug Design. *Aust. J. Chem., 57*, 1029-1037.

[27] Lyne, P. D. (2002). Structure-Based Virtual Screening: An Overview. *Drug Discov. Today, 7*, 1047-1055.

[28] Waszkowycz, B. (2002). Structure-Based Approaches to Drug Design and Virtual Screening. *Curr. Opin. Drug Discov. Dev., 5*, 407-413.

[29] Davis, A. M., Teague, S. J. and Kleywegt, G. J. (2003). Application and Limitations of X-Ray Crystallographic Data in Structure-Based Ligand and Drug Design. *Angew. Chem. Int. Ed., 42,* 2718-2736.

[30] Kasimova, M. R., Kristensen, S. M., Howe, P. W. A., Christensen, T., Matthiesen, F., Petersen, J., Sorensen, H. H. and Led, J. J. (2002). NMR Studies of the Backbone Flexibility and Structure of human Growth Hormone: A comparison of High and Low pH Conformations. *J. Mol. Biol., 318*, 679-695.

[31] Oshiro, C., Bradley, E. K., Eksterowicz, J., Evensen, E., Lamb, M. L., Lanctot, J. K., Putta, S., Stanton, R. and Grootenhuis, P. D. J. (2004). Performance of 3D-Database Molecular Docking Studies into Homology Model. *J. Med. Chem., 47,* 764-767.

[32] Berman, H. M., Westbrook, J., Feng, Z., Gilliland, G., Bhat, T. N., Weissig, H., Shindyalov, I. N., Bourne and P. E. (2000). The Protein Data Bank. *Nucl. Acids Res., 28*, 235–242.

[33] Good, A. (2001). Structure-Based Virtual Screening Protocols. *Curr. Opin. Drug Discov. Devel., 4*, 301-307.

[34] Vangrevelinghe, E., Zimmermann, K., Schoepfer, J., Portmann, R., Fabbro, D. and Furet, P. (2003). Discovery of a Potent and Selective Protein Kinase CK2 Inhibitor by High-Throughput Docking. *J. Med. Chem., 46,* 2656-2662.

[35] Xu, H. and Agrafiotis, D. K. (2002). Retrospect and Prospect of Virtual Screening in Drug Discovery. *Curr. Top. Med. Chem., 2*, 1305-1320.

[36] Alonso, H., Bliznyuk, A. A. and Gready, J. E. (2006). Combining Docking and Molecular Dynamic Simulations in Drug Design. *Med. Res. Rev., 26*, 531-568.

[37] Pozzan, A. (2006). Molecular Descriptors and Methods for Ligand Based Virtual High Throughput Screening in Drug DIscovery. *Curr. Pharm. Design., 12*, 2099-2110.

[38] Laggner, C., Schieferer, C., Fiechtner, B., Poles, G., Hoffmann, R. D., Glossmann, H., Langer, T. and Moebius, F. F. (2005). Discovery of High-Affinity Ligands of □1 Receptor, ERG2, and Emopamil Binding Protein by Pharmacophore Modeling and Virtual Screening. *J. Med. Chem., 48,* 4754-4764.

[39] Mestres, J. and Knegtel, R. M. A. (2000). Similarity *Versus* Docking in 3D Virtual Screening. *Persp. Drug Discov. Des., 20,* 191-207.

[40] Willett, P. (2003). Similarity-Based Approaches to Virtual Screening. *Biochem. Soc. Trans., 31,* 603-606.

[41] Willett, P. (2005). Searching Techniques for Databases of Two- and Three-Dimensional Chemical Structures. *J. Med. Chem., 48,* 4183-4199.

[42] Good, A. C., Mason, J. S. and Pickett, S. D. (2000). Pharmacophore Pattern Application in Virtual Screening, Library Design and QSAR. In: H. J. Böhm and G. Schneider (Eds.), *Virtual Screening for Bioactive Molecules: Methods and Principles in Medicinal Chemistry* (vol. 10, pp. 131–156). Wiley.

[43] Good, A. C., Krystek, S. R. and Mason, J. S. (2000). High-Throughput and Virtual Screening: Core Lead Discovery Technologies Move Towards Integration. *Drug Discov. Today, 5,* S61-S69.
[44] Mason, J. S., Good, A; C. and Martin, E. J. (2001). 3-D Pharmacophores in Drug Discovery. *Curr. Pharm. Des., 7,* 567-597.
[45] Doddareddy, M. R., Choo, H., Cho, Y. S., Rhim, H., Koh, H. Y., Lee, J., Jeong, S. and Pae, A. N. (2007). 3D Pharmacophore Based Virtual Screening of T-Type Calcium Channel Blockers. *Bioorg. Med. Chem., 15,* 1091–1105.
[46] Klebe, G. (2000). Virtual Screening: An Alternative or Complement to High Throughput Screening?, Berlin, Kluwer Academic Publishers.
[47] Mestres, J. (2002). Virtual Screening: A Real Screening Complement to High-Throughput Screening. *Biochem. Soc. Transact., 30,* 797-799.
[48] Klebe, G. (2005). Virtual Screening: Scope and Limitations. In: J. Alvarez and B. Shoichet (Eds.), *Virtual Screening in Drug Discovery* (pp. 3–4). New York, Taylor and Francis.
[49] Stahura, F.L. and Bajorath, J. (2004). Virtual Screening Methods that Complement HTS. *Comb. Chem. High Throughput Screen., 7,* 259-269.
[50] Oprea, T. I. (2002). Virtual Screening in Lead Discovery: A Viewpoint. *Molecules, 7,* 51-62.
[51] Lengauer, T., Lemmen, C., Rarey, M. and Zimmermann, M. (2004) Novel Technologies for Virtual Screening. *Drug Discov. Today, 9,* 27-34.
[52] Jenkins, J. L., Kao, R. Y. T. and Shapiro, R. (2003). Virtual Screening to Enrich Hit Lists From High-Throughput Screening: A case Study on Small-Molecule Inhibitors of Angiogenin. *Proteins, 50,* 81-93.
[53] Raaka, B. M., Samuels, H. H. and Abagyan, R. (2001). *In Silico* Discovery of Novel Retinoic Acid Receptor Agonist Structures Matthieu Schapira. *BMC Structural Biology,* 1, 1.
[54] Singh, J., Chuaqui, C. E., Boriack-Sjodin, P. A., Wen-Cherng, L., Pontz. T., Corbley, M. J., Cheung, H. K., Arduini, R. M., Mead, J. N., Newman, M. N., Papadatos, J. L., Bowes, S., Josiah, S. and Ling, L. E. (2003). Successful Shape-Based Virtual Screening: The Discovery of a Potent Inhibitor of the Type I TGF□ Receptor Kinase (T□RI). *Bioorg. Med. Chem. Lett., 13,* 4355-4359.
[55] Brenk, R., Naerum, L., Gradler, R., Gerber, H., Garcia, G. A., Reuter, K., Stubbs, M. T. and Klebe, G. (2003). Virtual Screening for Submicromolar Leads of tRNA-Guanine Transglycosylase Based on a New Unexpected Binding Mode Detected by Crystal Structure Analysis. *J. Med. Chem., 46,* 1133-1143.
[56] Rella, M , Rushworth, C. A., Guy, J. L., Turner, A. J., Langer, T. and Jackson, R. M. (2006). Structure-Based Pharmacophore Design and Virtual Screening for Novel Angiotensin Converting Enzyme 2 Inhibitors. *J. Chem. Inf. Model., 46,* 708-716.
[57] Desai, P.V., Patny, A., Sabnis, Y., Tekwani, B., Gut, J., Rosenthal, P., Srivastava, A. and Avery, M. (2004). Identification of Novel Parasitic

Cysteine Protease Inhibitors Using Virtual Screening. The ChemBridge database. *J. Med. Chem.*, *47*, 6609-6615.

[58] Zhao, L. and Brinton, R. D. (2005). Structure-Based Virtual Screening for Plant-Based ER□-Selective Ligands as Potential Preventative Therapy against Age-Related Neurodegenerative Diseases. *J. Med. Chem., 48,* 3463-3466.

[59] Evers, A. and Klebe, G. (2004). Successful Virtual Screening for a Submicromolar Antagonist of the Neurokinin-1 Receptor Based on a Ligand-Supported Homology Model. *J. Med. Chem., 47,* 5381-5392.

[60] Mohan, V., Gibbs, A. C., Cummings, M.D., Jaeger, E.P., and DesJarlais, R. L. (2005). Docking: Successes and Challenges. *Curr. Pharm. Design*, 11, 323-333.

[61] Smith, R., Hubbard, R. E., Gschwend, D. A., Leach, A. R and Good, A. C. (2003). Analysis and Optimization of Structure-Based Virtual Screening Protocols (3). New Methods and Old Problems in Scoring Function Design. *J. Mol. Graph. Model., 22*, 41-53.

[62] Schneider, G., Schneider, P. and Renner, S. (2006). Scaffold-Hopping: How Far Can You Jump? *QSAR Comb. Sci., 25,* 1162-1171.

[63] Rester, U. (2006). Dock Around the Clock - Current Status of Small Molecule Docking and Scoring. *QSAR Comb. Sci., 25*, 605-615.

[64] Bissantz, C., Bernard, P., Hibert, M. and Rognan, D. (2003). Protein-Based Virtual Screening of Chemical Databases. II. Are Homology Models of G-Protein Coupled Receptors Suitable Targets?. *Proteins, 50,* 5-25.

[65] McGovern, S. l. and Shoichet, B. K. (2003) Information Decay in Molecular Docking Screens Against Holo, Apo, and Modeled Conformations of Enzymes. *J. Med. Chem., 46,* 2895-2907.

[66] Martin, E. J. and Critchlow, R. E. (1999). Beyond Mere Diversity: Tailoring Combinatorial Libraries for Drug Discovery. *J. Comb. Chem., 1*, 32-45.

[67] Bleicher, K. H., Bohm, H. J., Muller, K. and Alanine, A. I. (2003). A Guide to Drug Discovery: Hit and Lead Generation: Beyond High-Throughput Screening. *Nat. Rev. Drug Discov., 2*, 369-378.

[68] Irwin, J. J. and Shoichet, B. K. (2005). ZINC - A Free Database of Commercially Available Compounds for Virtual Screening, *J. Chem. Inf. Model., 45,* 177-182.

[69] Walters, W. P. and Murcko, M. A. (1999). Recognizing Molecules With Drug-Like Properties. *Curr. Opin. Chem. Biol., 3,* 384-387.

[70] Walters, W. P. and Murcko, M. A. (2002). Prediction of 'Drug-Likeness', *Adv. Drug. Deliver Rev., 54,* 255-271.

[71] Oprea, T. I., Davis, A. M., Teague, S. J. and Leeson, P. D. (2001). Is There a Difference Between Leads and Drugs? A Historical Perspective. *J. Chem. Inf. Comput. Sci., 41,* 1308-1315.

[72] Lipinski, C. A., Lombardo, F., Dominy, B. W. and Feeney, P. J. (1997) Experimental and Computational APproaches to Estimate Solubility and Permeability in Drug Discovery and Development Settings. *Adv. Drug Delivery Rev., 23,* 3-25.

[73] Lipinski, C. and Hopkins, A. (2004). Navigating Chemical SPace for Biology and Medicine. *Nature, 432*, 855–861.

[74] Stahura, F. L. and Bajorath, M. (2005). New Methodologies for Ligand-Based Virtual Screening. *Curr. Pharm. Des., 11,* 1189-1202.

[75] Steindl, T. and Langer, T. (2005). Docking Versus Pharmacophore Model Generation: A Comparison of High-Throughput the Search of Human Rhinovirus Virtual Screening Strategies for Coat Protein Inhibitors. *QSAR Comb. Sci., 24,* 470-479.

[76] Langer, T. and Hoffmann, R. D. (2006). Pharmacophore Modelling: Applications in Drug DIscovery. *Expert Opin. Drug Discov., 1*, 261-267.

[77] Alvarez, J. C. (2004). High-throughput Docking as a Source of Novel Drug Leads, *Curr. Op. Chem. Biol., 8,* 365-370.

[78] Verkhivker, G. M., Bouzida, D., Gehlhaar, D. K., Rejto, P. A., Arthurs, S., Colson, A. B., Freer, S. T., Larson, V., Luty, B. A., Marrone, T. and Rose, P. W. (2000). Deciphering Common Failures in Molecular Docking of Ligand-Protein Complexes. *J. Comput. Aided Mol. Design, 14,* 731-751.

[79] Brooijmans, N. and Kuntz, I. D. (2003). Molecular Recognition and Docking Algorithms. *Annu. Rev. Biophys. Biomol. Struct., 32*, 335-373.

[80] Bissantz, C., Folkers, G. and Rognan, D. (2000). Protein-Based Virtual Screening of Chemical Databases. 1. Evaluation of Different Docking/Scoring Combinations, *J. Med. Chem., 43*, 4759-4767.

[81] Stahl, M. and Rarey, M. (2001). Detailed Analysis of Scoring Functions for Virtual Screening. *J. Med. Chem., 44,* 1035-1042.

[82] Salaniwal, S., Manas, E. S., Alvarez, J. C. and Unwalla, R. J. (2007). Critical Evaluation of Methods to Incorporate Entropy Loss Upon Binding in High-Throughput Docking. *Proteins, 66*, 422-435.

[83] Warren, G. L., Andrews, C. W., Capelli, A. M., Clarke, B., LaLonde, J., Lambert, M. H., Lindvall, M., Nevins, N., Semus, S. F., Senger, S., Tedesco, G., Wall, I. D., Woolven, J. M., Peishoff, C. E. and Head, M. S. (2006). A Critical Assessment of Docking Programs and Scoring Functions. *J. Med. Chem., 49*, 5912-5931.

[84] Coupez, B. and Lewis, R. A. (2006). Docking and Scoring - Theoretically Easy, Practically Impossible?. *Curr. Med. Chem., 13*, 2995-3003.

[85] Kellenberger, E., Rodrigo, J., Muller, P. and Rognan, D. (2004). Comparative Evaluation of eight Docking Tools for Docking and Virtual Screening Accuracy. *Proteins, 57*, 225-242.

[86] Perola, E. and Charifson, P. S. (2004). Conformational Analysis of Drug-Like Molecules Bound to Proteins: An Extensive Study of Ligand Reorganization upon Binding. *J. Med. Chem., 47,* 2499-2510.

[87] Barril, X. and Morley, S. D. (2005). Unveiling the Full Potential of Flexible Receptor Docking Using Multiple Crystallographic Structures. *J. Med. Chem.,* 48, 4432-4443.

[88] Carlson, H. A. and McGammon, J. A. (2000). Accommodating Protein Flexibility in Computational Drug Design. *Mol. Pharmacol., 57,* 213-218.

[89] FlexX/FlexE, BioSolveIt GmbH, http://www.biosolveit.de/; Tripos Inc., http://www.tripos.com/

[90] Glide/Prime, Scrçdinger Inc., http://www.schrodinger.com/

[91] GOLD, The Cambridge Crystallographic Data Centre, http://www.ccdc.cam.ac.uk/

[92] QXP/Flo+, Thistlesoft Inc., http://www.floqxp.com/

[93] ICM/IFREDA, MolSoft L. L.C., http://www.molsoft.com/

[94] Tame, J. R. H. (1999). Scoring Functions: A View From the Bench. *J. Comput. Aided Mol. Des., 13,* 99-108.

[95] Halperin, I., Ma, B. Y., Wolfson, H. and Nussinov, R. (2002). Principles of Docking: An Overview of Search Algorithms and Guide to Scoring Functions. *Proteins, 47,* 409-443.

[96] Bursulaya, B. D., Totrov, M., Abagyan, R. and Brooks III, C. L. (2003). Comparative Study of Several Algorithms for Flexible Ligand Docking. *J. Comput. Aided Mol. Des., 17,* 755-763.

[97] Kontoyianni, M., Sokol, G. S. and McClellan, L. M. (2005). Evaluation of Library Ranking Efficacy in Virtual Screening. *J. Comput. Chem., 26,* 11-22.

[98] Bender, A. and Glen, R. C. (2005). Discussion of Measures of Enrichment in Virtual Screening: Comparing the Information Content of Descriptors With Increasing Levels of Sophistication. *J. Chem. Inf. Model., 45,* 1369-1375.

[99] Charifson, P. S., Corkery, J. J., Murcko, M. A. and Walters, W. P. (1999). Consensus Scoring: A Method for Improved Hit Rates From Docking Databases of Three-Dimensional Structures Into Proteins. *J. Med. Chem., 42,* 5100-5109.

In: Drug Design Research Perspectives
Editor: Stanley P. Kaplan, pp. 275-306

ISBN: 978-1-60021-823-1

Chapter X

Structure Prediction of Proteins with Very Low Homology: A Comprehensive Introduction and a Case Study on Aminopeptidase

***Thomas Scior*[*1] *and Habibah A. Wahab*[2,3]**
[1] Department of Pharmacy, Fac. Ciencias Químicas, BUAP, 72570 Puebla, Mexico
[2] School of Pharmaceutical Sciences, Universiti Sains Malaysia, 11800 Penang, Malaysia.
[3] Laboratory of Biocrystallography and Structural Bioinformatics, Universiti Sains Malaysia, 11800, Penang, Malaysia

Abstract

The fact that protein sequences largely outnumber their structures gives rise to a steadily increasing demand for homology-based, atom-scale modeling of protein structures applying computational methods. Pertaining to the three-dimensional (3D) model generation of proteins, low amino acid sequence identity of the target sequence to known template structures imparts a twilight zone with uncertainties about the either homologous (evolutionary related) or coincidental (by chance) nature of relationships between target (known only by its sequence) and templates (known 3D structures). As a consequence, fully automated 3D model predictions tend to suffer from inappropriate candidate selections and/or sequence alignments. As an introduction for beginners we present a technical report on mainly web-based computational approaches how to convert protein sequences into 3D models. Using the gene registered at *GenBank* (Acc. No. AY364634), this chapter illustrates the necessary steps to obtain an atomic-scale model: Upon domain recognition (280 residues out of 898 total sequence length) which belongs to the alpha/beta hydrolase fold superfamily we exploit the highly conserved fold

[*] Correspondence: PhD THOMAS SCIOR; Department of Pharmacy, Facultad De Ciencias Químicas, Benemérita Universidad Autónoma De Puebla, Ciudad Universitaria C.U. Edificio 139, 14 Sur Con Avenida San Claudio, C.P. 72570 Colonia San Manuel, Puebla, Pue., Mexico; Email: tscior@siu.buap.mx

similarities to the cytosolic aminopeptidase family (EC 3.4.11.x) to find template structures. The resulting peptidase is theoretically involved in protein processing, a very plausible function in the light of the experiments in rat ovaries during cellular differentiation. In order to fill in the structural gap for this protein class the new aminopeptidase was deposited at PDB model repository under entry code 2B62 by AutoDep at RCSB (http://pdb.rutgers.edu/adit/).

Keywords: Protease; aminopeptidase; structure prediction; homology modeling; tutorial.

Abbreviations / Glossary

3D-	Three-dimension(al)
Aa	Amino acid(s)
AP	Aminopeptidase(s)
H	Helices
E	beta strand or sheets
C	Coils
L	Loops
Meta server	central user interface with local access to tools which are invisibly ("meta") run on remote machines
EC	Enzyme Class
HMM	Hidden Markov Models
NVT/E	ensembles in MD with N: Avogadro number, V: volume, T: temperature, / E: energy
HoMo	Homology Model(ing)
RMSD	root means square distance, measurement of geometrical resemblance
MD	Molecular Dynamics
SA	Simulated Annealing routine in MD
low-complexity	low compositional variation, repetitions of amino acid(s)
FF	Force Field

Why Protein Structure Prediction?

The scientific progress in experimental molecular biology aimed at gene and genome discovery has resulted in phenomenal accumulation of nucleotide data. Such strings or chains of nucleotides - albeit do not necessarily constitute identified genes - have been computationally translated into amino acid sequences or proteins at best. Hence, a larger number of the former have not been structurally and functionally defined and remain unknown. Nevertheless, all sequences enter the respective databanks for nucleotides (*GenBank[1]*, EMBL[2]) or protein sequences (*SwissProt[3]*). The latest release of EMBL Nucleotide sequence database[2] contains 83,666,567 sequence entries comprising 150,163,403,742 nucleotides (http://www.ebi.ac.uk/ embl/Documentation/ Release_notes/

current/ relnotes.html, last access 22 February 2007). The translation of these nucleotide sequences from EMBL[2] into the protein sequences enters another database at UniProtKB/TrEMBL[3]. Its current register recorded 3,826,359 entries, comprising 1,244,402,523 amino acids (last visit 20-Feb-2007). As these numbers continues to grow at an exponential rate, a new scientific discipline, bioinformatics, emerges and lays the theoretical groundwork for electronic, web-based knowledge enrichment embedding both nucleotide and amino acid codes in an interpretable textual context with computed data as well as expert-edited annotations.[11]

In this regard in the field of *in silico* (computational) biomolecular research the term *protein* has three implications:

1. the protein sequence, i.e. the biopolymer chain with its amino acids lined up as residues which are linked by peptide bonds. The residues are numbered in direction from its N-terminal to its C-terminal point. It was translated from the nucleotide sequence;
2. the protein structure, i.e. the three-dimensional (3D) biomolecule with its backbone of peptide linkage and side chains of its amino acids; and finally
3. the protein function, e.g. the biochemical role as antibody, enzyme, receptor, ligand or substrate, also ion channel or membrane pore and so forth.

Nowadays it is considered common knowledge that protein sequence determines its structure and the structure in turn, determines its function. Yet, one of the most important objectives of post-genomic studies henceforth is to determine the structure of proteins and thereby its molecular mechanism and biochemical function. It is noteworthy that while the sequences generated reached millions in number, the number of experimentally determined structures which are deposited in the protein data bank (PDB[44]) grows at a much slower pace with only 41,814 records (www.pdb.org, last access 22 February 2007). This databank of protein structures represents just a fraction of structurally unknown protein sequences (*SwissProt[3]* with 259,034 curated and annotated protein sequences, last visit 22 February 2007). This means that only about a sixth part of currently translated protein sequences are experimentally characterized by their structures (SwissProt[3]). The major reason for this is that elucidating structures of proteins (x-ray crystallography, NMR-studies) remains not only time consuming and expensive but also quite often is accompanied with some unsolved technical problems.

It is generally accepted that *in silico* methods complement more expensive and time-consuming *in vitro* or *in vivo* experiments and occasionally may also substitute them (see table 1). As the distinction matters the computed functional and structural models are edited with annotations and qualified as “hypothetical”, “putative” or “electronically predicted”. Web-based retrieval systems allow intuitive online access to selected data sources of nucleotide or amino acid sequences and structures (EMBL[2] SwissProt[3], PDB[44]). In addition annotated data can be retrieved for special tasks which are explained later in this chapter; for instance: amino acid patterns, motifs or families (PRÓSITE[7], PRINTS, BLOCKS, SUPFAM[8], MEROPS[9]).

Table 1. Data sources and services visited during this study

Data Sources	Web side
1. Online comparative or homology modeling servers	
3D-JIGSAW	http://www.bmm.icnet.uk/servers/3djigsaw/
CPH models	http://www.cbs.dtu.dk/services/CPHmodels/
EsyPred	http://www.fundp.ac.be/urbm/bioinfo/esypred/
FFAS03	http://ffas.ljcrf.edu/ffas-cgi/cgi/ffas.pl
ModBase	http://modbase.compbio.ucsf.edu/modbase-cgi-new/index.cgi
PALI	http://pauling.mbu.iisc.ernet.in/~pali/
PANAL Web Server	http://mgd.ahc.umn.edu/panal
Pedant	http://pedant.gsf.de
Pix	http://www.hgmp.mrc.ac.uk/Registered/Webapp/pix/
PredictProtein	http://www.embl-heidelberg.de/predictprotein/predictprotein.html http://cubic.bioc.columbia.edu/predictprotein/
PredictProtein	http://cubic.bioc.columbia.edu/predictprotein/index.html
SAM-T02/T99	http://www.soe.ucsc.edu/research/compbio/HMM-apps/HMM-applications.html
SCMF	http://bioweb.pasteur.fr/seqanal/interfaces/homology.html
SCWRL	http://www1.jcsg.org/scripts/prod/scwrl/serve.cgi
SDSC	http://cl.sdsc.edu/hm.html
Swiss-Model	http://www.expasy.ch/swissmod/SWISS-MODEL.html
2. Off-line modeling programs	
Arguslab	http://www.planaria-software.com/
Deepview	http://www.expasy.org/spdbv/
Fold X	http://foldx.embl.de/
FUGUE	http://www-cryst.bioc.cam.ac.uk/fugue/
LOOPP	http://msdlocal.ebi.ac.uk/docs/loopp/
Modeller	http://salilab.org/modeller/modeller.html
Molden	http://www.cmbi.ru.nl/molden/molden.html
MolIDE	http://dunbrack.fccc.edu/molide/
NCBI – pKb ftp	ftp://ftp.ncbi.nih.gov/pub/pkb/
PrISM	http://honiglab.cpmc.columbia.edu/
Rasmol	http://www.umass.edu/microbio/rasmol/index2.htm
SCWRL	http://dunbrack.fccc.edu/SCWRL3.php
TRITON	http://ncbr.chemi.muni.cz/triton/triton.html
VMD	http://www.ks.uiuc.edu/Research/vmd/
WHAT IF	http://swift.cmbi.kun.nl/whatif/
Yasara and many more links	http://www.nyu.edu/pages/mathmol/software.html http://www.edinformatics.com/mathmol/mm_software.htm
3. Online general structure prediction servers	
3D-PSSM	http://www.sbg.bio.ic.ac.uk/~3dpssm/
AGADIR (only helices)	http://www.embl-heidelberg.de/Services/serrano/agadir/agadir-start.html
BioInfoBank Meta Server	http://bioinfo.pl/meta/
CBS	http://www.cbs.dtu.dk/services/
CBSU - LOOPP	http://cbsuapps.tc.cornell.edu/loopp.aspx

Table 1. (Continued).

CMGM	http://cmgm.stanford.edu/WWW/www_predict.html
DALI	http://www.ebi.ac.uk/dali/
FAMS	http://www.pharm.kitasato-u.ac.jp/fams/
FunCat SCHEME	http://mips.gsf.de/projects/funcat
Geno3D	http://geno3d-pbil.ibcp.fr/cgi-bin/geno3d_automat.pl?page=/GENO3D/geno3d_home.html
GO	http://www.geneontology.org/
Data Sources	Web side
HMMTOP	http://www.enzim.hu/hmmtop/
META	http://www.embl-heidelberg.de/predictprotein/doc/meta_intro.html
Meta-PP	http://www.cmbi.kun.nl/bioinf/predictprotein/submit_meta.html
NNPREDICT	http://www.cmpharm.ucsf.edu/%7Enomi/nnpredict.html
Pred2ary	http://www.cmpharm.ucsf.edu/%7Ejmc/pred2ary/
Predator	http://bioweb.pasteur.fr/seqanal/interfaces/predator-simple.html
PredictProtein	http://www.embl-heidelberg.de/predictprotein/predictprotein.html
PredictProtein	http://cubic.bioc.columbia.edu/pp/
PROF	http://www.aber.ac.uk/%7Ephiwww/prof/
Protscale	http://us.expasy.org/tools/
PSCAN	http://www.sbc.su.se/%7Earne/pscan/seqalign2.html
PSI	http://www.structuralgenomics.org
PSIPRED	http://bioinf.cs.ucl.ac.uk/psipred/
RPFOLD	http://www.imtech.res.in/raghava/rpfold/
SAWTED	http://www.bmm.icnet.uk/%7Esawted/
SOSUI	http://sosui.proteome.bio.tuat.ac.jp/sosui_submit.html
THREADER	http://bioinf.cs.ucl.ac.uk/threader/threader.html
TM-Finder	http://www.ccb.sickkids.ca/tools/tmfinder/html/login.html
TMHMM	http://www.cbs.dtu.dk/services/TMHMM/
TMSTAT	http://bioinfo.mbb.yale.edu/tmstat/
TopPred	http://bioweb.pasteur.fr/seqanal/interfaces/toppred.html
4. Simple, multiple, local or global sequence alignments	
ALION	http://motif.Stanford.EDU/alion/
Clustal W (advanced)	http://npsa-pbil.ibcp.fr/cgi-bin/npsa_automat.pl?page=npsa_clustalw.html
ClustalW (simplified)	http://www.ebi.ac.uk/clustalw/
DOTLET	http://www.isrec.isb-sib.ch/java/dotlet/Dotlet.html
DOTMATCHER	http://bioweb.pasteur.fr/seqanal/interfaces/dotmatcher.html
DOTTER	http://www.cgr.ki.se/cgr/groups/sonnhammer/Dotter.html
DOTTUP	http://bioweb.pasteur.fr/seqanal/interfaces/dottup.html
FRAMESEARCH	http://www.dna.affrc.go.jp/swsrch/swx.pl
Global DBClustal	http://igbmc.u-strasbg.fr:8080/DbClustal/dbclustal.html
MATRIXPLOT	http://www.cbs.dtu.dk/services/MatrixPlot/
Needleman & Wunsch	http://bioweb.pasteur.fr/seqanal/interfaces/needle.html
PipeALign	http://igbmc.u-strasbg.fr/PipeAlign/
POLYDOT	http://bioweb.pasteur.fr/seqanal/interfaces/polydot.html
Smith & Waterman	http://bioweb.pasteur.fr/seqanal/interfaces/water.html
TMAP	http://bioinfo.limbo.ifm.liu.se/tmap/

Table 1. (Continued).

5. Sequence alignment editors	
Boxshade	http://www.ch.embnet.org/software/BOX_form.html
CINEMA	http://www.biochem.ucl.ac.uk/bsm/dbbrowser/CINEMA2.02/
TeXshade	http://homepages.uni-tuebingen.de/beitz/biotex.html
6. Retrieval systems for sequences, templates and families, domains, motifs, fingerprints	
BLAST - DDBJ	http://www.ddbj.nig.ac.jp/search/blast-e.html
BLAST2	http://bioweb.pasteur.fr/seqanal/interfaces/psiblast.html
BLAST-NCBI10	http://www.ncbi.nlm.nih.gov/BLAST/
Blocks	http://blocks.fhcrc.org/blocks/
DAS	http://www.sbc.su.se/%7Emiklos/DAS/
DbEST	http://www.ncbi.nlm.nih.gov/dbEST/index.html
DbSNP	http://www.ncbi.nlm.nih.gov/SNP/
Data Sources	Web side
DDBJ	http://www.ddbj.nig.ac.jp/searches-e.html
EMBL-EBI	http://www.ebi.ac.uk/embl.html
ENTREZ	http://www.ncbi.nlm.nih.gov/Entrez/
FASTA – DDBJ	http://www.ddbj.nig.ac.jp/search/fasta-e.html
FASTA - EMBL	http://www.ebi.ac.uk/fasta/index.html
FASTA - Log Statistics	http://www.dna.affrc.go.jp/statistics/fasta.html
FASTA - Virginia	http://fasta.bioch.virginia.edu/
Genome	http://www.ncbi.nlm.nih.gov/entrez/query.fcgi?db=Genome
Getentry	http://getentry.ddbj.nig.ac.jp/getstart-e.html
HMMPfam	http://hmmer.wustl.edu/
InterPro	http://www.ebi.ac.uk/interpro/
MEROPS	http://merops.sanger.ac.uk/
Molphy	http://bioweb.pasteur.fr/seqanal/interfaces/molphy.html
NCBI Databases	http://www.ncbi.nlm.nih.gov/Database/
Pfam Sanger	http://www.sanger.ac.uk/Software/Pfam/
Phylip - protdist	http://bioweb.pasteur.fr/seqanal/interfaces/protdist.html
PIR PSD	http://pir.georgetown.edu/pirwww/search/textpsd.shtml
Prints	ftp://ftp.bioinf.man.ac.uk/pub/fingerPRINTScan/
ProfileScan	http://www.isrec.isb-sib.ch/profile/profile.html
Protein Domains at NPD	http://npd.hgu.mrc.ac.uk/domains.html
PSIBLAST10	http://helix.nih.gov/docs/gcg/psiblast.html
RCSB	http://pdb.rutgers.edu/adit/
ScanRegExp & PPsearch	ftp://ftp.ebi.ac.uk/pub/software/unix/
Sequenzanalyse tutorial–EBI	http://www.ebi.ac.uk/2can/index.html
Sequenzanalyse tutorial-NCBI	http://www.ncbi.nlm.nih.gov/About/primer/
SRS–DDBJ	http://srs.ddbj.nig.ac.jp/index-e.html
SRS–EBI	http://srs6.ebi.ac.uk/
SRS–EMBL	http://srs.embl-heidelberg.de:8000/srs5/
SRS-LION	http://downloads.lionbio.co.uk/publicsrs.html
SSEARCH	http://watson.genes.nig.ac.jp/homology/ssearch-e.shtml

Table 1. (Continued).

SUPFAM43	http://pauling.mbu.iisc.ernet.in/~supfam
SWISS PROT	http://www.expasy.ch/sprot/
TIGRFAMs	http://ftp.tigr.org/pub/data/TIGRFAMs/
Trace Archive	http://www.ncbi.nlm.nih.gov/Traces/trace.cgi?
UCLA-DOE	http://www.doe-mbi.ucla.edu/people/frsvr/frsvr.html
UniGene	http://www.ncbi.nlm.nih.gov/entrez/query.fcgi?db=unigene
WU Blast EMBL	http://www.ebi.ac.uk/blast2
7. Secondary data source (derived from primary)	
Boehringer Mann-heim Bio-chemical Pathways	http://www.expasy.ch/cgi-bin/search-biochem-index
BRENDA pathways	http://www.brenda.uni-koeln.de/
FINGERprintsSCAN	http://www.ebi.ac.uk/printsscan/
InterProScan - EBI	http://www.ebi.ac.uk/InterProScan/
LIGAND	http://www.genome.jp/kegg/ligand.html
PFAM-Wustl	http://pfam.wustl.edu/
PRATT	http://www.ebi.ac.uk/pratt/
PRATT Quellcode	http://www.ii.uib.no/~inge/Pratt.html
PRINTS	http://bioinf.man.ac.uk/dbbrowser/PRINTS/PRINTS.html
PRODOM	http://protein.toulouse.inra.fr/prodom/current/html/home.php
Data Sources	Web side
SCANPROSITE	http://au.expasy.org/tools/scanprosite/
8. Databases and Information servers for protein prediction	
Amino acid	http://www.russell.embl-heidelberg.de/aas/
BCM-PSSP	http://searchlauncher.bcm.tmc.edu/seq-search/struc-predict.html
BIOINBGU	http://www.cs.bgu.ac.il/~bioinbgu/
CAMD	http://www.netsci.org/Science/Compchem/feature01.html
CASP	http://predictioncenter.org/
CATH	http://www.biochem.ucl.ac.uk/bsm/cath/
Comparative protein modeling	http://swissmodel.expasy.org/course/text/chapter6.htm
ENZYME Enzyme nomenclature database	http://ca.expasy.org/enzyme/
EVA	http://pdg.cnb.uam.es/eva/doc/explain_method.html
ExPasy	http://ca.expasy.org/
FORESST	http://abs.cit.nih.gov/foresst2/
GeneBank	http://www.ncbi.nlm.nih.gov/Genbank/GenbankSearch.html
GOR	http://molbiol.soton.ac.uk/compute/GOR.html
Guide to structure prediction	http://www.russell.embl.de/gtsp/index.html
ICM	http://www.molsoft.com/icm/index.html
Interpro-EBI	http://www.ebi.ac.uk/interpro/databases.html
IUBMB	http://www.chem.qmw.ac.uk/iubmb/
Jpred	http://www.compbio.dundee.ac.uk/~www-jpred/
KEGG	http://www.genome.jp/kegg/
NPS@-IBCP-PBIL	http://npsa-pbil.ibcp.fr/cgi-bin/npsa_automat.pl?page=/NPSA/npsa_seccons.html

Table 1. (Continued).

PDB	http://www.rcsb.org/
CMS Protein data	http://restools.sdsc.edu/biotools/biotools9.html
RCSB	http://www.rcsb.org/databases.html
SCOP	http://scop.mrc-lmb.cam.ac.uk/scop/
Sisyphus Homology	http://www.embl-heidelberg.de/~rost/Papers/sisyphus.html
UniProtKB/TrEMBL	http://www.ebi.ac.uk/trembl/index.html
Zink finger	http://npd.hgu.mrc.ac.uk/domains.html#Z
9. Protein model evaluation	
ANOLEA	http://protein.bio.puc.cl/cardex/servers/anolea/index.html
ERRAT	http://nihserver.mbi.ucla.edu/ERRAT/
GeneCensus	http://bioinfo.mbb.yale.edu/genome
PROCHECK	http://www.biochem.ucl.ac.uk/~roman/procheck/procheck.html
ProSa II and ProSa 2003	http://www.came.sbg.ac.at/
VERIFY3D	http://www.doe-mbi.ucla.edu/Services/Verify_3D/
WHAT_CHECK	http://www.cmbi.kun.nl/swift/whatcheck

Homology for Protein Structure Prediction

Surprisingly, the spatial features (secondary and tertiary structures) or topology (3D fold information) of proteins are often more conserved in divergent evolution than the linear amino acid sequences (1D genetic information) of their respective protein chains.[4] In consequence, proteins can be grouped together into families with spatial resemblance despite missing amino acid sequence identity. The amino acids of the family members diverge in certain segments where they are dispensable, i.e. residue changes are not detrimental to structure or function. The evolutionary relatedness of proteins to a common ancestor is called *homology* and measured in alignment studies as similarity or identity scores (%). If the amino acid sequences are compared against each other (sequence alignments) the amino acids in certain segments can be found either identical or homologous (evolutionarily related) or not similar at all (mutation), and sometimes even lost (deletion, gaps). Therefore, low or remote homology means poor resemblance of the amino acids in equivalent positions of the aligned sequences. But even remote homology often conserves both spatial folds and biochemical function.[5] Homologous with membership to a particular protein family or structural patterns (fingerprints, signatures, motifs, domains or folds) are unambiguously detectable for higher sequence identity only. Below a certain limit of sequence similarity we enter the twilight zone of uncertainty where proteins may or not form a so-called superfamily, i.e. they share some hidden structural and functional common ground.[6] The identity threshold depends on the sequence length. Its average value ranges from 25 to 30% for a typical domain (functional or structural protein subunit) or cytosolic protein with a sequence length of 200 to 250 residues (100%) .

In 2004, Ng and Tan[10] summarized the idea of homologous proteins: "a widely accepted approach of assigning function to newly sequenced genes is to compare a new sequence with those of annotated proteins in other species. If the new gene or protein's sequence bears significant similarity to the sequence of a gene or protein in an annotated database of another species, it can be assumed that the two proteins are either of the same genetic instantiation, or at the very least, share very similar properties and functions.[10]

Sequences sharing identity greater than 30% are considered homologous sequences, which stem from the same ancestor. The higher this identity score, the more likely that the proteins might be evolutionary related and probably share the same fold and a similar biological function. This principle has to be fulfilled in order to conduct homology modeling on a rational basis.

Our prediction study on a aminopeptidase demonstrate that (i) despite extremely low total sequence identity (below 15%) computational homology modeling of a structurally unknown protein (FXNA, *GenBank* Acc. No.AY364634) against 3D template structures is feasible, (ii) provided reliable relationships between structurally unknown target and experimentally determined template structure exist, i.e. "despite remote evolutionary relationships the overall spatial resemblance has been conserved".

Selecting the Methods for Protein Structure Prediction

There are two ways on how one can start the process of finding related structures (templates) of a structurally unknown protein (target): either by database searching or automatic homology modeling.

Protein structure prediction is best started by database searching to determine any structural similarity or to identify homologous sequence, i.e. selecting a template structure that will constitute the core building block for the backbone of the protein structure one wishes to predict (target). The former is of paramount importance in order to decide which method of structure prediction to be employed.

There are basically three general different methods to predict the target protein structure:

1. *Homology modeling* uses observed 3D structures as templates with sequence similarity to the protein being modeled (target).
2. *Threading* is based on known 3D structures, while
3. *ab initio* methods use just the sequence of the protein as input.

A search which returns that the query sequence shares structural identity of 30% or better indicates the structures could be predicted using fully automated homology modeling, whereas lower percentage scores suggest the use of threading or - tentatively - *ab initio* methods. The latter stays in an earlier stage of development. The fact that some fold processes forming the secondary and tertiary structure depend not only on the intrinsic information provided by the inherited nucleotide chain for that protein but also on the molecular assistance and multimolecular rearrangements at a cytosolic level defies present ab initio work. Of the many different approaches to estimate partial or total structure from scratch we would like to name the newer developments. They are referenced in table 1 and encompass fold recognition (Fugue) or folding pathways predictions (FoldX) or combined approaches (*PROFcon*) or others like PETRA, FREAD, CODA, *Elan-Prot*, Isites and improve steadily. Other knowledge-based methods already have experienced reliability: conservative replacing residues on the template's backbone with the new side chains of the query according to the sequence alignment. A convenient option is the SCWRL server thanks to its side chain rectifications (see table 1). Side chain conformations are usually curated by means of molecular mechanics with empirical energy refinements towards local minima.

On the other hand molecular dynamics (NAMD/VMD, GROMOS, YASARA/*Whatif*, referenced in table 1) was capable to cross kinetic barriers in a temperature-dependant manner and therefore explores larger conformational space, simulating certain conditions like simulation time, temperature, volume, pressure and solution. Special force fields exist for either biopolymers (AMBER, CHARMM, GROMOS) or both polypeptides and metallorganic compounds (MMFF94). When dynamics starts the system (all the molecules: solute, solvent) is heated during the initial phase (equilibration stage, NVE) reaching the thermal equilibrium state (data collection stage, NVT) and temperature becomes related to the kinetic energy of the molecule(s). A typical conformational space search takes some tens of pico-seconds and higher temperature (350-1000K). Cycles of temperature raising and lowering (simulated annealing) end up in ("freeze out") more meaningful and stable structures. They can be analyzed for flexible and rigid regions at room temperature during longer simulation periods with shorter time steps (0.5-1 fs).

Detection of Homologues and Templates from Protein Sequence Information

Besides searching for the template(s), database search is also important for functional analysis where the sequences can be aligned to detect any patterns that might cluster the proteins into *families* or *domains* (fold units). Alignments with protein family members can identify any residues that are conserved throughout the family. The conserved regions will give hints on the important sites, like binding sites or on the sequence that might involve in catalytic activity (active site).

The most popular database search program is the basic local alignment search tool (BLAST[29]), and to a similar extent is FASTA[57]. These two search programs compare the query sequence with each sequence form the selected databases using pair-wise sequence comparison.

One should expect that when dealing with protein with low homology, direct BLAST[29] search against the PDB[44] source usually results in poor answers (few hits). Thus, related sequences are best collected from a PSI-BLAST[29] against SwissProt[3] (see table 2). In SwissProt repository, each listed sequence for which a PDB entry exists is labeled "pdb" and linked to its web-source. Annotations allow for keyword queries about the putative domain, family or enzyme class, the user is assisted by a "Sequence Retrieval System" (SRS, see table 1) which explores expert edited (or secondary) data sources, e.g. the family database of structurally similar proteins from PDB, called "Fold classification based on Structure-Structure alignment of Proteins" (FSSP).[36] On the other note, PDB too, allows access to its entries by keyword search, and its virtual visitor is wise to collect "structural neighbors" by just a mouse-click.

Table 2. PSI-BLAST[29] output summery with 26 hits, i.e. matches to query found among 178,545 sequences stored in SwissProt (last visit 2005-08-27)

Acc. No.	Annotation	Length, Identity
Q09216	Hypothetical protein B0495.7	204/282 (72%)
Q18600	Putative zinc metallopeptidase	146/246 (59%)
094479	Probable metallopeptidase C1919	138/261 (52%)
P38244	Hypothetical protein YBR074W	106/208 (50%)
P25152	Hypothetical peptidase ywaD precursor	79/187 (42%)
P83913	Leupeptin-inactivating enzyme 2	43/100 (43%)
Q9M1S8	Probable glutamate carboxypeptidase	79/194 (40%)
Q01693	Bacterial leucyl aminopeptidase precursor	34/71 (47%)
P80561	Aminopeptidase (SGAP)	53/112 (47%)
10 to 26	hydrolases (EC3), transferases (EC 2)	lower values

The first step is to find identical or similar sequences with textual annotations about their structure and function to our query. The basic local alignment search tool (BLAST) filters selects data sources to find sequence files with resemblance to partial or total query length (target).[29] In our study, after analyzing the first results further examination was restricted to the N-terminus (see figure 1). In order to reduce the amount of retrieved hits it is wise to switch from translated nucleotide libraries (BLAST-label "nr", no longer non-redundant!)[1] to the curated (expert edited and non-redundant) *SwissProt* database.[3] When initial Swiss-Model prediction fails far-distant homology can be expected, PSI-BLAST[29] is especially suited either with lower BLOSUM 45 or higher PAM 250 mutation matrices. The method applies profiles with numerical information (20 scores) about which amino acid type (20) change (mutation) is reasonable (insertions, deletions, substitutions) at a certain position during this position-specific iteration (PSI) process. The evolutionary distance unit is measured as 1PAM, i.e. one accepted point mutation per 100 amino acids. Such probability matrices evaluate aligned amino acid pairs on the basis of the observed mutation frequency of one amino acid for the other. Within a particular protein family, gap-free blocks (conserved regions without deletions) exist and can be treated separately with blocks substitution matrices (BLOSUM), i.e. amino acid tables with scores for substitutions in aligned residue blocks.

```
MEWSSESAAVRRHRGTAERREGQAAASHPQREASAQEDARGGGRHRGRTE
SGGESRGAKTALSEARTALALALYLLALRALVQLSLQRLVLSRTSGLQGE
FDARQARVYLEHITAIGPRTTGSAENEILTVQYLLEQITLIEEQSNSLHR
ISVDVQRPTGSFSIDFLGGFTSYYDNITNVVVKLEPQDGAKYAVLANCHF
DSVANSPGASDDAVSCAVMLEVLRVMAASPEPLQHAVVFLFNGAEENVLQ
ASHGFITQHPWASLIRAFINLEAAGVGGKELVFQTGPENPWLVQAYVSAA
KHPFASVVAQEVFQSGIIPSDTDFRIYRDFGNIPGIDLAFIENGYIYHTK
YDTADRILIDSIQRAGDNILAVLKYLATSDMLASSSEYRHGSMVFFDVLG
LLVIAYPSRVGSIINYMVVMAVVLYLGRKLLRPNHSNSNYVRDFLCGLGI
TFISWFTSLVTVLIIAVFVSLIGQSLSWYNYFYIAVCLYGTATVAKIILI
HTLAKRFYYVNASDLYLGELFFDTSLFVHCGFLVALTAQGFCSAFMSAVW
VAFPLLTKLCVYKDFKKHGAKGRFIALYLLGMFIPYLYGLYLIWAVFEMF
TPILGRSGSEIPPDVVLASILAVCVMILSSYFITFIYLVNSTKKTILTLI
LVCAVTFLLVCSGAFFPYSSNPDSPKPKRVFLQHVSRTFHNLEGSVVKRD
SGIWINGFDYTGMSHVTPHIPEINDTIRAHCEENAPLCGFPWYLPVHFLI
RKNWYLPAPEISPRNPAHFRLVSKEKMPWDSIKLTFEATGPSHMSFYVRT
HKGSTLSQWSLGNGIPVTSRGGDYFVFYSHGLQASAWQFWIEVQVSEEQP
EGMVTVAIAAHYLSGENKRSSQLDALKEKFPDWSFPSAWVSTYSLFVF
```

Figure 1. The entire 898 amino acids-long target sequence "FXNA" (GenBank access number NM_184050) and the query input (bold letters) to retrieve homologous sequences with known domains and motifs. The final homology 3D model comprises a domain of 280 underlined amino acids at the N-terminus of alleged protein precursor.

Detection of Structural and Functional Patterns

A pattern is a string of amino acids with special wild cards: "X" standing for any amino acid type, and "[…]" standing for exactly one residue out of a defined group, for instance "[Val, Leu, Ile]" is either valine, leucine or isoleucine. A profile includes discontinuity (large gaps) and more dynamic expressions (variables, repetitions) to find more divergent protein families (with very old ancestors) or domains that cannot be detected using regular patterns. Contreras-Moreira and Bates suggest domain fishing to improve the template collecting steps.[30] PHD is a tool box based on threading techniques for remote homology search below 25% identities using BLAST, *ProSite* or *ProDom*, while PHDsec predicts secondary structures for sequences after multiple alignments.[22,31] Moreover, PHD[31] was accessed through a so-called meta-server which passes the requests to hidden remote machines, here: MetaPP (see table 1).

InterPro[32] was launched indirectly through the PHD option of MetaPP. *InterPro* itself is a suite of useful pattern recognition tools: *ProDom* is an acronym for Protein Domains.[33] They are automatically compiled from homologous domains. PROSITE constitutes an electronic domain and protein family collection to inform about catalytic sites, patterns like zinc fingers or profiles.[7] Pfam library covers many common families of curated protein domains, i.e. sequences of homologues in each family are aligned accurately. SUPERFAMILY[34] currently contains 1447 superfamilies which represent all proteins of known structure following the Structural Classification Of Proteins (SCOP) definitions.[35]

Groups of hidden *Markov* models (HMM) form the retrievable profiles, i.e. structural predefined patterns are within amino acid sequences.[34]

Identification of Structural Template Candidates from PDB

In order to identify possible 3D templates from PDB, few considerations must be made. First of all, the templates that show higher overall sequence identity with the target sequence frequently yield a better a better model. However, this is not always the case and sometimes the best ranking template doesn't always necessarily reflect the most suitable candidates, on the contrary they could be false positives. To overcome this problem, results from the threading programs must be carefully examined and compared so that the target sequence has good compatibility against the proposed templates. The second consideration is the quality of the template must be high, for example, the resolution and the R-factor of the crystallographic structure must meet certain accuracy; and the third is that there should be at least a remote evolutionary relation between the target and the template that can be analyzed using phylogenetic tree.[59]

One good approach for selecting the template is to compare the results from the various threading methods; templates that scored high in all or most methods possibly represent the best templates for the target sequence. Once the template has been selected, the user could conduct manual homology modeling to build the 3D model based on the template's 3D structure.

Automated Homology Modeling

The most straightforward way of prediction structure is by employing the automatic homology modeling method. Computer assisted protein model building by homology is launched either off-line or on-line, e.g. *Swiss-Model[13]*, EsyPred[15], CPH[16], SDSC-1[17] (see table 1). The former has become very popular because it can be handled as a black box delivering ready-to-use answers by e-mail.

In principle, for a fully automated procedure, the user inputs the amino acid sequence of the spatially unknown target, and this is termed as the query sequence. The program uses the query sequence to find closely related three-dimensional structures, which are called templates, as bases to build the target 3D model. The templates are provided either by the user or by the program itself (fully automated solution). The built-in template selection criteria are based on residual resemblance (in a larger sense) between target and template candidates and is not yet resolved satisfactorily. Thusly, a great amount of methods exist to uncover similarities. Simpler functions measure pair-wise identities in percents or homologies following the supposition of genetic mutation during protein evolution.[18] Sequences with alignment scores lying below a 35 to 25 % identity threshold are better treated with overlaying techniques, i.e. threading the query into the template backbone to detect hidden resemblance.[19]

There are many more automated servers available on-line; 3D-PSSM, *GenThreader* and PHD[31] are some of the examples. PHD is a tool box based on threading techniques for remote homology search below 25% identities using BLAST, *ProSite* or *ProDom*; and can be accessed through a so-called meta-server which passes the requests to hidden remote machines, here: MetaPP (see table 1).

3D-PSSM, or nowadays updated as PHYRE, evaluates the match of the query sequence to both the sequence of the template and the solvent accessibility pattern of the template structure. *GenThreader* is another popular threading method which works by threading a sequence into different structures; and those structures that produces the lowest energy fold is said to be the best threading template for the target sequence.

In general, the automated such as *Swiss-Model[13]* fails or does not generate any useful 3D models for sequences with a Poisson unlikelihood probability P(N) lower than 10-5 (BLAST).[16] Should this happens, user should consider manual homology-based structure modeling. In addition, sequences with alignment scores lying below a 35 to 25 % identity threshold are better treated with overlaying techniques, i.e. threading the query into the template backbone to detect hidden resemblance.[19] For a deeper look at issues on model accuracy and reliability, please refer to the model evaluation section in table 1, BioInfo.pl, CASP and EVA or special literature.[20, 21, 22, 23, 24, 25, 26]

Although fully automated structure prediction (e.g. Swiss-Model[13]) examines structural resemblance between available 3D template(s) and unknown target (query) its failure due to poor similarities and has often been misunderstood as limitation until now. Sometimes not all building assumptions are realistic, resulting in various solutions that lend straightforward insight into otherwise experimentally determined function and/or structure under higher costs of time and resources.

Threading Methods

Using threading method, unlike homology modeling, two proteins need not be evolutionarily related but exploiting convergent evolution, they can share similar backbone structures, too. *Threading* works by passing the amino acid sequence of the target protein through each position of the amino acids in the known 3D structure(s) found in the structural databases, i.e. comparing the target sequence to the 3D template, then evaluating how they both fit based on the potential energy scores. In the first step of threading, one would need to carry out thorough sequence analysis to determine the function of the protein target prior to the search of functionally converging proteins in an unconventional way. Eventually, one detects the target's remote homologues of divergent evolution as potential templates during threading. Subsequent to identification of all possible template candidates, the user can then utilize manual homology modeling to build the 3D model of the target.

Manual Homology Modeling

Manual 3D model generation follows the same fundamental steps: search for PDB structure candidates related to the amino acid sequence of the query, template selection, alignment of the target sequence with one or more structures, model building and it is evaluation. Molecular editing and dynamics can be conducted by either commercial or academic programs (*ArgusLab, NAMD/VMD, Molden, PyMol, Rasmol, Swiss-PDB-Viewer/Deep View[27], What_If[28]*, see table 1). In addition, a universal file converter (Babel) is often required to combine the packages which do not house all the necessary features:

i. identify all possible homologous protein sequences, not just protein structures (PDB entries);
ii. analyzing memberships to possible functional and structural groups called domains and superfamilies in case of very low homology;
iii. find appropriate 3D template candidates, preferably crystal structures of proteins;
iv. select final 3D templates after analyzing edited multiple sequence alignment and known and/or predicted secondary structures;
v. change the side chains on the template's backbone according to the target's sequence and add the variable loop segments manually (e.g. Swiss-PdbViewer Loop Database, see below), optimising the geometry/conformations of backbone/side chains and validate the probability of achieving a realistic 3D model compared to statistical evidence.

Target Backbone Based on a Multiple Template Approach

They are all used to assign the new coordinates to the target in two complementary ways. Firstly, the template with the best (largest, fewest gaps, highest id) overlap is the one to build the target model from. Secondly, multiple template combination means each template covers a different part (domain, motif) of the target. The likelihood of existence of such a new backbone and side chain conformations composed of natural components depends on the magnitude of changes the template(s) suffered from. Bad side chain contacts are repaired by conformational search tools, mostly by molecular mechanics: force-field energy minimization, molecular dynamics or knowledge-based (e.g. SCWRL[60]). In cases of mayor changes the model remains speculative until experiments assert these new overall folds.

Missing or Superfluous Structural Information: Loops

After internal aligning the sequences, less conserved regions (loops) and highly conserved segments for its structural and functional role become evident.

The principle tenet for comparative modeling states that the overall structure (typically a domain with 90-120 aa) remains completely unaltered above 70 % identity and the relative mean square deviation (rmsd) of the alpha-carbon co-ordinates probably lies around 0.1nm or better. Below this threshold deviations of the conserved backbone are mostly due to so-called loops, i.e. deletions or insertions with missing or superfluous segments, respectively. In contrast, parts with helices, sheets and turns keep their fold architecture. Below 25-35 % fold resemblance is only found within the same subfamily. Loop search tools may be found in commercial software. Loops as long as 15 amino acids can be retrieved from LiP.[43]

Special care has to be taken in cases of different sequence lengths, i.e. when residue deletions (gaps) occur in an equivalent segment of the successfully aligned sequences. Two situations emerge: the target segment is longer than the template segment or vice versa. In the final model superposition to template the longer segment protrudes like a handle of a cup. If the template is longer than the target is easily modeled taking a short-cut between the edges of the conserved neighbors. In the opposite case, the loop to be inserted has to be searched in databases of structures (*Swiss-PdbViewer* Loop Database) or guessed.

Case Study: Structure Prediction of FXNA

In 2003, our initial attempts to structure prediction (see figure 4) by fully automated homology modeling server *Swiss-Model[13]* did not generate any useful 3D models because it does not consider sequences with a Poisson unlikelihood probability P(N) lower than 10-5 (BLAST)[16]. Since that time we have launched prediction attempts to follow the progress. The predictive power or success shows a twofold dependence: on the one side the need of higher sequence identity between target and the 3D template(s) whereas, on the other side, the programs enhance to detect lower homologies between target and 3D template(s). Since neither better templates nor programs have emerged, our study on aminopeptidase constitutes a general guideline for all those cases where fully automated tools do not solve the prediction task. Hence, the know-how described in this chapter about manual homology-based structure modeling becomes a priceless asset. As can be inferred from our ongoing submissions, progress has been made either by new templates (1z8l) or algorithms (3D-JIGSAW, see table 3). However, crystallization of membrane-bound or cytosolic hydrolases remains a daunting technical endeavor. Until now, suitable crystal or NMR structures with higher sequence identity (> 25 to 35%, for an average domain length of 80 to 120 residues) have not been available at the Brookhaven protein database (PDB[44]).

Table 3. Automated model generation for query

Homology Modeling Method	Year: 2003 to 2004; # residues describes the model with number of amino acids	Year 2005
ESyPred3D (Based on consensus and neural network-based alignments. MODELLER program build the 3D-model)	No model returned; when 1F2P template was forced it shared 0.0% identity with query.	253 residues from hydrolase 1z8l, release 2005, EC 3.4.17.21 glutamate carboxypeptidase 2
CPH models (version 2)	84 residues (27 % id) 1amp with two short alpha-helices, two beta-strands around the Cys-Cys-bridge	97 residues from transferase 2AFO, release 2005, EC 2.3.2.5, hydrolase; glutaminyl cyclase
3D-JIGSAW (domains search with SCOP, Pfam for PDB-templates to generate molecular models).	Not applied	211 residues from hydrolase 1F2O, release 2001; EC 3.4.11.-*Streptomyces griseus* AP
SWISS-Model	No model returned; when 1F2P template was forced 42 residues around no Cys-Cys bridge built	No model returned, ditto.
SDSC version 1	No model returned	No model returned
MOE release 2004	No model returned with its built-in Protein Family Database	(Not licensed)
ModBase	Not applied	273 aa, 1rtq, chain A, EC 3.4.11.10; 367 aa, 1cg2, chain A, EC 3.4.17.11; 353 aa, 1de4, chain C, CATH 3.40.630.10
Manual comparative model under multiple template recombination for conserved backbone and loop insertions	280 residues from 1QQ9, 1LOK with a binuclear zinc finger and catalytic H_20 molecule.	282 residues including a dipeptide substrate

The sequences with remote homology to query were obtained with the PSI-BLAST search tool against PDB (see table 2). The latter currently contains over 120 aminopeptidase-like structures (last visit 2005-08-26). The initial PSI-BLAST against PDB retrieved half a dozen (today15) hits from two different enzyme classes: transferases (EC2) and aminopeptidases (EC3). In 2005 the picture becomes clear up, SRS against the FSSP database yields ten chains of cytosolic aminopeptidases (AP) (EC 3.4.11.) out of 330 predefined structural families (see table 4). *InterPro* server links to eleven AP's (EC3) while the *SuperFamily* consistently lists six M28 domain-bearing PDB files (see table 5). *InterPro* assigns a peptidase M28 domain (IPR007484) to the N-terminus of full length query (see figure 2). Some details are briefly deferred in the following: *ProDom* finds N-terminal part of query belongs to the peptidases subfamily M28 (PD043272).[33]

Table 4. Inspected template candidates after eliminating false hits by secondary structure analysis. Percentage identities (%id) rise after editing Clustal W[42] alignments. For instance, the unedited multiple alignments for the two templates, 1FT7 and 1F2P, showed 7% vs. 11%. The former corresponds to 1LOK[49], the latter to 1QQ9[48]

PDB	%id	Aminopeptidase (AP)	Enzyme Class (EC)
1LOK	15	Aeromonas Proteolytica Leucyl AP (Aap)	3.4.11.10
*1FT7	15	Aap Complexed With L-Leucine phosphonic Acid	3.4.11.10
1RTQ	15	Aeromonas Proteolytica Aminopeptidase (AP)	3.4.11.10
*1F2P	11	Streptomyces Griseus AP (Sgap)	3.4.11.
2MAT	10	E. Coli Methionine AP	3.4.11.18
1B6A	10	Human Methionine AP 2	3.4.11.18
1GYT	09	Escherichia coli leucine AP A (pepa)	3.4.11.1
1LCP	07	Bovine lens leucine AP	3.4.11.1
1AMP	06	Aeromonas Proteolytica AP	3.4.11.10
1CP6	06	Aeromonas Proteolytica AP	3.4.11.10

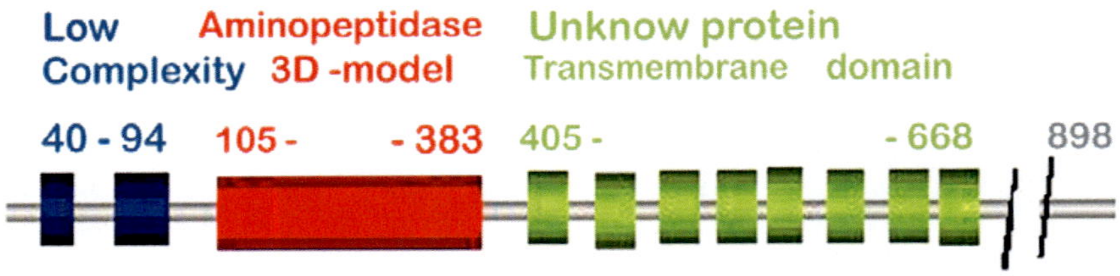

Figure 2. Schematic representation of the three inspected sequence sections and the location of the target 3D-model. The numbers indicate the amino acid positions in the "FXNA" protein with 898 residues length.

ProSite predicts localization of predefined patterns, called motifs like sites of phosphorylation, N-myristoylation or N-glycosylation[7] while another neural network routine (DISULFIND) mistakenly rejects existence of one disulphide bridge between the two unique cysteines in the mid-section separated by 17 residues.[45] *SuperFamily* characterizes a Zn-dependent excpeptidase superfamily (SSF53187), typical of the sort of phosphorylase/hydrolase-like fold group. It forms a alpha and beta (a/b) protein based on predicted 30% alpha-helical, and 17% beta-sheet composition therein practically complying with the a/b SCOP class definition (a >30%, b >20%).[34] In addition to the (super-) family and domain search by PHD/*InterPro* the BLOCKS database was explored to find blocks along the total query chain, i.e. conserved segments without gaps. It contains homologous groups of proteins represented in the PROSITE database. The search confirms the existence of a "M28 peptidase family" with 14% identity for 50 over 341 residues, or 26% homology for 91 allowed mutations over 341 residues. Upon visiting cross-linked *InterPro/ProDom* this database reveals M28 domains belong to metallopeptidases, while MEROPS imparts a "M28 family, aminopeptidase Y"[9] The Pfam code is PF04389, CATH no. 3.40.630.10.1, 3.40.630.10.4[46], and SCOP c.56.5.4.[35]

The M28 family contains aminopeptidases and carboxypeptidases alike, and peptidases with M28 domains possess co-catalytic zinc ions. Among the peptidase-like structures (some are just older versions or complexes, so redundant) two crystallographic structures of bacterial leucine aminopeptidases (EC 4.4.10.) were hand-selected with highest sequence homology (40%) after identification and dissection of loops. Both templates (see table 4) were exchanged against better crystallographic versions (see table 5). They serve as templates for the conserved segments of alpha helices, beta-sheets and their turns. Herein the Clustal W[42] alignments (see figure 3) were edited guided by some statistical rules for the preferred location of amino acids (one letter code): RNDQEHK at the surface, ILMFV buried inside, GTMRKHF starting helices, SNDELWP ending them, PGY breaking them, AMLEQK or less frequently VIFW forming the helical body, while beta sheets are formed by VIMLTFW, broken by DEKGP and turns formed of GSDNP and suppressed by AVLIMHFWC.

Table 5. PDB-retrieval[44] of all available versions of the two bacterial aminopeptidase 3D-templates to examine data quality, missing atoms, crystallographic resolution (in nm), R-factors etc. (LEU-AP stands for leucyl aminopeptidase). The asterisks mark the two selected templates

PDB	EC	Aminopeptidase (AP)	Year	nm
1RTQ	3.4.11.10	Aeromonas proteolytica LEU-AP	2003	0.09
*1LOK	3.4.11.10	Aeromonas proteolytica LEU-AP	2002	0.12
1AMP	3.4.11.10	Aeromonas proteolytica LEU-AP	1994	0.18
1CP6	3.4.11.10	Aeromonas proteolytica LEU-AP	1999	0.19
1TXR	3.4.11.10	Aeromonas proteolytica LEU-AP	2004	0.20
1FT7	3.4.11.10	Aeromonas proteolytica LEU-AP	2000	0.22
1IGB	3.4.11.10	Aeromonas proteolytica LEU-AP	1996	0.23
*1QQ9	3.4.11.-	Streptomyces Griseus AP	2000	0.15
1CP7	3.4.11.-	Streptomyces Griseus AP	2000	0.15
1F20	3.4.11.-	Streptomyces Griseus AP	2001	0.17
1F2P	3.4.11.-	Streptomyces Griseus AP	2000	0.18
1XJO	3.4.11.-	Streptomyces Griseus AP	1997	0.18
1Z8L	3.4.17.21	Glutamate Carboxypeptidase 2	2005	0.35

When we started our study (2003) less supportive data was available with only a minority of protein sequences characterized structurally. Back then, total identities of hits ranged from as low as 6% up to 12% affecting all project stages: detection of remote homology; significance of alignments, fidelity to overall fold. Secondary structure profiles (DSSP[37]) had to be used to filter out false candidates.[30] Despite the fact that the predictions of secondary structures result more reliably than of other spatial features they are not foolproof and as a last resort several tools were applied to find coincidences (consensus): PHDsec (see above), PSIPRED[38] a neural network program analyses the position specific scoring matrix output obtained from PSI-BLAST, and Jpred/JNET[39] the consensus method over a built-in set of tools were utilized. They were accessed through the MetaPP service. Besides, two more tools were used: offline Predator[40] and web-based Agadir.[41] Number and positions of helices (but not exact lengths for its ambiguity) were screened because alpha-helical prediction is safest mainly due to known correlations on contacts between intrahelical neighbors (every third or fourth position), in contrast to a beta sheet where (unforeseeable) distant residues in strands on one or more chains have contact due to loops and turns (see also web information on amino acids in table 1).

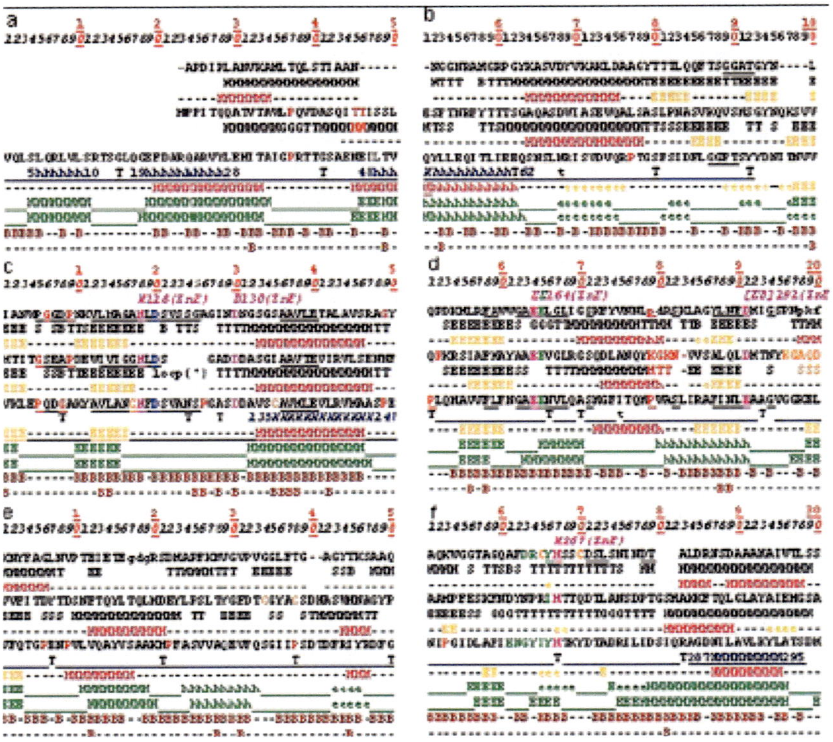

Legend
Line 1: Decimal leap numbering
Line 2: Block-wise amino acid numbering from one to ten
Line 3: Zn Finger (ZnF) Motif
Line 4: 1qq9: aligned to match predicted secondary structures of query
Line 5: 1qq9: H= alpha helix; B, E=extended beta sheet/strand; T= turn; S=bend
Line 6: 1qq9: JPRED indicates Helical structures (H) and Extended sheets (E)
Line 7: 1lok: aligned to match predicted secondary structures with query
Line 8: 1lok: secondary structure from PDB entry, see line 5
Line 9: 1lok: JPRED indicates Helical structures (H) and Extended sheets (E)
Line10: Query / target sequence of 3D-model
Line11: Query: AGADIR: Helix prediction (H), and EMBL beta-turn prediction (T)
Line12: Query: JPRED: Helical structure (H) and Extended sheets (E) for query
Line13: Query: PREDATOR: Helical (H), Extended (E) and coiled (__) segments
Line14: Query: Jnet_25 % Buried (B) location with 25 % probability of error
Line15: Query: Jnet_5 % Buried, i.e. inside location with 5 % probability of error

Figure 3. Final alignments editing yields 40 % homology in conserved segments after identifying and extracting variable loops. To satisfy the requirements of sequence length for these manual arrangements the template 1qq9[48] was shortened at L188: LPSADR becomes LPDR and a little further down at N197 NpGYFVYDDDPVIEkTf becomes Npkf. The template 1lok[49] was adopted to the query, too: a longer loop TIGSHTNEQSVAP between D120 and D130 was deleted. At position Q177 the helical end

QYKSEGKNVV was shortened to QYKGKNVV resp. NYKGSAQD to NYKGAQD starting at K196. Critical regions are underlined. Whereas capital letters in the query line symbolize straightforward replacements with template 1lok, local inconsistencies are marked up in lower case letters and solved individually by local geometry optimization keeping the reminder rigid. Each template bears two unique cysteins forming one disulfide bridge each (1qq9: C265-C270; 1lok: C234-C238; query: C117-C135). We identified a nucleophilic elbow with E165. The catalytic site encompasses a double zinc finger motif on 1qq9: H085 D097 E131 E132 D160 H247, and on 1lok: H097 D117 E151 E152 D179 H256, and then was copied to query: H096 D098 E142 E143 E169 H245. (N.B. sequence tails are cut out.).

In order to compare the template candidates multiple sequences alignment studies were performed. Clustal W[42] generates phylogenetic trees with relative distances in "time" based on the available members of clans, families and subfamilies. Sometimes, editing the alignment helps to identify conserved regions where the backbone remains constant with typical folds of alpha-helical and beta-sheet elements. If the edited multiple sequence alignments are re-run after eliminating potential structurally variable regions (loops, coiled coils, low complexity, i.e. residue repetitions) the scores will improve when the fold or protein class is correctly represented. The priorities of the following criteria listed for selecting templates depend on the particular case: The best candidate may be hand-selected based on the highest identity scores but if they lie almost in the same value range templates with some degree of spatial differences within one fold, domain, family class should be considered for a multiple template combination. If a larger number represents the same crystal structure several choices are to be made: evolutionary distance, host organism, atomic resolution and R-factors, missing atoms or residues, co-crystallization with water or ions, and complex with ligands. Even a lower scoring candidate might be appropriate, reconsidering that a poor overall similarity along the total length may be compensated by a higher local homology at interesting segments, like binding site, zinc finger motifs, lipophilic core, as well as the catalytic triad. Gap penalties determine the number and length of gaps to be tolerated. In the same sense fitness to the secondary structural profile and number and length of loop insertions may be decisive. The personal preferences taken in this step mark the difference to the automated 3D modeling.

The manual construction of the model was conducted with MOE.[47] Due to the higher degree of conservation, especially with respect to the amino acids forming a lipophilic core of the protein, the backbone of 1lok[49] was mainly threaded (see figure 4). The side chain conformations of identical amino acids were adopted from this template. Thereafter, any amino acid, not present with respect to 1qq9[48] but conserved in the 1lok was modeled according to the conformation in that structure. Certain amino acids (D, E, K, R) are possible salt bridge partners due to charge and position at the surface. The correlated occurrences were detected by multiple sequence alignment. A binuclear zinc finger was copied from the template 1qq9. Two unique cysteines with surface access were bridged, whereas the location is not conserved (1qq9:C245-C250; 1lok: C223-227, query: C117-C135).

By keeping the zinc finger, backbone and conserved side chains rigid, all the other side chains were treated by geometry refinement with the AMBER 94 force field implemented in the modeling software MOE[47] then a recommended simulated annealing (SA) protocol[50] was adapted to MOE. The structure was solvated in MOE's predefined water box and the geometry fully optimized by 5000 steps conjugate gradient energy minimization before running SA under standard MMFF94 conditions (0-450K, 250 ps, constant dielectric epsilon

= 1). Temperature equilibrium was reached and a dipeptide substrate placed into the shallow binding groove. The stable model was subjected to a final conformational stability test removing the solvents and all rigid restrains, further a dynamics run was carried out in vacuum without periodic boundary conditions (distance-dependent dielectric epsilon = 40). The force field choice was found appropriate, since both metal ions were properly held in placed by the six surrounding side chains and the over-all shape of the fold kept through long period (100 ps, a more detailed report is beyond our scope). Only the catalytic water and the dipeptide ligand probe were lost. The quality of the target structure was positively evaluated by all statistical criteria related to modeling by *Gerd Vriend*'s protein evaluation package What_Check[51] (http://nihserver.mbi.ucla.edu/SAVS/) (see figure 5).[52]

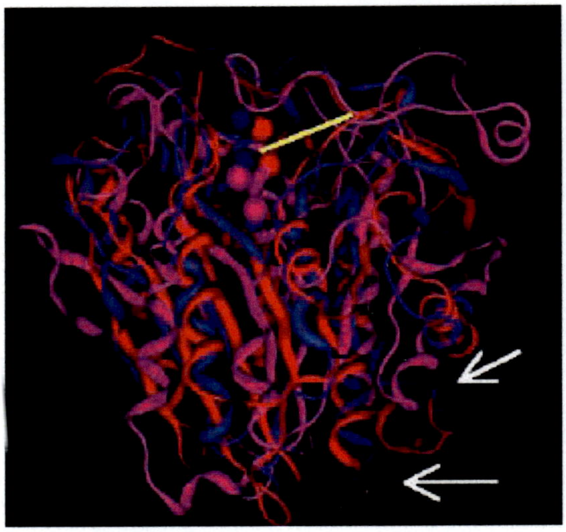

Figure 4. Superposition of final 3D model of FXNA (blue) upon its two nearest, hand-selected templates 1lok (red) and 1qq9 (magenta, RMSD 18). The root means square distance (RMSD) serves as a measurement of geometrical resemblance. The RMSD values of FXNA against 1lok or 1qq9 is 12, or 18, respectively. This means that 1lok is closer to target even in comparison with a RMSD of 20 between both templates. Displaying the backbone by colored ribbons as well as the zinc pairs by atomic volumes allowing a visual comparison of structural similarities and variations (mostly loops). N- and C-termini (upper vs. lower arrow) are situated at the bottom to the right, and a short ribbon (yellow) locates a hypothetical dipeptide substrate at the shallow active site groove.

According to our results, remote homology does not impede the generation of atom-scale models of aminopeptidases (see figure 2 and 4), in which case conclusive evidence for both its successful computational evaluation and its useful biological interpretation can be drawn (see table 6):

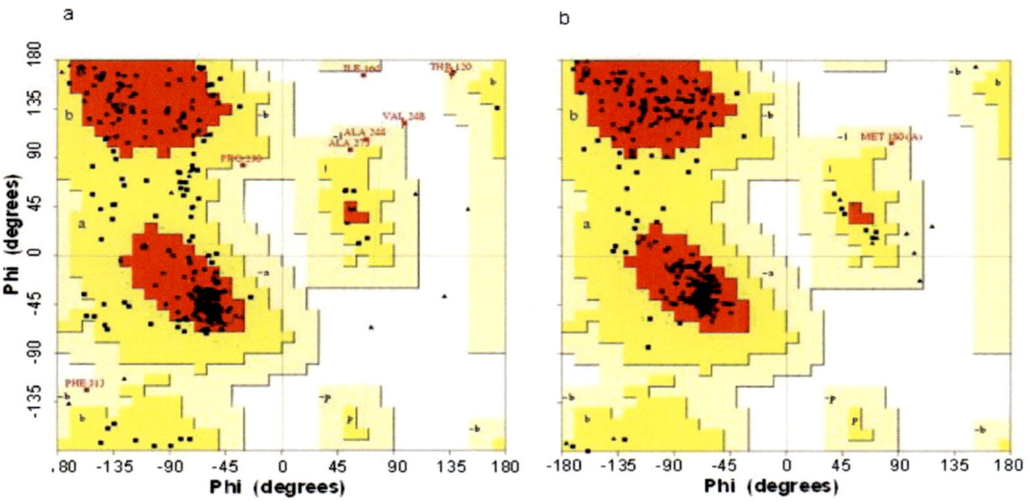

Figure 5. A *Ramachandran* plot generates a figure of two dihedral (torsion) angles, called phi (x-axis: atoms C-N-Calfa-C) and psi (y-axis: atoms N-Calfa-C-N) to evaluate numerically the orientation of the side chains on a protein. Allowed phi-psi combinations lie in the red to yellow zones and forbidden zones are left white (Procheck plots found at http://nihserver.mbi.ucla.edu/SAVS/).[53]

1. eight stranded, mostly parallel alpha/beta secondary structure comprising eight alpha-helices and eight beta-sheets;
2. probably a double zinc finger coordinated by two pairs of three typical side chains (HIS199, ASP211, GLU272 and HIS348, ASP212, GLU246) although cobalt can not be ruled to form the binuclear motif, however zinc is the more common case and the divalent cations are normally separated by 3.6 Å, here by 3.7 Å;[54]
3. our target protein exhibits hydrolase activity since a concert of important amino acids was identified (GLU245, SER320, TYR347, ASP212, ASN343). The catalytic triad in target and templates is conserved: a general base (glutamic anion), nucleophilic water (or stronger hydroxyl anion) with one zinc ion and histidine. The metal ion activates the bound water molecule for catalysis. Despite considerable diversity among the 30 subfamilies of metalloproteases the mechanism is well understood, e.g. serine proteases conserve an Asp-His-Ser catalytic triad, too;
4. the catalytic triad is composed of two glutamic acids (GLU245, GLU272) and a histidine residue (HIS348) including probably one Zn-associated water molecule/hydroxide; GLU245 sits on the nucleophilic elbow, next to conserved GLU246 which coordinates Zn; tyrosine (TYR347) assists to stabilize amino groups on substrate binding; its adjacent residues infer substrate specificity because they are significantly distinct compared to 1lok[49] and 1qq9[48];
5. hydrophobic lip for ligand binding on target: ILE138, VAL203, ALA204, VAL248, LEU249, PHE313, ILE317, ILE318, PHE340, ILE341, ILE346;
6. not conserved prolines (P207, P319, underlined in figure 2) may play important roles in substrate recognition, situated at the edge of the shallow groove, such prolines have been reported to bind to a substrate at the surface containing aromatic residues.[55] We examined the no polar contact zone around the active site with the double zinc finger. Unspecific APs (1qq9[48], 1lok[49]) remove wide range of

different types of hydrophobic amino acids starting at the N-terminus of the substrates. The query AP is even more lipophilic in nature (see GRAVY in table 6), its shallow groove is contoured by the hydrophobic lip. Therefore we inserted an Ala-Ala dipeptide to mark the location (MOE docking);

7. other prolines (P118, P158, P230, P260, P287, P303) are known to break up helices, so they are located near the surface at either end of the helices;
8. a unique disulfide bridge is located at the bottom of the shallow cleft forming the binding site with two unique cysteins (C117, C135) between beta-sheet 4 and helix 4;
9. a well conserved and highly hydrophobic core;
10. all positive charges are located at the surface;
11. on residue exchange on the template the new side chains accommodate easily and avoid bad contacts;
12. after editing the alignments the predicted secondary structure of the query matches the observed secondary structure of templates up to 70 % for a residues length of 280;
13. insertions, gaps, loops and other evolutionary changes, like the analogous shift of the disulfide bridge. Statistics say 67 % can be exchanged still preserving structure;
14. the manual modeling of a dZnF, as well as the catalytic triad with an acid, nucleophile and histidine gives evidence for a realistic mechanism to cleave peptide bonds at amino-terminus. Surrounded by a crown of lipophilic amino acids the peptide-binding groove is not so pronounced as with other exopeptidases from eukaryotes or bacterial thermolysins (endopeptidases) assuming other type of substrates [41 54];
15. The new protein's mechanism is most likely that of *Streptomyces Griseus* AP and not *Aeromonas proteolytica* LEU-AP because target's TYR266 is found homologous to TYR246 in template 1qq9 but not in template 1lok. So it can be inferred that the target may recognize a wide range of hydrophobic N-terminal amino acids in substrates on binding.[56] As can be seen in figure 4 the target follows template 1lok with higher structural fidelity but multiple template procedure is needed to explain the substrate recognition difference;
16. our 3D model is consistent with known facts about aminopeptidases: they specifically cleave the amino-terminal residue from polypeptide chains and are involved in the metabolism of biologically active peptides. The family includes zinc-dependent enzymes possessing either one or two zinc ions per active site. The structure is stabilized by disulfide bridges and metal ligands, paralleled helices, extensive hydrogen net spanning over the eight strands, and through a pronounced hydrophobic core;

Table 6. Final statistical data about formal protein composition, structure and function of target and templates

	FXNA	1LOK[49]	1QQ9[48]
Superfamily	Zn-dependent exopeptidases		
Class	Alpha and beta proteins (a/b), i.e. mainly parallel beta sheets (beta-alpha-beta units)		
Fold	Phosphorylase/hydrolase-like		
Source	*Rattus Norvegicus*	*Aeromonas Proteolytica*	*Streptomyces Griseus*
Protein function	Aminopeptidase	Leucyl Amino-peptidase	Amino-peptidase
Synonyms	Rnap(?)	Aap, Lap	Sgap
Classification	Hydrolase	Hydrolase	Hydrolase
EC Number	3.4.11.?	3.4.11.10	3.4.11.
CATH	3.40.630.10.?	3.40.630.10	3.40.630.10.4
Protein structure	alpha/beta globular	alpha/beta globular	alpha/beta globular
Number of helices, content of helices	8, (+2 minor fragments) 36 %	8, (+2 minor fragments) 37 %	9, (no fragments at C-term) 38 %
Number and content of sheet	8 and 16 %	8 and 16 %	8 and 17 %
	FXNA	**1LOK[49]**	**1QQ9[48]**
Known substrate type by complexed ligands	Putative hydrophobic N-terminus of peptides	L-leucine-phosphonic acid	Met, Phe, Leu
Catalytic site	Bridged putative bimetallic amino-peptidase (double Zn finger), hydroxyl anion	Proteolytic C-N cleaving, bridged, double-zinc exo-peptidase removes the N-terminal residues from a peptide	Proteolytic C-N cleaving, bridged, double-zinc exo-peptidase; zinc-bound water, catalytic Glu131, Tyr246
Metal ion for dinuclear zinc finger	2 Zn; mammalian	2 Zn; bacterial	2 Zn; bacterial
Disulfide bridge	Cys117 - Cys135	Cys223 - Cys227	Cys245 - Cys250
GRAVY hydropathicity	+0.068 lipophilic	-0.326 hydrophilic	-0.303 hydrophilic
Number of aa	280	291	278
Molecular mass	30613	31313	29164
Functional unit	monomer	monomer	monomer
Total similarity	7 % vs. 1lok 11% vs. 1qq9	7 % vs. query 16% vs. 1qq9	11 % vs. query 16% vs. 1lok
Partial similarity without loops	100 % vs query	29% vs. query	22% vs. query

Table 6. (Continued).

Amino acid (aa)	Partial content percentage % (total number)		
Cys (C)	0.7 (2 aa)	0.7 (2 aa)	0.7 (2 aa)
His (H)	2.6 (8 aa)	2.4 (7 aa)	2.5 (7 aa)
Pro (P)	3.9 (12 aa)	4.1 (12 aa)	4.7 (13 aa)
Negatively charged Asp (D), Glu (E)	34	31	26
Positively charged Arg (R), Lys (K)	19	14	21
Isoelectric point: pI	4.83	4.54	5.77

Conclusion

In the present introduction to mainly web-based computational techniques to predict the structure of a protein with low homology with all solved structures in the PDB, practical hints are deferred with a down-to-earth practical flavor for both the biomolecular novice as well as the computationally inexperienced biologist in a platform-independent general manner. The application of molecular modeling programs requires knowledge in the field of computer assisted molecular design (CAMD) and we recommend the reading of the basic concepts and suggested references by Richon[12] for further reading (see table 1).

While this chapter is hoped to serve as a starting point, the reader may dig deeper into the protein prediction field from the CASP-project "Critical Assessment of techniques for protein Structure Prediction" (see table 1) or the enzyme activity prediction, e.g. POPS.[14]

The present 3D model demonstrates low homology modeling is possible with acceptable statistics for protein backbone and side-chain geometries (see figure 5). Since aminopeptidase structures are not yet found in PDB (last visit February 2007) it was deposited as a hypothetical contribution under entry code 2B62 by AutoDep at RCSB (http://pdb.rutgers.edu/adit/).[58]

Acknowledgement

Adit co-worker *Massy Rajabzadeh* at RCSB Protein Data Bank, Rutgers, New Jersey, USA *Felix Luna* at BUAP, Puebla, Mexico for delivering and discussing the FAXNA gene and figure drawings.

References

[1] Benson DA, Karsch-Mizrachi I, Lipman DJ et al. (2002). GenBank. *Nucleic acids research*, 30:17-20.

[2] Baker W, van den Broek A, Camon E et al. (2000). The EMBL nucleotide sequence database. *Nucleic acids research*, 28:19-23.

[3] Boeckmann B, Bairoch A, Apweiler R et al. (2003). The Swiss-Prot protein knowledgebase and its supplement TrEMBL in 2003. *Nucleic acids research*, 31:365-370.

[4] Chothia C and Lesk AM (1986). The relation between the divergence of sequence and structure in proteins. *The EMBO journal*, 5:823-826.

[5] Chothia C and Gerstein M (1997). Protein evolution. How far can sequences diverge?. *Nature*, 385:579-581.

[6] Murzin AG, Brenner SE, Hubbard T et al. (1995). SCOP: a structural classification of proteins database for the investigation of sequences and structures. *Journal of molecular biology*, 247:536-540.

[7] Bairoch A, Bucher P and Hofmann K (1997). The PROSITE database, its status in 1997. *Nucleic acids research*, 25:217-221.

[8] Pandit SB, Bhadra R, Gowri VS et al. (2004). SUPFAM: a database of sequence superfamilies of protein domains. *BMC Bioinformatics*, 5:28.

[9] Rawlings ND and Barrett AJ (1995). Evolutionary families of metallopeptidases. *Methods in enzymology*, 248:183-228.

[10] Ng SK and Tan SH (2004). Discovering protein-protein interactions. *Journal of bioinformatics and computational biology*, 1:711-741.

[11] D'Alfonso G, Tramontano A and Lahm A (2001). Structural conservation in single-domain proteins: implications for homology modeling. *Journal of structural biology*, 134:246-256.

[12] Richon AB (1994). An introduction to molecular modeling, *Mathematech*, 1:83.

[13] Schwede T, Kopp J, Guex N et al. (2003). SWISS-MODEL: An automated protein homology-modeling server. *Nucleic acids research*, 31:3381-3385. Site (URL) http://www.expasy.org/swissmod/SWISS-MODEL.html

[14] Boyd SE, Pike RN, Rudy GB et al. (2005). PoPS: a computational tool for modeling and predicting protease specificity. *Journal of bioinformatics and computational biology*, 3:551-585.

[15] Lambert C, Leonard N, De Bolle X et al. (2002). ESyPred3D: Prediction of proteins 3D structures. *Bioinformatics*, 18:1250-1256.

[16] Lund O, Frimand K, Gorodkin J et al. (1997). Protein distance constraints predicted by neural networks and probability density functions. *Protein engineering*, 10:1241-1248.

[17] Shindyalov IN and Bourne PE (2000). Improving alignments in HM protocol with intermediate sequences. In: "Forth meeting on the critical assessment of techniques for protein structure prediction", p. A-92. online http://cl.sdsc.edu/hm.html

[18] Barton GJ (1998). Protein sequence alignment and database scanning. In MJE Sternberg (ed), Protein structure prediction: A Practical Approach. IRL Press at Oxford University Press. pp 312.

[19] Meller J and Elber R (2001). Linear programming optimization and a double statistical filter for protein threading protocols. *Proteins*, 45:241-261.

[20] Wallner B and Elofsson A (2005). All are not equal: a benchmark of different homology modeling programs. *Protein science*, 14:1315-1327.

[21] Rost B and O'Donoghue S (1997). Sisyphus and prediction of protein structure. *Computer applications in the biosciences*, 13:345-356.

[22] Park J, Karplus K, Barrett C et al. (1998). Sequence comparisons using multiple sequences detect three times as many remote homologues as pairwise methods. *Journal of molecular biology*, 284:1201-1210.

[23] Johnson MS, Srinivasan N, Sowdhamini R et al. (1994). Knowledge-based protein modeling. *Critical reviews in biochemistry and molecular biology*, 29:1-68.

[24] Rost B and Sander C (1996). Bridging the protein sequence-structure gap by structure predictions. *Annual review of biophysics and biomolecular structure*, 25:113-36.

[25] Morris AL, MacArthur MW, Hutchinson EG et al. (1992). Stereochemical quality of protein structure coordinates. *Proteins*, 12:345-364.

[26] Rodriguez R, Chinea G, Lopez N et al. (1998). Homology modeling, model and software evaluation: three related resources. *Bioinformatics*, 14:523-528.

[27] Guex N and Peitsch MC (1997). SWISS-MODEL and the Swiss-PdbViewer: an environment for comparative protein modeling. *Electrophoresis*, 18:2714-2723.

[28] Vriend G (1990). WHAT IF: a molecular modeling and drug design program. *Journal of molecular graphics*, 8:52-56.

[29] Altschul SF, Madden TL, Schaffer AA et al. (1997). Gapped BLAST and PSI-BLAST: a new generation of protein database search programs. *Nucleic acids research*, 25 (17):3389-3402.

[30] Contreras-Moreira B and Bates PA (2002). Domain fishing: a first step in protein comparative modeling. *Bioinformatics*, 18:1141-1142.

[31] Rost B (1996). PHD: predicting one-dimensional protein structure by profile-based neural networks. *Methods in enzymology*, 266:525-539.

[32] Mulder NJ, Apweiler R, Attwood TK et al. (2005). InterPro, progress and status in 2005. *Nucleic acids research*, 33:D201-D205.

[33] Corpet F, Gouzy J and Kahn D (1998). The ProDom database of protein domain families. *Nucleic acids research*, 26:323-326.

[34] Madera M, Vogel C, Kummerfeld SK et al. (2004). The SUPERFAMILY database in 2004: additions and improvements. *Nucleic acids research*, 32:D235-D239.

[35] Murzin AG, Brenner SE, Hubbard T et al. (1995). SCOP: a structural classification of proteins database for the investigation of sequences and structures. *Journal of molecular biology*, 247:536-540.

[36] Holm L and Sander C (1998). Touring protein fold space with Dali/FSSP. *Nucleic acids research,* 26:316-319.
[37] Kabsch W and Sander C (1983). Dictionary of protein secondary structure: pattern recognition of hydrogen-bonded and geometrical features. *Biopolymers,* 22:2577-2637.
[38] Jones DT (1999). Protein secondary structure prediction based on position-specific scoring matrices. *Journal of molecular biology,* 292:195-202.
[39] Cuff JA and Barton GJ (1999). Evaluation and improvement of multiple sequence methods for protein secondary structure prediction. *Proteins,* 34:508-519.
[40] Frishman D and Argos P (1996). Incorporation of non-local interactions in protein secondary structure prediction from the amino acid sequence. *Protein engineering,* 9:133-142.
[41] Muñoz V and Serrano L (1997). Development of the multiple sequence approximation within the AGADIR model of alpha-helix formation: comparison with Zimm-Bragg and Lifson-Roig formalisms. *Biopolymers,* 41:495-509.
[42] Thompson JD, Higgins DG and Gibson TJ (1994). CLUSTAL W: improving the sensitivity of progressive multiple sequence alignment through sequence weighting, position-specific gap penalties and weight matrix choice. *Nucleic acids research,* 22:4673-4680.
[43] Michalsky E and Goede AR (2003). Preissner, Loops In Proteins (LIP)-a comprehensive loop database for homology modeling. *Protein engineering,* 16:979-985.
[44] Berman HM, Westbrook J, Feng Z et al. (2000). The Protein Data Bank. *Nucleic acids research,* 28:235-242.
[45] Ceroni A, Frasconi P, Passerini A et al. (2003). Predicting the disulfide bonding state of cysteines with combinations of kernel machines. *Journal of VLSI SIgnal Processing,* 35:287-295.
[46] Bray JE, Todd AE, Pearl FM et al. (2000). The CATH Dictionary of Homologous Superfamilies (DHS): a consensus approach for identifying distant structural homologues. *Protein engineering,* 13:153–165.
[47] Molecular Operating Environment (MOE), software, Chemical Computing Group Inc., 1010 Sherbrooke Street West, Suite 910, Montreal, Canada H3A 2R7, http://www.chemcomp.com, versions 2003.02, 2004.03.
[48] Gilboa R, Greenblatt HM, Perach M et al. (2000). Interactions of Streptomyces griseus aminopeptidase with a methionine product analogue: a structural study at 1.53 A resolution. *Acta crystallographica. Section D, Biological crystallography*, 56:551-558.
[49] Desmarais WT, Bienvenue DL, Bzymek KP et al. (2002). The 1.20 Å resolution crystal structure of the aminopeptidase from Aeromonas proteolytica complexed with tris: a tale of buffer inhibition. *Structure,* 10:1063-1072.

[50] Spadola L, Novellino E, Folkers G et al. (2003). Homology modeling and docking studies on Varicella Zoster Virus thymidine kinase. *European journal of medicinal chemistry,* 38:413-419.

[51] Hooft RW, Vriend G, Sander C et al. (1996). Errors in protein structures. *Nature,* 381:272.

[52] Structure Analysis and Verification Server, online: *http://www.doe-mbi.ucla.edu/Services/SV/http://shannon.mbi.ucla.edu/DOE/Services/SV/*; with Ramachandran plots and What_Check and ProCheck options.

[53] Laskowski RA, MacArthur MW, Moss DS et al. (1993). PROCHECK: program to check the stereochemical quality of protein structures. *Journal of Applied Crystallography,* 26:283-291.

[54] Weston J (2005). Mode of action of bi- and trinuclear zinc hydrolases and their synthetic analogues. *Chemical reviews*, 105:2151-2174.

[55] Hopfner KP, Karcher A, Shin DS et al. (2000). Structural biology of Rad50 ATPase: ATP-driven conformational control in DNA double-strand break repair and the ABC-ATPase superfamily. *Cell,* 101: 789-800.

[56] Fundoiano-Hershcovitz Y, Rabinovitch L, Langut Y et al. (2004). Identification of the catalytic residues in the double-zinc aminopeptidase from Streptomyces griseus. *FEBS letters,* 571:192-196.

[57] W.R. Pearson (1990) Rapid and sensitive sequence comparison with FASTP and FASTA. *Methods Enzymol*., 183:63-69.

[58] ADIT at RCSB Protein Data Bank, Rutgers The State University of New Jersey, USA *http://pdb.rutgers.edu/adit/*

[59] private communication on phylogenetic tree (Wahab et al., 2006)

[60] A.A. Canutescu, A.A. Shelenkov, R.L. Dunbrack Jr, A graph-theory algorithm for rapid protein side-chain prediction, *Protein Sci.* 12, 2001(2003).

[61] Wahab HA, Ahmad Khairudin NB, Samian MR, Najimudin N. , Sequence analysis and structure prediction of type II Pseudomonas sp. USM 4-55 PHA synthase and an insight into its catalytic mechanism. *BMC Struct. Biol.* 2006 Nov 1;6:23.

[62] A.A. Canutescu, A.A. Shelenkov, R.L. Dunbrack Jr, A graph-theory algorithm for rapid protein side-chain prediction, *Protein Sci.* 12, 2001(2003).

[63] Gouet P, Robert X, Courcelle F (2003) ESPript/ENDscript: extracting and rendering sequence and 3D information from atomic structures of proteins. *Nucleic Acids Res.* 31: 3320-3323.

[64] Deleage G, Combet C, Blanchet C, Geourjon C (2001) ANTHEPROT: an integrated protein sequence analysis software with client/server capabilities. *Comput. Biol. Med.* 31: 259-267.

[65] Rost B, Yachdav G, Liu J (2004) The PredictProtein server. *Nucleic Acids Res.* W321-W326.

[66] Wojcik J., Mornon J-P., Chomilier J. WLoop: New efficient statistical sequence dependent structure prediction of short to medium-sized protein loops based on an exhaustive loop classification. *J. Mol. Biol.* (1999) 289:1469-1490

[67] Gautier R., Camproux A-C., SCit: web tools for protein side chain conformation analysis Tuffery P. *Nucleic Acids Res.* (2004) 32: W508-W511.

Appendix

The following description for the inexperienced beginner proposes a straight forward procedure step by step for a typical homology modelling situation using only Internet-based resources or free tools for downloading to the local computer. The bottom line for such studies is that the three-dimensional structure of the target is unknown, but the amino acid sequence has already been identified. Sometimes, however, only a part of the gene or nucleotide sequence will be related to a known protein structure as illustrated in details in our example of an aminopeptidase. Recently, a more sophisticated procedure has been published as a free full text article at *Pub Med Entrez* (http://www.ncbi.nlm.nih.gov/entrez) [59].

1. Preparation of the Input Data Files

1a.

The primary structure of the target can be retrieved from the *EMBL database* [2]. This amino acid sequence of the peptide chain should be converted in *FASTA, PIR* and/or free format according to the file specification of the read-in operation of the alignment software in use. All three are simple ASCII text format and can be edited with any text editor.

1b.

The template(s) are in PDB format taken from the Brookhaven protein structure database [44]. The amino acid sequence(s) of the template(s) can be generated from the PDB files and stored in FASTA, PIR or free text formats.

The original PDB files contain information about the structure, data generation, reliability and inconsistencies. To read it the user opens the file with a text editor. The PDB sequences should be compared with the (nucleotides to amino acids) translated sequences obtained from the *GeneBank* [1,2].

2. Sequence Alignment

The sequence analysis is conducted by aligning the amino acid chains for comparison. The target and template sequences are aligned and then compared using the required input file format mentioned above. The recommended multiple sequence alignment program is *Clustal W* (http://www.ebi.ac.uk/clustalw) [42]. The amino acids are positioned to find maximal identities and homologies between the sequences. Another program builds alignment based on secondary structure information: *Espript* software (http://espript.ibcp.fr/ESPript/cgi-bin/ESPript.cgi)[61].

3. Prediction of Secondary Structure

The secondary structure can readily be predicted by several algorithms including the following programs:

Antheprot (http://antheprot-pbil.ibcp.fr)[62],

PSIPRED (http://bioinf.cs.ucl.ac.uk/psipred/psiform.html)[38],

Jpred (http://www.compbio.dundee.ac.uk/~www-jpred/submit.html)[39] and PROF which can be downloaded and installed on SGI machines and on PC with Linux (http://www.aber.ac.uk/~phiwww/prof)[63].

4a. Fully Automated Generation of 3D Target Model

The target that has not been structurally known hitherto is send as a sequence in FASTA format by e-mail to the Swiss-Model Server (http://swissmodel.expasy.org) [13]. The server replies automatically and normally sends back the results with the 3D model of the target protein within hours.

4b. Manual Generation of 3D Target Model

The use selects a protein structure and substitutes all side chains along the backbone. The substitution of all the residues by other amino acid types will result in building the target structure. The loop segments, i.e. insertion and deletion have to be treated separately using the Loop Database delivered with *Swiss-PdbViewer* [27] or predicted with *WLoop* [64]. To clear the side chain geometries the user can use online tools for protein side chain conformation prediction: SCWRL [60]. Another program offers residue substitution and positions side chains (SCs) under standard conformations. The geometries can be analyzed and manipulated by a built-in editor of SCit (http://bioserv.rpbs.jussieu.fr/Help/SCitHelp.html) [65].

5. 3D Model Validation

The predicted structures can be verified upon request of the Web page of the Swiss-Model Server user, and the online program send a *WhatCheck* report along with the 3D model in PDB format [51].

6. 3D Model Visualization

DeepView (Swiss PDB Viewer SPDBV) is a tool to manipulate the model on the screen. For instance, atoms of backbone or side chains can be displayed in different style and colors. Also surfaces can be colored according to the calculated electrostatic potential loading predefined atom charges for the residues (http://ca.expasy.org/spdbv)[27].

Index

B

C

D

E

F

G

H

I

J

K

L

M

P

Q

R

S